Edexcel GCSE
Geography B
Evolving Planet

Student Book

David Flint • Lindsay Frost • Simon Oakes • Andy Palmer • Michael Witherick • Phil Wood • Nigel Yates
Series editor: Nigel Yates

A PEARSON COMPANY

Published by Pearson Education Limited, a company incorporated in England and Wales, having its registered office at Edinburgh Gate, Harlow, Essex, CM20 2JE. Registered company number: 872828

Edexcel is a registered trade mark of Edexcel Limited

© Pearson Education 2009
The right of David Flint, Lindsay Frost, Simon Oakes, Andy Palmer, Michael Witherick, Phil Wood and Nigel Yates to be identified as the authors of this work has been asserted by them in accordance with the Copyright, Designs and Patents Act 1988.

First published 2009

British Library Cataloguing in Publication Data
A catalogue record for this book is available from the British Library.

ISBN 978 1 846905 01 8

Edited by Graham Bradbury
Original illustrations © Oxford Designers and Illustrators 2009
Illustrated by Oxford Designers and Illustrators
Picture research by Louise Edgeworth
Printed and bound in Great Britain at Scotprint, Haddington

Acknowledgements
The publisher would like to thank the following for their kind permission to reproduce their photographs:

(Key: b-bottom; c-centre; l-left; r-right; t-top)
Alamy Images: Aerial Archives 188; AfriPics.com 91; Alvey & Towers Picture Library 202; Bailey-Cooper Photography 2 204; David Ball 167; Brett Baunton 120b; bjp images 265, 280r; Sara Blancett 11; David Burton 193; Thomas Cockrem 7b, 145; Brandon Cole Marine Photography 112; CuboImages srl 53b; dmark 88b; Chad Ehlers 5t, 269t; Julio Etchart 95; Greg Balfour Evans 218tl; Leslie Garland Picture Library 77t; geogphotos 77b; Les Gibbon 70b; Chris Gomersall 53t; Melvin Green 82; Tom Hanley 218b; David Hoffman Photo Library 253; Images&Stories 21; James Kubrick 131, 179; Frans Lemmens 118; Barry Lewis 144; Steven May 256; Gianni Muratore 148; nagelestock.com 261; Chris Pancewicz 89; Andrew Parker 212r; Photofusion Picture Library 168r; Purple Marbles 98; David Robertson 200l; Andrew Rubtsov 208; Sami Sarkis 269b; Skyscan Photolibrary 74; Doug Steley B 128; Homer Sykes Archive 147; The Photolibrary Wales 168l; Peter Titmuss 280l; Janine Wiedel Photolibrary 217; Worldwide Picture Library 47b, 47t; Ariadne Van Zandbergen 221; Justin Kase ztenaz 205; Art Directors and TRIP photo Library: Chris Kapolka 203; Jane Sweeney 186r; Chris Wormald 227; Corbis: Bilderbuch/Design Pics 225; Carlos Cazalis 206; Sherwin Crasto/Reuters 260; Bennett Dean; Eye Ubiquitous 212l; Walter Geiersperger 207; Jose Goitia 178; Jason Hawkes 72; Lindsay Hebberd 176; Jana Renee Cruder/Digital Nikon D200 196; Kendra Luck/San Francisco Chronicle 130; Jose Fuste Raga 124; Rickey Rogers/Reuters 48; Michael T. Sedam 16; Skyscan 75, 259; George Steinmetz 182; Strauss/Curtis 126; Jim Sugar 19b; Rudy Sulgan 61; Olivier Pitras/Sygma 122; William Taufic 246; The Irish Image Collection 218tr; Elizabeth Whiting & Associates 215b; DK Images: 33r; Kim Sayer 70t; Ecoscene: Simon Grove 7t, 103; GeoScience Features Picture Library: 129l, 129r; Geoslides/Geo Aerial Photography: 88t; Getty Images: Daniel Berehulak 92; DEA/C.BEVILACQUA/De Agostini Picture Library 13b; Rafael Duran/AFP 186l; Jack Dykinga/Stone 120t; Jason Edwards/National Geographic 59; Christopher Furlong 93; Ramzi Haidar/AFP 67; Jyrki Komulainen/Gorilla Creative Images 215t; Banaras Khan/AFP 19t; Dan Kitwood 209; Peter Macdiarmid 200r; Haywood Magee/Hulton Archive 146; Simon Maina/AFP 213; Randy Olson/National Geographic 123; Sam Panthaky/AFP 44; Alan Smith/Stone 172; Solar and Heliospheric Observatory 29c; Charles Thatcher/Stone 143; Gandee Vasan/Photographer's Choice 4b, 187; Art Wolfe/Stone 29b; iStockphoto: Andrew Howe 51 (willow warbler); Lubomir Jendrol 51 (turtle dove); Lurii Konoval 51 (yellow wagtail); Jupiter Unlimited: 51 (garden warbler); Brand X Pictures 304bl, 307; Goodshoot 304tl; Stockxpert 288, 304r; NASA: Goddard Space Flight Center. Scientific Visualization Studio 108b, 108t; Image courtesy of MODIS Rapid Response Project at NASA/GSFC 240; Natural History Museum Picture Library: Michael Long 33l; No Trace: 175; Simon Oakes: 65, 110; Lucia Oritiz: 236; PA Photos: AP Photo/Andy Wong 153; Andy Palmer: 71; Photofusion Picture Library: Dorothy Burrows 5b, 285; Photolibrary.com: Digital Vision/StockTrek 4t, 29t; Guildhall Library & Art Gallery/ImageState 31; Jacques Jangoux 125; Practical Action/ZUL: Steve Fisher 242; Reuters: Phil Noble 111; Rex Features: Dobson Agency 76; Andrew Drysdale 250; Eye Ubiquitous 13t, 58; Sipa Press 50; Jane Sweeney/Robert Harding 62; Skyscan Photolibrary: 99; Southampton Science Park: 255; Andrew Stacey: 81; Visage by Barratt: 171; Phil Wood: 289, 298

All other images © Pearson Education
Also see pages 327-328 and the backcover for figures, tables and text.

Every effort has been made to trace the copyright holders and we apologise in advance for any unintentional omissions. We would be pleased to insert the appropriate acknowledgement in any subsequent edition of this publication

The websites used in this book were correct and up to date at the time of publication. It is essential for tutors to preview each website before using it in class so as to ensure that the URL is still accurate, relevant and appropriate. We suggest that tutors bookmark useful websites and consider enabling students to access them through the school/college intranet.

Disclaimer
This Edexcel publication offers high-quality support for the delivery of Edexcel qualifications. Edexcel endorsement does not mean that this material is essential to achieve any Edexcel qualification, nor does it mean that this is the only suitable material available to support any Edexcel qualification. No endorsed material will be used verbatim in setting any Edexcel examination/assessment and any resource lists produced by Edexcel shall include this and other appropriate texts.

Copies of official specifications for all Edexcel qualifications may be found on the Edexcel website - www.edexcel.com

Contents: delivering the Edexcel GCSE Geography B Evolving Planet specification

Welcome to Edexcel GCSE Geography B Evolving Planet

Why should I choose GCSE Geography?

Because you will:

- learn about and understand the world that you live in
- develop skills that will help you in other subjects and your future career
- get to complete practical work away from the classroom
- learn how to work as a team
- learn by investigating, not just listening and reading

What will I learn?

You only have to switch on the news or pick up a newspaper to see that we live in a fast-pace, ever-changing world. GCSE Geography gives you the chance to learn about those changes: from those on your own doorstep to those of global proportions. There are four units:

Unit 1: Dynamic planet

Have you ever wondered...

- How and why climate has changed in the past, and the impact this will have on the future?
- Why water is important to the health of the planet?
- Why conflict occurs on the coast and how these conflicts can be managed?
- What the challenges of extreme climates are?

In this unit you will get a chance to investigate geological processes, ecosystems, the atmosphere and climate, and the hydrological cycle. These topics are interlinked and, although you may study them separately, the unit is designed to show you how physical geography combines to create a 'life support system' for the planet.

Unit 2: People and the planet

Have you ever wondered...

- How and why the population is changing in different parts of the world?
- What the ingredients of good living spaces are?
- What the environmental issues facing cities are?
- How countries might develop sustainably in the future?

This unit focuses on human geography. In a similar way to Unit 1, it links together to build an overall understanding of how humans interact with the planet. You will study how populations grow and change, where people live and work, and how they exploit and use resources.

There are also options in Units 1 and 2 so you will choose to study some topics in more depth such as rivers or coasts, cities or the countryside, development or economic geography, and oceans or extreme climates.

Unit 3: Making geographical decisions

This is a decision-making exercise, where you get to stand in the shoes of a real geographer. You will study a specific topic, such as Antarctica, in detail. This is designed to teach you how to make decisions about a specific topic, based on the evidence studied. The skills you will learn in this topic will be valuable in all aspects of this GCSE in Geography, and in the rest of your life.

Unit 4: Researching geography

This is the unit where you can really get stuck in! It will involve undertaking research, carrying out fieldwork and then writing it up. The research and fieldwork can be undertaken out of class, but the writing up will all be in class time. This means you have to spend less time at home doing your geography coursework!

How will I be assessed?

The great thing about the course is that the three 1-hour exams for Units 1, 2 and 3 can be spread over your two-year GCSE course. For unit 4, you will need to write up your fieldwork task in the classroom under controlled conditions while supervised by your teacher.

- Higher and Foundation examination papers are available.
- Units 1 and 2 exams are resource based. You will have a booklet containing maps, photographs and diagrams to help you answer the questions.
- Units 1 and 2 exam questions will range from short questions up to larger extended-writing questions.
- Unit 3 is a decision-making exercise based on pre-released resources. These will consist of a colour resource booklet on a geographical issue or location. Your teacher will work with you on the resources. Questions will assess your understanding of the resources in relation to environmental issues and sustainability.
- Unit 4 is the controlled assessment unit. You will complete fieldwork and data collection for this unit, and analyse and write up your results in class.

About this book

Objectives provide a **clear overview** of what you will learn in the section. Objectives increase in difficulty from ● to ◉

Clear and accessible diagrams **highlight key concepts and enable skills practice**.

ResultsPlus features combine real exam performance data with examiner insight to give **guidance on how to achieve better results**.

Activities provide extra **support** to ensure understanding and opportunities to **stretch** your knowledge.

Key terms are highlighted in the text, summarised at the end of each chapter, and are detailed in full in the glossary at the end of the book to enable you to **develop your geographical language**.

Chapter 7 Oceans on the edge

102

Objectives

- Describe the impacts of human activities on marine ecosystems such as mangrove forest.

◉ Explain how marine food webs work, and why they become damaged.

◉ Understand that climate change brings new and often unpredictable stresses to oceans.

ResultsPlus
Build Better Answers

Study the distribution of mangrove swamp shown in Figure 2. Identify the main populated areas where mangrove swamp grows. **(3 marks)**

■ **Basic answers** (0–1 marks)
Name only one or two areas, or are very imprecise, writing things like 'there is a lot on the right hand side of the map'.

● **Good answers** (2 marks)
Correctly identify Asia and Central America as named locations.

▲ **Excellent answers** (3 marks)
Also provide more specific details of Asian locations (naming the west coast of India and Indonesia) or mentioning north-west and south-east Africa.

How and why are some ecosystems threatened with destruction?

The term ecosystem describes a grouping of plants and animals that is linked with its local physical environment – through use of soil nutrients, for example. The oceans, covering two-thirds of our planet, are home to distinctive **marine ecosystem** communities composed of fish, aquatic plants and sea birds – as well as tiny but very important organisms such as krill and plankton.

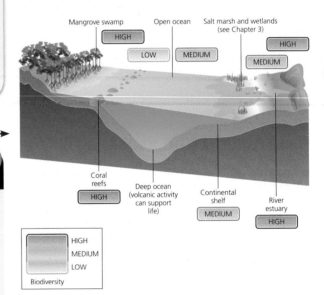

Mangrove swamp — HIGH
Open ocean — LOW / MEDIUM
Salt marsh and wetlands (see Chapter 3) — HIGH / MEDIUM
Coral reefs — HIGH
Deep ocean (volcanic activity can support life) — HIGH
Continental shelf — MEDIUM
River estuary — HIGH

HIGH
MEDIUM
LOW
Biodiversity

Figure 1: Biodiversity in the oceans' ecosystems

Coral reef ecosystems have an incredible level of **biodiversity** (their variety of plant and animal species). Although they cover less than 1% of the Earth's surface, they are home to 25% of all marine fish species. Biodiversity is also high in waters close to the edges of the world land masses – above the **continental shelf**. There, species enjoy shallow warm water, enriched with silt nutrients from river **estuaries**. Even in the deep ocean, where light cannot penetrate, unique ecosystems are found. Underwater volcanic activity can create densely populated sites of plant and animal life. Here, often at great depth, life has evolved that can survive truly extreme conditions of heat and pressure.

Activity 2

(a) Use the Internet to find another location that suffers from coastal flooding.

(b) Research how this area tries to predict or prevent coastal flooding.

Investigating how mangrove swamp is used

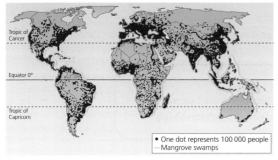

Figure 2: The distribution of mangrove swamp and populated areas

• One dot represents 100 000 people
— Mangrove swamps

Mangroves are areas of swampy forest, found in estuaries and along marine shorelines in around 120 tropical and subtropical countries. Many are densely populated, especially in coastal Asia (Figure 2). Mangrove plants have evolved to tolerate daily tidal flooding and high salinity. Long twisting roots anchor the trees against a constantly ebbing and flowing tide (Figure 3). The roots trap mud, making a **habitat** for lobster and prawn.

These species used to be fished in **sustainable** ways that preserved the **mangrove swamp**. However, prawns can be more intensively harvested from ponds dug in mud on cleared land. Global demand for prawns (worth £30 billion a year) has caused widespread removal of mangrove vegetation. Mangrove swamp naturally covered 200,000 km² of the Earth's surface but only half that area now remains, often because of this prawn **aquaculture**. Vietnam's Mekong Delta is a typical site where mangrove trees have been removed, leaving a flat muddy plain studded with blue plastic-lined ponds.

Mangrove removal creates problems of coastal erosion and loss of nursery grounds for fish. Crocodiles, snakes, tigers, deer, otters, dolphins and birds all lose an important habitat. Carbon dioxide stored over centuries in the rich mud beneath the swamp is released when the trees are removed.

Mangrove swamp is also nature's defence against tsunamis. The enormous ocean wave that struck the coast of south-east Asia in 2004 killed 230,000 people. Where they were still in existence, mangrove trees shielded lives and property. In Thailand, recent conversion of mangrove habitat into prawn farms and tourist resorts close to Phuket contributed significantly to the catastrophic losses experienced there.

Prawn aquaculture can also cause serious **pollution** problems. Antibiotics and pesticides used in the prawn pools frequently leak into the delicate ecosystem of neighbouring areas.

Figure 3: Mangrove forest

Quick notes (Mangrove swamp):

• Human use of this environment was originally sustainable.
• Because of commercial pressures, **unsustainable** exploitation now takes place.
• Vegetation removal and pollution damage local ecosystems and settlements.

Skills Builder 1

Study Figure 3

(a) Name one
helps the
environme

(b) Describe
mangrove
activities.

Case Study: Singapore's 'Have three or more' policy

Since the mid-1960s, the Singapore government has controlled the size of its population. First, it wanted to reduce the rate of population growth, because it was worried that the small island would soon become overpopulated. This policy was so successful that in the mid-1980s the government was forced to completely reverse the policy. The old family planning slogan of 'Stop at two' was replaced by 'Have three or more – if you can afford it'. Instead of penalising couples for having more than two children, they now introduced a whole new set of incentives to encourage them to do just that. These include:

• Tax rebates for the third child and subsequent children
• Cheap nurseries
• Preferential access to the best schools
• Spacious apartments.

Pregnant women are offered special counselling to discourage 'abortions of convenience' or sterilisation after the birth of one or two children.

Case study quick notes:
• Governments are able to control population numbers in a variety of ways.
• Control is usually achieved by a 'stick and carrot' approach.

examzone

A dedicated suite of revision resources for **complete exam success.**

We've broken down the six stages of revision to ensure that you are prepared every step of the way.

Zone in: How to get into the perfect 'zone' for your revision.

Planning zone: Tips and advice on how to effectively plan your revision.

Know zone: All the facts you need to know and exam-style practice at the end of every chapter.

Chapter overview: Outlines the key issue that the chapter examines. Keep this issue in mind as you work through the Know Zone pages.

Key terms: A matching exercise to ensure that you can **understand and apply important geographical terminology**.

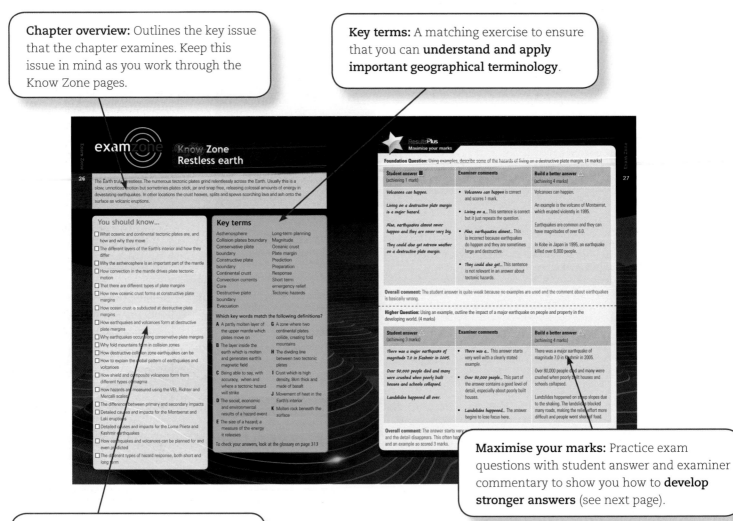

You should know: A check-yourself list of the concepts and facts that you should know before you sit the exam. Use this list to **identify your strengths and weaknesses** so you can plan your revision wisely.

Maximise your marks: Practice exam questions with student answer and examiner commentary to show you how to **develop stronger answers** (see next page).

Don't panic zone: Last-minute revision tips for just before the exam.

Exam zone: Some exam-style questions for you to try, an explanation of the assessment objectives, plus a chance to see what a real exam paper might look like.

Zone out: What do you do after your exam? This section contains information on how to get your results and answers to frequently asked questions on what to do next.

ResultsPlus

These features are based on the actual marks that students have achieved in past exams. They are combined with expert advice and guidance from examiners to show you **how to achieve better results**.

ResultsPlus
Exam Question Report

The residents of Keswick have conflicting views about the increasing numbers of tourists. Suggest reasons why some are for and some are against the increase in tourist numbers. (4 marks)

How students answered

Some students only stated reasons for or against the increase in tourist numbers and failed to provide reasons for these views.

■ 11% (0–1 marks)

Most students identified groups that would oppose and support the increase, but the reasons were more limited.

■ 52% (2 marks)

Many students could identify the economic advantages to some groups, and the social and economic costs to others. The best answers defined their groups carefully, such as recent retirees to the area opposing development, whereas young locals needing employment opportunities welcoming it.

■ 37% (3–4 marks)

Exam question report: These show previous exam questions with details about how well students answered them.

- Red shows the number of students who scored low marks (less than 35% of the total marks)
- Orange shows the number of students who did okay (scoring between 35% and 70% of the total marks)
- Green shows the number of students who did well (scoring over 70% of the total marks).

They explain how students could have achieved the top marks so that you can make sure that you answer these questions correctly in future.

ResultsPlus
Build Better Answers

Explain one natural cause of climate change in the past. (2 marks)

■ **Basic answers** (0 marks)
Are about the impact humans have had on climate, instead of natural causes.

● **Good answers** (1 mark)
Correctly describe a natural cause of climate change, such as volcanic activity, but do not explain why this causes climate to change.

▲ **Excellent answers** (2 marks)
Not only describe a natural cause, such as volcanic activity, but also explain that ash clouds ejected into the atmosphere can cut out the Sun's rays, causing temperatures to fall.

Build better answers These give you an opportunity to answer some exam-style questions. They contain tips for what a basic ■, good ● and excellent ▲ answer will contain.

Exam tip: These provide examiner advice and guidance to help improve your results.

Watch out! These warn you about common mistakes and misconceptions that examiners frequently see students make. Make sure that you don't repeat them! The ■, ● and ▲ symbols highlight the severity of the error.

Maximise your marks These are featured in the Know Zone pages at the end of each chapter. They include an exam-style question with a student answer, examiner comments and an improved answer so that you can see how to build a better response.

ResultsPlus
Maximise your marks

Foundation Question: Using examples, describe some of the hazards of living on a destructive plate margin. (4 marks)

Student answer ■ (achieving 1 mark)	Examiner comments	Build a better answer ▲ (achieving 4 marks)
Volcanoes can happen.	• *Volcanoes can happen* is correct and scores 1 mark.	Volcanoes can happen.
Living on a destructive plate margin is a major hazard.	• *Living on a...* This sentence is correct but it just repeats the question.	An example is the volcano of Montserrat, which erupted violently in 1995.
Also, earthquakes almost never happen and they are never very big.	• *Also, earthquakes almost...* This is incorrect because earthquakes do happen and they are sometimes large and destructive.	Earthquakes are common and they can have magnitudes of over 6.0.
They could also get extreme weather on a destructive plate margin.	• *They could also get...* This sentence is not relevant in an answer about tectonic hazards.	In Kobe in Japan in 1995, an earthquake killed over 6,000 people.

Overall comment: The student answer is quite weak because no examples are used and the comment about earthquakes is basically wrong.

Unit 1 Dynamic planet

Your course

This unit investigates geological processes, ecosystems, the atmosphere and climate, and the hydrological cycle. These topics are interlinked and show you how physical geography combines to create a 'life support system' for the planet. There are three sections:

Section A topics are compulsory and introduce you to the main areas of our planet: the geosphere, atmosphere, biosphere and hydrosphere. You will study **all** topics:

- Topic 1 (Chapter 1): Restless Earth
- Topic 2 (Chapter 2): Climate and change
- Topic 3 (Chapter 3): Battle for the biosphere
- Topic 4 (Chapter 4): Water world

Section B will cover how aspects of our planet work on a small scale and you will study **one** topic:

- Topic 5 (Chapter 5): Coastal change
- Topic 6 (Chapter 6): River processes

Section C will cover how aspects of our planet work on a large scale and you will study **one** topic:

- Topic 7 (Chapter 7): Oceans on the edge
- Topic 8 (Chapter 8): Extreme climates

Your assessment

- You will sit a 1-hour written exam worth a total of 50 marks

- There will be a variety of question types: short answer, graphical and extended answer, which you will practice throughout the chapters that you study. You will answer **all** the questions in Section A, **one** question from Section B and **one** question from Section C.

- **Section A** contains questions on the compulsory topics.

- **Section B** contains questions on the two small-scale topics.

- **Section C** contains questions on the two large-scale topics.

Remember to answer the questions for the topics that you have studied in class!

Study the photograph of a desert in Australia.

(a) Explain why droughts may become more frequent in some countries.

(b) Describe two impacts of drought on the economy of a country.

Chapter 1 Restless Earth

12

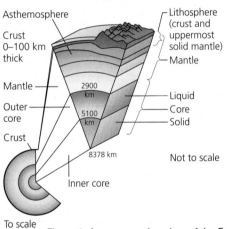

Figure 1: A cross-section view of the Earth

Skills Builder 1

Study Figure 1. Calculate the thickness of the mantle.

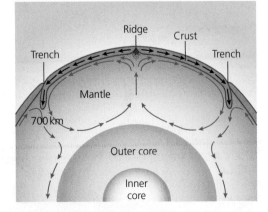

Figure 2: The convectional currents in the mantle

How and why do the Earth's tectonic plates move?

The structure of the Earth

The Earth is made up of a series of distinctive layers that are sometimes compared with onion skins (or even a Scotch egg). The solid crust is made up of relatively low density, allowing it to rest on the more flexible and fluid mantle. A real 'journey to the centre of the Earth' would not get very far because temperatures rise quite quickly, reaching more than 200°C at the boundary with the upper mantle.

As you can see in Figure 1, the Earth's crust is very thin compared to the diameter of the planet. The crust is made up of sections, called plates, ranging from the very small to the size of an entire continent. **Continental crust** is much thicker than **oceanic crust** but is made of material that is less dense than ocean crust. The Earth's surface and its interior are both constantly moving, but you will never notice this movement – unless you are unlucky enough to be caught up in an earthquake.

Beneath the Earth's crust lies the mantle, the upper part of which is solid, just like the crust. The crust and this solid top layer of the mantle are together known as the 'lithosphere'. Below the lithosphere the sticky, viscous dense rock of the mantle slowly moves in great convectional currents (see Figure 2). This process is bit like a pot of thick soup that is heated to boiling. The heated soup rises to the surface, spreads and begins to cool, and then sinks back to the bottom of the pot, where it reheats and rises again. In the upper part of the mantle – called the **asthenosphere** – the flows affect the lithosphere above, causing earthquakes and creating volcanoes.

Because of the heat and pressure that builds up beneath the surface, the crust is constantly being stressed, which breaks it up. These large-scale processes within the Earth's crust are known as plate tectonics. The processes were suspected for many years but they were only finally accepted as geological fact about fifty years ago. When two plates are being pulled away from each other, deep cracks are opened through the crust. This allows magma to rise to the surface and then, when it cools, it forms new crust in the shape of a ridge (Figure 2).

Plate size can vary greatly, from a few hundred to thousands of kilometres across. The Pacific and Antarctic plates are two of the largest.

How do these massive slabs of solid rock remain on the Earth's surface? Why don't they sink into the mantle? The answer lies in the composition of the rocks. Continental crust is composed of granitic rocks which are made up of relatively low density minerals such as quartz and feldspar (Figure 3). By contrast, oceanic crust is composed of basaltic rocks, such as gabbro, which are much denser (Figure 4). When a tectonic plate composed mostly of oceanic material meets a plate composed mostly of continental material, it is the denser oceanic plate that is forced downwards. The variations in plate thickness are nature's way of partly compensating for the imbalance in density of the two types of crust. Because continental rocks are less dense, the crust under the continents is much thicker (as much as 100 km) whereas the crust under the oceans is generally only about 7 km thick.

The **convection currents** in the mantle are themselves driven by the heat of the **core**. That heat is partly created by the pressure of overlying material but also by the radioactivity of the core material itself. As long as there is a temperature difference with depth, there will be a cycle of rising and sinking material. (A lava lamp is a perfect illustration of convection.)

Temperatures in the core are probably much the same as on the surface of the Sun. The other basic facts about the core are:

- It is mostly made up of iron and nickel

- Just over half the diameter of the Earth

- One-sixth of the volume of the Earth

- One-third of the Earth's mass

- The outer core is liquid

- The inner core is solid

- The currents in the outer core generate the Earth's magnetic field.

The four types of boundary between tectonic plates

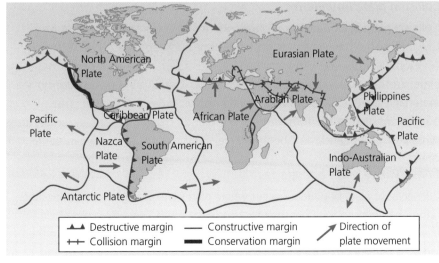

Figure 5: The Earth's tectonic plates and their boundaries

Figure 3: Granite is made up of large crystals. The magma which formed these granitic rocks cooled slowly, which explains the very large crystals.

Figure 4: Basalt is fine-grained with no visible crystals. It ranges from black to dark grey in colour. Most basalts occur in lava flows and sheets.

Activity 1

Describe the differences between the rocks that dominate continents and those that dominate oceanic crust.

Activity 2

1. State two differences between oceanic and continental crust.

2. Explain how tectonic plates move.

Skills Builder 2

Study Figure 5.

1. Describe the distribution of plate boundaries.

2. Identify the type of boundary or boundaries found between: (a) The Nazca plate and the South American plate (b) The Indo-Australian plate and the Eurasian plate (c) The Eurasian plate and the North American plate.

There are four types of boundaries between tectonic plates – **destructive**, **constructive**, **collision** and **conservative**. Each of them is associated with characteristic tectonic events and landforms.

Destructive plate boundaries

Destructive plate boundaries are found where two plates are moving together and oceanic plate material is destroyed (Figure 6). The boundary between the Nazca plate and the South American plate would be an example.

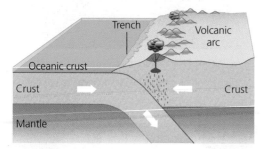

Figure 6: A destructive plate boundary

Destructive plate boundaries are associated with frequent earthquakes and volcanoes. The collision of the two plates buckles the leading edge of the continental plate forming fold mountains – and causing earthquakes. The denser basaltic oceanic plate is dragged downwards below the lighter granitic continental plate – a process known as 'subduction'. This creates an ocean trench and the melting of this material creates molten material (magma) as it is dragged deeper and deeper into the upper mantle. This magma rises through weaknesses in the overlying continental crustal material, some of which is inevitably melted by the rising magma, forming volcanoes on the surface. This is what has happened at the boundary between the Nazca and South American plates.

Constructive plate boundaries

Constructive plate boundaries (Figure 7) are found where new basaltic material rises to the surface, forcing plates apart (for example the mid-Atlantic boundary between the Eurasian and North American plates).

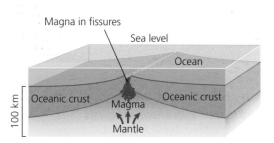

Figure 7: A constructive plate boundary

Rising convection currents in the mantle cool and spread outwards as they near the surface. This pulls the crust apart and creates fissures and faults through which molten magma can reach the surface. The vast majority of this creation of new crust takes place in the oceans, forming a ridge and chains of submarine volcanoes. But sometimes, as in Iceland, these volcanoes reach the surface to form islands. It is believed that constructive plate boundaries evolve from rising magma splitting apart continental crust and creating 'new' oceans. This process is thought to be happening in the modern Red Sea and the African Rift valley.

Collision plate boundaries

Collision plate boundaries are found where two continental plates move towards each other. Neither is destroyed but buckling takes place (Figure 8). An example would be between the Indo-Australian plate and the Eurasian plate. Earthquakes are very common (e.g. Pakistan 2005) but because no material is being subducted and melted no volcanoes are formed. The buckling has led to the formation of the world's biggest mountain range, the Himalayan chain, and the Tibetan plateau to the north of it.

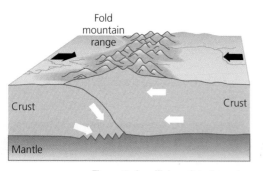

Figure 8: A collision plate boundary

Conservative plate boundaries

Conservative plate boundaries occur where plates are 'sliding' past one another, rather than moving together or away from each other. There is no plate being created or destroyed – because there is no magma and no subduction. But the sliding is not smooth – there is friction between the two plates, and extreme stresses build up in the crustal rocks.

When this is eventually released the result is an earthquake. The best example of this is along the west coast of North America, where a series of faults mark the boundary between the North American plate and the Pacific plate (Figure 9). The size of the earthquakes relates to the frequency of movement, so if there is a long period with no earthquake activity the pressure builds up and the eventual movement will be much greater, generating much more energy. On this boundary the plates are actually moving in the same direction but at different rates. Some of the faults are visible on the surface but many are not – and the results can be devastating when there is a sudden movement, as there was along the previously unknown Northridge fault in 1994.

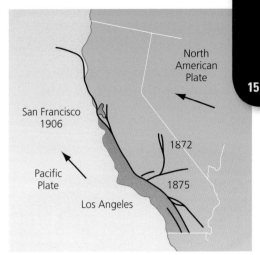

Figure 9: A conservative plate boundary

Earthquakes and volcanoes

Figures 10 and 11 show the global distribution of earthquakes and volcanoes. By comparing these with Figure 5 it is obvious that there is a very close relationship between **plate margins** and tectonic activity. However there are some important exceptions.

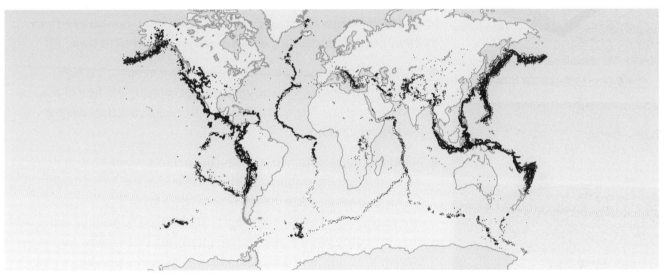

Figure 10: Global distribution of earthquakes (as shown by the red dots)

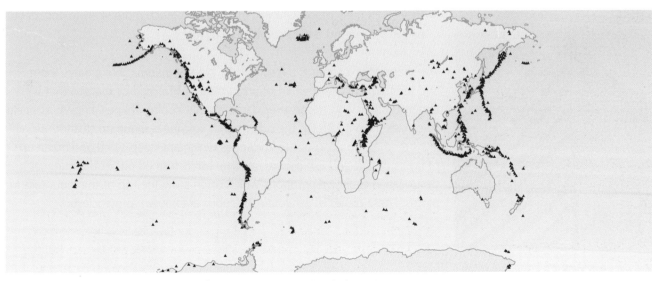

Figure 11: Global distribution of volcanoes (as shown by the red triangles)

Activity 5

Study Figure 9.

Describe the movements of the North American and Pacific plates.

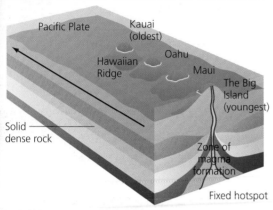

Figure 12: The formation of the Hawaiian chain of islands, according to the hotspot theory

Figure 13: The shield volcano, Mauna Loa, Hawaii – the largest volcano in the world

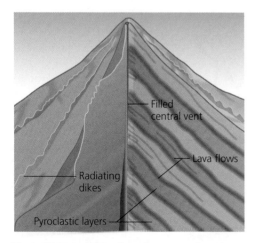

Figure 14: A cut-away view of a composite volcano

The Hawaiian Islands, which are entirely of volcanic origin, have formed in the middle of the Pacific Ocean – more than 3,200 km from the nearest plate boundary. This exception to the usual pattern has been explained by the 'hotspot theory' – the theory that there are fixed spots in the mantle where magma rises to the surface. As the crust is moved over these fixed spots, volcanoes are created, eventually forming a chain of volcanoes, as shown in Figure 12. There are many other hotspots on the Earth, some of which have been extremely destructive in the past and pose threats for the planet in the future (probably a very long time in the future). The best known example of this is the so-called 'supervolcano' under Yellowstone National Park in the USA.

There are many different types of volcano. Experts recognise 539 volcanoes that have erupted in recorded history. These are classified as 'active'. There are a further 529 volcanoes that have not erupted in historic times, but which exhibit clear evidence of the capacity to do so again. These are classified as 'dormant'. The rest are classified as 'extinct'.

In general, the shape and structure of volcanoes – and the explosive threat that they pose – is related to the type of magma that created them:

● Basaltic magma is usually found on constructive margins. It is high temperature, very low in silica, with low gas content. This type of magma produces fluid lava flow with relatively little explosive activity when it reaches the surface.

● Andesitic magma is formed at destructive margins where continental rocks are melted by rising magma. (It gets its name from the Andes mountains.) It is lower in temperature, has more silica and a lot of dissolved gases. As a result, this magma is much less fluid than basaltic magma, and is more likely to explode when it reaches the surface. (It's much harder for the gas bubbles to escape from a viscous magma than from a less viscous one.)

● Granitic magma is relatively low temperature, with a high silica content. These magmas are usually so viscous and 'sticky' that they get stuck before they reach the surface. Granitic magmas are much more likely to cool deep below the ground and become igneous rocks.

The shape of volcanoes is a result of the type of magma that creates them and the frequency of their eruptions. Volcanoes formed by magma that is largely basaltic tend to form very large, gently sloping shapes, known as shield volcanoes (Figure 13). On the other hand, volcanoes made up of andesitic lavas tend to form composite volcanoes which are steep-sided and made up of layers of lava and ash (Figure 14). The ash is formed out of the material destroyed in explosive eruptions which often blow the top of the mountain to pieces. These pieces of ash and rock debris are known as pyroclastics.

Mauna Loa, Hawaii, a shield volcano

Mauna Loa is the largest volcano on the planet. Its long sides descend to the sea floor to add an additional 5,000 metres to its above sea-level height of 4,170 metres. This makes it well over 9,000 metres in height in total – more than Mount Everest. This enormous volcano forms half of the Big Island in the Hawaiian Islands (and by itself its volume is almost as much as all the other Hawaiian Islands combined). Mauna Loa is shaped like a shield, with very gentle slopes (Figure 13), because its basaltic lava is extremely fluid (it has low viscosity). This lava means that eruptions are rarely violent – the most common form being lava 'fountains' feeding lava flows. Typically, at the start of an eruption, a long rift up to several kilometres long opens towards the summit of the volcano with lava fountains occurring along its length in a 'curtain of fire'. After a few days, activity normally becomes concentrated at one or two fountains only.

Mauna Loa is one of the Earth's most active volcanoes, having erupted 33 times since its first recorded eruption in 1843. Its most recent eruption was in 1984. Mauna Loa is certain to erupt again, and the volcano is constantly monitored for warning signs.

Mount Pinatubo, Philippines, a composite volcano

In the Philippines, volcanoes provide one of several threats to the population. The volcanoes here are almost all composite because the magma is made up of sticky andesitic lava. This frequently blocks the vent, leading to a build up of pressure. Eventually, the top of the mountain may be blown off in a large explosion.

In 1991 one of the most dangerous Philippine volcanoes erupted, for the first time in 500 years. This eruption produced at least 10 km³ of erupted material – ten times the size of the Mount St Helens eruption of 1980 – putting it at level 6 on the 'Volcanic Explosivity Index' (Figure 15). The eruption had been predicted, and thousands of people had been evacuated from the surrounding areas. Many lives were therefore saved, but the region was damaged by flows of lava and ash.

Large eruptions have an effect on the climate because huge quantities of dust and gases are ejected into the atmosphere. Pinatobu's eruption led to a drop in global temperatures of about 0.5 °C.

Activity 6

Study Figure 13.

(a) Identify the main characteristics of shield volcanoes.

(b) Explain how the type of magma affects the shape that you have described.

ResultsPlus
Build Better Answers

Describe the main features of composite volcanoes. (2 marks)

■ **Basic answers** (0 marks)
Confuse composite volcanoes with shield volcanoes.

● **Good answers** (1 mark)
Identify one feature, such as steep sides.

▲ **Excellent answers** (2 marks)
Clearly identify two or more features, including their appearance and their structure.

		VEI	Volume of erupted tephra	Examples
Small	Non-explosive	0		
		1	0.00001 km³	
		2	0.001 km³	
Moderate		3	0.01 km³	
		4	0.1 km³	
Large	Very large	5	1 km³	Mount St. Helens, May 18 1980 (1 km³)
		6	10 km³	Pinatubo, 1991 (10 km³)
		7	100 km³	Tambora, 1815 (100 km³) Long Valley Caldera, 760 000 years ago (600 km³)
		8		Yellowstone Caldera, 600 000 years ago (1000 km³)

Figure 15: The Volcanic Explosivity Index (VEI). The erupted volumes of the examples are estimates.

Objectives

- Be able to identify the various impacts of tectonic hazards on people and property.

- Describe differences in the impact of tectonic hazards in developed and developing countries.

- Understand the role of various agencies in relieving the social and economic impact of various hazards.

Activity 7

1. Define the term 'vulnerability'.

2. Explain why some areas are more vulnerable to hazards than others.

What are the effects and management issues resulting from tectonic hazards?

The effects and impacts of volcanic hazards

Earthquakes and volcanoes are good examples of **tectonic hazards**. Hazards pose a threat to us, but not all hazardous events are disasters. That depends on a number of other factors:

- The type of hazard

- The place's vulnerability to hazards

- The ability or 'capacity' to cope and recover from a hazardous event.

Not all natural hazards are equally devastating. The size and the 'type' of event is crucial. An eruption of Mauna Loa, for example, is seldom threatening to life because the volcano is not explosive. This is not the case for Mount Pinatubo.

Not all of the Earth's inhabitants are at equal risk from natural hazards. Despite the exceptions referred to on page 16, unless people live close to a plate boundary, it is very rare for them or their property to be damaged by earthquakes – they are not 'vulnerable' to earthquakes.

The idea of vulnerability can also be applied to volcanoes. If you do not live close to an active volcano, then you are not likely to be threatened by lava flows. However, you may very well be affected by clouds of volcanic ash, which can significantly alter the climate of places many miles, even continents, away from their point of origin.

'Capacity' refers to the ability of a community to absorb, and ultimately recover from, the effects of a natural hazard. The Japanese increase their capacity to cope with the effects of earthquakes by regularly practising how to respond. In theory, this means that they will have a better chance of coping with a large earthquake if one happens. Compare this capacity to cope with the situation in a sprawling slum in the developing world, where dwellings have been hastily constructed from poor-quality materials, and where there is neither the time nor money to commit to a large-scale community training programme, let alone improve the quality of the buildings.

It is important to distinguish between the primary impacts and the secondary impacts of disasters. Primary impacts are those that take place at the time of the event itself, and are directly caused by it. Secondary impacts are those that follow the event, and are indirectly caused by it. The social and economic impact of volcanic eruptions can be very considerable. In extreme cases it might mean the entire **evacuation** of the area.

What are the effects and management issues resulting from tectonic hazards?

19

The Kashmir earthquake, 2005

This earthquake occurred in one of the most remote regions in the world – one that is difficult to reach even in normal conditions. This poor, largely agricultural, mountainous region is disputed between Pakistan and India, but is currently administered by Pakistan. Mountainous areas are fragile environments and earthquakes are likely to cause very considerable damage because of landslides and falling rock. The earthquake happened on a Saturday morning – a normal school day – and many of the dead and injured were children. What made it worse was that because it was during Ramadan (the Muslim period of fasting during daytime) others were sleeping after getting up early to have a pre-dawn meal.

A considerable relief effort by the Pakistani government and international relief agencies was able to prevent the secondary disaster that threatened to overwhelm the region. With winter drawing in and conditions becoming more and more difficult, it became a race against time. But, using helicopters, the agencies managed to fly blankets, tents, basic provisions and medical supplies into the area.

The Loma Prieta earthquake, California, 1989

The Loma Prieta earthquake (named after a local mountain) is sometimes known as the 'San Francisco 'quake of '89'. It was caused by a slip of several metres on the San Andreas fault and the other faults that mark the boundary between the Pacific plate and the North American plate. In common with the rest of the Californian coastal area, earthquakes are expected in this area – but **prediction** of time and **magnitude** is not possible. The earthquake took place during the evening rush hour, and offices were mostly empty or emptying. Purely by chance it was an exceptionally quiet rush hour because the two local baseball teams were competing for the World Series in Candlestick Park in San Francisco. Many people were either at the match or had made an early trip home to watch this local derby on television. The earthquake caused more property damage in the Bay Area than its strength and location suggested likely. This was because the clay soils of that area shook so much that they liquefied, causing properties to sink, gas mains to burst as they broke and fires to break out. (This was a smaller-scale reminder of the great San Francisco earthquake of 1906, when it was the fire after the earthquake that caused most of the destruction of the city.)

Figure 17: The collapsed Cypress Street Overpass – the scene of 41 deaths in the Loma Prieta earthquake

**Quick notes
(The Kashmir earthquake):**
- Date: Saturday 8 October 2005
- Magnitude: 7.6 on the Richter Scale
- Epicentre: Muzaffarabad, the capital of Pakistani-administered Kashmir
- Death toll: 75,000
- Injured: 75,000
- Homeless: 2.8 million
- Property cost: $440 million.

Figure 16: In Patikka, 17 km from the epicentre, villagers have to cross this ruined bridge to get to the Red Cross emergency unit, where they can pick up tents and blankets.

**Quick notes
(The Loma Prieta earthquake):**
- Date: Tuesday 17 October 1989, at 17.04
- Magnitude: 6.9 on the Richter Scale
- Epicentre: in a mountainous part of Santa Cruz County, 90 km south-east of San Francisco
- Death toll: 63
- Injured: 3,757
- Homeless: 12,000
- Property cost: $10 billion.

Montserrat, 1995: fleeing the volcano

Montserrat is a tiny Caribbean island of about 100 km² that is a very small leftover of the British Empire. The Chances Peak volcano in the south of the island was dormant – it had not erupted since the seventeenth century. This was because its very viscous lava had blocked the vent but, as with any dormant volcano, from time to time there can be violent eruptions. And between 1995 and 1997 it erupted huge quantities of lava, ash and extremely dangerous pyroclastic flows – high-speed avalanches of hot gases, ash and rock fragments which moved at speeds of 100–150 km/h.

Before the 1995 eruption, Montserrat had a population of about 10,500. Now it has about 4,000. Throughout the twentieth century, Montserratians migrated from their island because of a lack of employment opportunities. Recently, however, since the eruption of Chances Peak approximately two-thirds of the island's population have left the island for very different reasons. They have been evacuated by the government for their own protection. Meanwhile, three-quarters of the remaining population in the south have had to relocate to the north of the island. Most of those who left the island altogether came to the UK and others went to neighbouring islands such as Antigua. Such drastic mass movements have had, and continue to have serious social, economic, political and cultural effects upon all those involved.

The management of tectonic hazards

The management of volcanic and earthquake hazards is expensive. But the costs of a disaster can be much worse. The pattern of volcanic and earthquake risk is fairly well known and, apart from the occasional totally unexpected events, communities that are located in tectonically active areas can develop management strategies to cope.

Preparedness – being ready

Preparation means that governments, communities and individuals are ready to respond rapidly when disaster strikes and cope with the situation effectively. These measures include the formulation of emergency plans, the development of warning systems and the training of personnel. The measures may include evacuation plans for areas that may be at risk from a disaster and training for search and rescue teams. Preparedness therefore encompasses those measures taken before a disaster event which are aimed at minimising loss of life, disruption of critical services, and damage when the disaster occurs.

Mitigation – reducing the impact

Mitigation measures are taken to reduce both the effect of the hazard and the vulnerability to it, in order to reduce the scale of a disaster. They can be focused on the hazard itself or on the elements exposed to the threat. Hazard-specific measures include relocating people away from the hazard-prone areas and strengthening structures – and using hazard-resistant design to reduce damage when a hazard occurs.

Quick notes (Montserrat):

- Eruption of a 'dormant' volcano in 1995
- Very viscous lava that gets 'stuck', causing violent eruptions
- Tiny island, so there was nowhere to 'hide'
- Two-thirds of population have emigrated
- Devastating social and economic impact.

Activity 8

Read the case studies on Montserrat and Laki.

(a) Describe the differences in the impact of these eruptions on the population.

(b) Explain why some volcanic eruptions cause much more loss of life than others.

Iceland's Laki eruption, 1783–84, and beyond

The Laki eruption was one of the most devastating eruptions in human history. Iceland lies on the mid-Atlantic ridge and its volcanoes pose a constant threat, although very few of them produce violent eruptions because the magma is usually basaltic and relatively free-flowing. In 1783–84, a major eruption from the Laki fissure poured out an estimated 14 km^3 of basaltic lava and clouds of poisonous compounds. The volcano is located in a remote part of Iceland and no one was killed by the event itself. However, the secondary effects were devastating because the poisonous cloud killed over half of Iceland's livestock population, leading to a famine which killed approximately a quarter of the population. At that time, there was no system of international relief in Iceland. The dust cloud (which was much larger than that caused by the eruption of Mount Pinatubo) is thought to have reduced temperatures in Europe for several years, causing poor summers, reduced harvests and, as a result, social unrest. Some historians believe that it helped trigger the French Revolution in 1789.

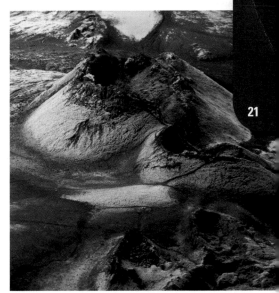

Figure 18: Laki volcano cones after the 1783 eruption

Today, despite its banking collapse in 2008, Iceland is now one of the richest countries on the planet and the impact of volcanic eruptions is very different. The recent eruptions of Mount Hekla illustrate this. It is very active, erupting most recently in 1970, 1980, 1981, 1991 and 2000. The last eruption started on 26 February 2000 and lasted for 12 days, producing lava that covers approximately 18 km^2. Mount Hekla is very carefully monitored by the Icelandic authorities, using seismometers to record any movement of magma. They also measure the gas that the volcano is constantly leaking, watching carefully for any sudden changes. The area is very sparsely populated but in the last eruption, 200 families were evacuated and the island ring-road was closed. A future eruption of Hekla might interfere with flights to and from Iceland which would impact on its tourist industry.

Some Icelandic volcanoes are under its ice-sheets and these pose an interesting secondary threat. During their under-ice eruptions they melt vast quantities of water that can be released suddenly in massive floods known as 'jokulhaups'.

Some of these measures are expensive and beyond the budgets of poor communities, unless they can get help from international aid agencies. This explains the difference in impacts of disasters in different parts of the world. A community's ability to cope with a disaster depends on how well prepared they were and how much they had been able to spend on mitigation. But lack of preparedness is not just a feature of poor countries. Democratically elected leaders frequently make promises to potential voters that they will reduce taxes and cut public spending – almost always a popular message. In San Francisco after the Loma Prieta earthquake there was a call for greater spending on the emergency services but, with memories fading, it isn't obvious that the city is ready for the next earthquake.

Quick notes (Laki eruption):

- Huge eruption in a remote area
- Population unable to escape
- No aid possible at that time (1783–84)
- No direct deaths but vegetation died, so animals died – and then people died
- Impact on the climate of Europe.

Activity 9

In relation to the management of tectonic hazards, define and illustrate the terms 'preparation' and 'mitigation'.

21

How ready is San Francisco?

San Francisco is bound to be struck by the 'Big One' – an earthquake many times more powerful than the Loma Prieta. Many experts say that the city is poorly prepared.

Although the city has firefighters and police who train regularly everything is not 'ready'. Experts say many critical structures in the city may also not have been adequately upgraded to withstand earthquakes, including:

● The city's biggest hospital, San Francisco General, where many of the injured would be taken, as well as many of the schools.

● The Bay Bridge Connecting Oakland and San Francisco damaged in 1989 and very vulnerable to a larger earthquake.

● The famous Golden Gate Bridge – if both bridges are damaged San Francisco becomes very isolated with many of its firefighters and hospital staff living on the wrong side of the bridges.

The city has evacuation plans and websites that aim to inform the citizens about the correct procedures. The Hurricane Katrina disaster forced many cities to take a long look at their plans in the light of what happened in New Orleans, which was widely though to be poorly organised, with the poor and disadvantaged being neglected.

As a result, the Red Cross, fire department and other agencies have trained many citizens in emergency **response**. They urge residents to prepare an emergency kit that will last them three days with food, water and other supplies, while bilingual students offer training to older foreign residents who may not have the language skills to understand the procedures.

San Francisco is famous for its old Victorian buildings, which add to the city's charm. But many are at risk in a major earthquake. High-rise buildings put up in the 1950s and 1960s are also at risk. Earthquake scientists say that many buildings are not being made safe because owners have little incentive to do so, and in a city with a desperate shortage of housing the last thing the authorities want is housing to become even less affordable.

Scientists say there is nearly a 'two in three' chance that another major earthquake will strike San Francisco by 2032, and that the only way to get through disaster is to be ready for it. Although popular attention is fixed on the San Andreas fault San Francisco is also threatened by the Hayward fault.

As Figure 19 shows the whole region is criss-crossed with faults. Residents can log-on to the seismic net online to check local earthquake activity. The San Andreas fault runs close to the coast across the bay, while the Hayward fault runs in the same south-east/north-west orientation but through Berkeley and Santa Rosa.

US Geological Survey Senior Seismologist Tom Brocher said: 'The Hayward Fault is a tectonic time bomb. (It's) the single most dangerous fault in the entire Bay Area, because it is ready to pop and because nearly 2 million people live directly on top of it.'

23

Figure 19: The San Francisco seismic net online

Figure 20: Earthquake-resistant design for tall buildings

Building design

One of the main ways of mitigating the impact of an earthquake is to improve building design. Engineers have developed ways to build earthquake-resistant structures – not just houses, but office blocks and bridges too. The methods range from extremely simple to complex. For small- to medium-sized buildings, the simpler reinforcement techniques include bolting buildings to their foundations and providing support walls, called 'shear walls'. Shear walls are made from concrete that has steel rods embedded in it to help strengthen the structure and help resist rocking forces. The centre of a building can be constructed to form what is called a 'shear core'. Walls may also be reinforced and supported by adding diagonal steel beams in a technique called cross-bracing (Figure 20).

Medium-sized buildings are constructed using devices that act like shock absorbers between the building and its foundation. These devices – known as 'base isolators' – are usually bearings made of alternating layers of steel and an elastic material, such as synthetic rubber. Base isolators absorb some of the sideways motion that would otherwise damage a building.

Skyscrapers need special construction to make them earthquake-resistant. The foundations need to be very deep. They need a reinforced framework with stronger joints than those used in an ordinary skyscraper. Such a framework makes the skyscraper strong enough and yet flexible enough to absorb the energy of an earthquake – flexibility is the key.

All heavy appliances, furniture and other structures should be fastened down to prevent them from falling. Gas and water lines must be specially reinforced with flexible joints to prevent breaking. Fire, fuelled by broken gas pipes, is a real risk after earthquakes.

In the developing world, all of these methods are used, especially for important buildings in the Central Business Districts (CBDs) and in the richer areas of cities. Elsewhere in the developing world, however, money is rarely available for expensive engineering solutions, and simpler ways of strengthening existing buildings and new buildings are used, as the tables show.

Strengthening of existing buildings

Problem	Strengthening methods used
Heavy roof	Removal of mud overlay on top.
Poor timber frame connections	Adding diagonal bracing to the frame, using timber (cheaper) or steel (may not be locally available).
Thick walls without 'through-stones' to bind walls together	Installation of through-stones. This needs training of local artisans (new skills) and must be performed very carefully.
Separation joint at wall corners	Strengthening of wall corners, using wire mesh and cement overlay (though welded wire mesh not always available in rural areas).
Walls moving outwards	Installation of a ring beam (band of concrete) at the roof level.
Shaking of stones/bricks from exterior walls	Pointing of exterior walls with cement mortar.

Strengthening of new buildings

Part vulnerable to earthquakes	Strengthening provisions used
Walls	Use cement/sand mortar and shaped stones (including through-stones) in construction. Construct concrete ring beam at the roof level.
Roof	Limit the thickness of mud overlay to 200 mm.
Timber frame	Install 'knee-braces' to reinforce the vertical/horizontal connections.

Tectonic disasters stimulate both **short-term relief** and **long-term planning** to improve preparedness and mitigation for any future events. Immediate aid is needed to keep people alive, especially if, as in Kashmir in 2005, there are few local resources to fall back on. The aid needed urgently usually includes:

- Tents
- Blankets
- Garbage bags
- Antibiotics
- Baby food
- Milk
- Canned food
- Generators
- Tranquilisers
- Jerry cans
- Prefab toilets
- Disinfectants
- Mobile field kitchens.

Volunteers are also required, especially from the professional rescue services and health services. Disaster relief is by no means always smooth. The Turkish authorities, for example, were widely criticised after the 1999 Izmit earthquake – not just for an absence of preparedness and mitigation in a country that is far from the poorest, but also for bureaucratic blunders after the shock. Relief efforts were hindered by a serious underestimation of the size of the earthquake (because of poor quality and inaccurate sensors) which led to the Red Crescent sending far too little aid. The government also seemed slow in responding to help from the aid agencies. Generous aid from the United States and Europe did pick up the slack from NGOs. But as the disaster ran its course, the government mishandled this aid as well. Foreign-aid workers were required to pay customs duties on equipment and wait days before entering the disaster area. Failure to provide maps, interpreters and information on Islamic burial practice also impeded foreign workers' efforts.

25

ResultsPlus
Build Better Answers

Describe two ways in which buildings in developing countries can be made more resistant to earthquakes. (2 marks)

■ **Basic answers** (0 marks)
Confuse developing with developed countries or discuss volcanic eruptions and not earthquakes.

● **Good answers** (1 marks)
Offer one method, such as bracing the frame..

▲ **Excellent answers** (2 marks)
Add a second method, such as installing a ring beam to prevent the walls moving outwards.

Decision-making skills

Imagine you were involved in assembling aid packs for after an earthquake.

Which of the thirteen bullet point items would be your essential choice for emergency aid? You may choose up to six. Provide clear instructions for local aid workers as to how your six could best be used.

Decision-making skills

Why are there sometimes conflicts between the following people after a major tectonic disaster has occurred:

- Local people
- Specialist rescue services
- Local leaders
- Local charities
- International charities
- Central government
- Armed forces

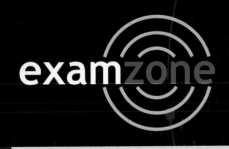

examzone

Know Zone
Restless Earth

The Earth truly is restless. The numerous tectonic plates grind relentlessly across the Earth. Usually this is a slow, unnoticed motion but sometimes plates stick, jar and snap free, releasing colossal amounts of energy in devastating earthquakes. In other locations the crust heaves, splits and spews scorching lava and ash onto the surface as volcanic eruptions.

You should know...

- [ ] What oceanic and continental tectonic plates are, and how and why they move
- [ ] The different layers of the Earth's interior and how they differ
- [ ] Why the asthenosphere is an important part of the mantle
- [ ] How convection in the mantle drives plate tectonic motion
- [ ] That there are different types of plate margins
- [ ] How new oceanic crust forms at constructive plate margins
- [ ] How ocean crust is subducted at destructive plate margins
- [ ] How earthquakes and volcanoes form at destructive plate margins
- [ ] Why earthquakes occur along conservative plate margins
- [ ] Why fold mountains form in collision zones
- [ ] How destructive collision zone earthquakes can be
- [ ] How to explain the global pattern of earthquakes and volcanoes
- [ ] How shield and composite volcanoes form from different types of magma
- [ ] How hazards are measured using the VEI, Richter and Mercalli scales
- [ ] The difference between primary and secondary impacts
- [ ] Detailed causes and impacts for the Montserrat and Laki eruptions
- [ ] Detailed causes and impacts for the Loma Prieta and Kashmir earthquakes
- [ ] How earthquakes and volcanoes can be planned for and even predicted
- [ ] The different types of hazard response, both short and long term

Key terms

Asthenosphere
Collision plate boundary
Conservative plate boundary
Constructive plate boundary
Continental crust
Convection currents
Core
Destructive plate boundary
Evacuation

Long-term planning
Magnitude
Oceanic crust
Plate margin
Prediction
Preparation
Response
Short-term emergency relief
Tectonic hazards

Which key terms match the following definitions?

A Where two tectonic plates slide past each other

B Circulating movements of magma in the mantle caused by heat from the core

C The upper part of the Earth's mantle, where the rocks are more fluid

D Tectonic plate margins where oceanic plate is subducted

E The part of the crust dominated by denser basaltic rocks

F The central part of the Earth, consisting of a solid inner core and a more fluid outer core, and mostly composed of iron and nickel.

G Tectonic plate margin where rising magma adds new material to the diverging plates.

H The part of the crust dominated by less dense granitic rocks

To check your answers, look at the glossary on page 313

ResultsPlus
Maximise your marks

Foundation Question: Using examples, describe some of the hazards of living on a destructive plate margin. (4 marks)

Student answer ■ (achieving 1 mark)	Examiner comments	Build a better answer △ (achieving 4 marks)
Volcanoes can happen. Living on a destructive plate margin is a major hazard. Also, earthquakes almost never happen and they are never very big. They could also get extreme weather on a destructive plate margin.	• *Volcanoes can happen* is correct and scores 1 mark. • *Living on a...* This sentence is correct but it just repeats the question. • *Also, earthquakes almost...* This is incorrect because earthquakes do happen and they are sometimes large and destructive. • *They could also get...* This sentence is not relevant in an answer about tectonic hazards.	Volcanoes can happen. An example is the volcano of Montserrat, which erupted violently in 1995. Earthquakes are common and they can have magnitudes of over 6.0. In Kobe in Japan in 1995, an earthquake killed over 6,000 people.

Overall comment: The student answer is quite weak because no examples are used and the comment about earthquakes is basically wrong.

- -

Higher Question: Using an example, outline the impact of a major earthquake on people and property in the developing world. (4 marks)

Student answer △ (achieving 3 marks)	Examiner comments	Build a better answer △ (achieving 4 marks)
There was a major earthquake of magnitude 7.0 in Kashmir in 2005. Over 80,000 people died and many were crushed when poorly built houses and schools collapsed. Landslides happened all over.	• *There was a...* This answer starts very well with a clearly stated example, which scores 1 mark. • *Over 80,000 people...* This part of the answer contains a good level of detail, especially about poorly built houses, so scores another mark. • *Landslides happened...* The answer begins to lose focus here but scores 1 mark.	There was a major earthquake of magnitude 7.0 in Kashmir in 2005. Over 80,000 people died and many were crushed when poorly built houses and schools collapsed. Landslides happened on steep slopes due to the shaking. The landslides blocked many roads, making the relief effort more difficult and people went short of food.

Overall comment: The answer starts very well, with some good, accurate detail. However, it becomes much more general and the detail disappears. This often happens when students have not revised in depth. The answer had some good impacts and an example so scored 3 marks.

Chapter 2 Climate and change

Objectives

- Recognise that the average temperature of the Earth has changed a great deal over time.

- Explain the natural causes of past climate changes.

- Understand that climate change challenges existed for people and ecosystems living in the past.

Skills Builder 1

Study Figure 1.

(a) State how many times temperatures have risen to be warmer than average over the past 450,000 years.

(b) When was the temperature lowest?

(c) Describe the changes during the past 200,000 years.

How and why has climate changed in the past?

Natural climate change over time

The average temperature of the Earth's atmosphere has changed a great deal in the past. For example, 100 million years ago – at the time of the dinosaurs – conditions were much hotter than they are today. There have also been many cold phases, called **ice ages**. Our last major cold period, the Pleistocene, started 1.8 million years ago and ended just 10,000 years ago. Since then, conditions have been warmer. This most recent 10,000 years is called the Holocene. The Pleistocene and the Holocene are part of the **Quaternary Period** of Earth history.

We know – from a range of evidence – that temperatures were colder during much of the Pleistocene. The most important evidence is found in ice cores extracted from polar ice caps. Snow has been falling and building up into thick ice there for many thousands of years. Ice cores allow us to look back in time – they are like cross-sections drilled down through the snow and ice. One core was cut down through 3 kilometres of ice, to where the ice was 500,000 years old. The ice was taken to a laboratory and melted, releasing bubbles of ancient air. Changes in air content were then analysed, showing scientists how temperatures have warmed and cooled over time (Figure 1).

Another important data source for science is the fossil record. This shows whether animals preferring warm or cool conditions were alive and thriving at different times in the Earth's past.

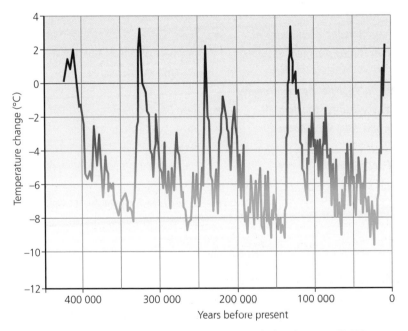

Figure 1: Changes in the Earth's average temperature during the past 450,000 years

Natural causes of climate change

The Earth's atmosphere is affected by changes in the lithosphere as well as by cosmic influences such as the Sun's activity or variations in the Earth's orbit. These are **natural causes** of climate change. As shown below, some changes take place over a short timescale and these can explain the particular years or decades of warming or cooling. Other changes are far more significant and can last for thousands, or even millions of years.

Long-term changes (may last for many centuries) ↓ Short-term changes (lasting just a few years)	**Orbital changes** Changes in how the Earth moves around the Sun are believed to cause ice ages. According to Milankovitch (a Serbian physicist), every 100,000 years or so the Earth's orbit changes from a circular to elliptical (egg-shaped) pattern. This changes how much sunlight we receive. He also identified that the Earth's axis moves and wobbles about, changing over 41,000 and 21,000 year cycles. This also affects how much sunlight is received. Put all of this together and the history of ice ages can be explained!	
	Solar output The Sun's output is not constant. Cycles have been detected that reduce or increase the amount of solar energy. The most well-known phenomenon is sunspot activity, when uneven temperatures develop on the Sun's surface. These can be seen as tiny black spots on photographs of the sun taken by experts (never try this yourself). Sunspots seem to come and go following an irregular cycle that lasts about 11 years. Interestingly, temperatures are greatest when there are plenty of spots – because it means other areas of the Sun are working even harder!	
	Volcanic activity Major volcanic eruptions lead to a brief period of global cooling, due to ash and dust particles being ejected high into the atmosphere, blanketing the earth. The 1883 explosion of Krakatoa is believed to have reduced world temperatures by 1.2 °C for at least one year afterwards. The most recent explosion to have a similar effect was Pinatubo (1991). Sunlight reaching earth was reduced by 10%. World temperatures fell by nearly half a degree in the following year.	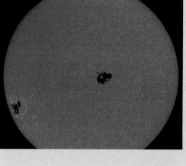

The impact of natural climate change on people and ecosystems

So far we have only considered the **climate changes** that took place many thousands of years ago. We now need to analyse more recent changes. In just the last few centuries, there have been noticeable changes in world temperatures (see Figure 2 on page 30).

The world warmed slightly about one thousand years ago. This was around the time of the Norman invasion of England. However, it began to cool again during the thirteenth and fourteenth centuries. By 1600, when Elizabeth the First was on the English throne, a notably cooler phase had begun. Icy winters became far more common in Europe. This period is called the **Little Ice Age**.

ResultsPlus
Build Better Answers

Explain one natural cause of climate change in the past. (2 marks)

■ **Basic answers** (0 marks)
Are about the impact humans have had on climate, instead of natural causes.

● **Good answers** (1 mark)
Correctly describe a natural cause of climate change, such as volcanic activity, but do not explain why this causes climate to change.

▲ **Excellent answers** (2 marks)
Not only describe a natural cause, such as volcanic activity, but also explain that ash clouds ejected into the atmosphere can cut out the Sun's rays, causing temperatures to fall.

Skills Builder 2

Study Figure 2.

(a) When did the Medieval Warm Period start and end?

(b) Describe one possible natural cause of the onset of the Little Ice Age.

(c) Describe one disadvantage and one advantage of the Little Ice Age for people alive at that time.

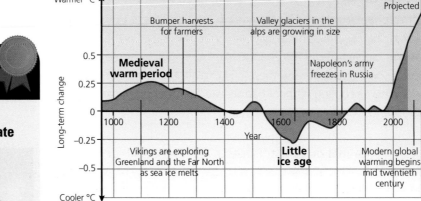

Figure 2: The Medieval Warm Period and the Little Ice Age

Of course, modern technology did not exist to accurately record these events when they were occurring. Worldwide temperature readings (using reliable thermometers) have only been recorded in the past couple of centuries, and photography only really got going in the late nineteenth century. Luckily, however, we do have all kinds of other evidence to help us build a case for climate change before modern scientific readings started to be made. Diaries, folklore, paintings, books and very old newspapers can all provide a pretty reliable picture of the Little Ice Age and the years that followed. Some scientists study old wood, examining tree rings (because a cold year brings less new growth and thinner rings). We also get clues from old coral reefs (because when seas were cooler, coral grew more slowly).

Evidence for the warmer world that preceded the Little Ice Age has also been pieced together. For instance, archaeological evidence shows that the Vikings sailed to many northerly places after they left their homes in Norway and Denmark around AD 900. They even settled and farmed in icy Greenland. Many historians believe that there must have been less Arctic sea ice at this time to allow the Vikings to move around so easily.

Modern **global warming** is widely believed to be the result of pollution caused by humans since 1800. So these much older climate changes must have been natural in origin. Sunspot activity, as we have already learned, can result in warmer or cooler temperatures on the Earth. One explanation for the Little Ice Age is that sunspot activity was much lower than today. Another is that it coincided with a more active period of volcanic activity.

The Little Ice Age

Charles Dickens wrote his novels as the Little Ice Age was drawing to a close. Many of his books, including *A Christmas Carol*, describe very cold and snowy winters in London. But these days, London and south-east England seldom experience a 'white Christmas'. Were winters actually colder back in the seventeenth, eighteenth and early nineteenth centuries? Plenty of data sources suggest so.

Scientists have examined roof timbers in houses built at this time. They have found that tree-ring growth was often slow during winter in the 1700s. Backing up this scientific fact-finding is evidence gathered by art and book historians. Some landscape paintings from the Middle Ages show a winter landscape strikingly different from today. There are, for example, many illustrations of the River Thames covered with ice. Very cold temperatures are required to freeze moving water, especially in a salty tidal river like the Thames. But it is clear that between 1607 and 1814 Londoners enjoyed 'ice fairs' and skating competitions on the frozen river – events never seen today (Figure 3).

There is evidence for the Little Ice Age from all over Europe. Farming records tell us that the thermal growing season was often a little shorter than today. In a cool year, with spring arriving late, planting was often delayed. When autumn came early, so too did the harvest. Some of the most useful and reliable records are grape harvest dates from winemaking regions. These have been used to reconstruct summer temperatures in Paris from 1370 to 1879, and a clear decline in temperature can be seen through much of the period after 1500.

Cold winters often played a part in wars at this time – which are well documented. During Napoleon's retreat from Russia, for example, thousands of his troops froze to death in the winter of 1812. Changes in ecosystems and animal populations are also recorded. Seal populations appear to have been badly affected by a fall in their fish food supplies during this period. Most importantly of all, surviving records tell us that our European ancestors adapted well to a cooler world (otherwise we would not be here today).

Figure 3: Paintings of the Little Ice Age include ice fairs on the River Thames

ResultsPlus
Watch out!

■ The changes associated with the Little Ice Age and Medieval Warm Period were not *major* events in the Earth's history. Look carefully at Figure 2 and you will see that variations in average annual temperature never exceed half a degree either way. This was enough to make winters colder, or to make sea ice melt a little – but was in no way comparable with the arrival or departure of proper ice ages.

Activity 1

1. How reliable are art and literature, compared with scientific data?

2. Do you trust the evidence shown in a painting (Figure 3) as much as a photograph or a thermometer reading?

Quick notes (The Little Ice Age):
- The world may be warming today but in the past it has sometimes been colder.
- Data showing such changes come from both scientific and artistic sources.
- Humans successfully adapted to these changes in their environment.

Geological climate events

Significant changes taking place over millions of years are called **geological climate events**. This is because they form major chapters in the Earth's history (and the study of the Earth's history is called 'geology'). You have already been introduced to one major geological era, the Quaternary Period (and its colder years, called the Pleistocene).

Whenever the Earth starts a new chapter of its history, enormous climatic changes are usually taking place. In the past, these had drastic knock-on effects for flora and fauna (plants and animals). The fossil record is what alerts scientists to big changes that took place in the distant past, such as the extinction of the dinosaurs 65 million years ago.

Humans have also been affected by big climate events and changes in the past. Although modern humans evolved just 200,000 years ago, we have been around in one shape or another for about 5 million years. One of our oldest ancestors (who scientists named 'Lucy') lived in Africa at the end of a geological period called the Pliocene, when the world was a little cooler and drier.

The world's thermometer has moved up and down many times since then. Part of humanity's story is our adaptation to the ongoing environmental changes around us. Sometimes our ancestors have had to migrate away from warming or cooling regions. It may be the reason why Lucy's descendants left Africa.

The extinction of megafauna

The most recent major change for plant and animal life was the **extinction** of certain large animals called **megafauna** at the end of the last Pleistocene ice age. At its coldest, climate change had left the UK covered in ice as far south as what is now the Midlands. It was not just the UK that was affected of course, as changes were global. What are now the central states of the US were ice-bound too. However, a warming of the Earth's climate was well under way by 10,000 years ago, probably linked to the orbit changes discussed earlier. The ice that covered the UK and much of Northern Europe retreated to its present-day Arctic limit.

During the Pleistocene ice ages, very large mammals lived in Europe and North America (Figure 4). These included woolly mammoths, sabre-toothed tigers, large wolves and giant beavers (the size of bears). However, within just a few centuries of the ice melting, these animals – as many as 135 species – were all extinct. Why?

Scientists think that they were unable to adapt to new conditions. Weather and plant life were changing, affecting whole food chains. Some places were left drier once the glaciers – a source of meltwater – retreated north. If food was scarce, some megafauna would have died out naturally.

There is an opposing argument that says that humans were to blame, rather than changes in the climate. This theory claims that towards the end of the ice ages, humans migrated around the world in large numbers and that wherever they went they hunted native megafauna species such as the mammoth – until they eventually became extinct. Remains of mammoths have been found at many settlement sites dating from this time.

Fast forward to today and it seems humans may be guilty of starting another mass extinction. This time it is linked to a new era of climate change – one that we are very probably responsible for. The Earth's temperature is rising due to the human addition of extra **greenhouse gases** to the atmosphere. By some estimates, a quarter of the world's species may not survive the worst predicted changes.

Figure 4: Megafauna that became extinct near the end of the last ice age – sabre-toothed tiger and a woolly mammoth

Watch out!

■ Remember that dinosaurs died out way before the most recent ice age began. The last dinosaur lived 65 million years ago. The final icy phase of the Pleistocene only began 100,000 years ago. Its ending, around 12,000 years ago, coincided with the extinction of giant *mammals* like mammoths – very different kinds of creature from dinosaurs.

Objectives

- Recognise that human activities produce greenhouse gases.

- Explain how this results in an enhanced greenhouse effect and a changing climate.

- Understand that people everywhere will face climate change challenges in the future.

Build Better Answers

Describe how human activities produce two different types of named greenhouse gas. (4 marks)

■ **Basic answers** (0–1 marks)
Name carbon dioxide but do not link its production with an activity such as deforestation or car driving.

● **Good answers** (2 marks)
Also name a second gas, such as methane. However, still do not say how human activities produce the gases.

▲ **Excellent answers** (3–4 marks)
Not only describe a second greenhouse gas, usually methane, but also accurately link the growth of this gas with more cattle being reared by people (the best answers even state that this was a response to rising demand in Asia).

What challenges might our future climate present us with?

The human causes of modern climate change

Greenhouse gases naturally help to warm our atmosphere and make the Earth habitable. However, if *extra* greenhouse gases are added then the Earth begins to get unnaturally warmer. What are these greenhouse gases? What kinds of human activities are boosting their supply?

Greenhouse gases and the activities that produce them

Carbon dioxide (CO_2) is a greenhouse gas that is naturally produced when humans and other animals respire (breathe out). Plants do the opposite, taking in carbon dioxide, while producing oxygen. For this reason, trees and other plants are called a carbon store. However, deforestation releases all that stored carbon kept locked up in the trees (see Chapter 3). Other human activities have also boosted CO_2 levels well beyond their natural level. The burning of fossil fuels, such as oil and gas, contributes greatly. Other significant activities that release CO_2 are cement making and steel manufacturing.

In addition to carbon dioxide, there are other greenhouse gases that human activities produce.

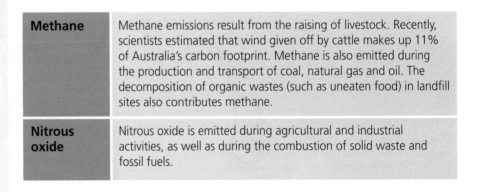

Methane	Methane emissions result from the raising of livestock. Recently, scientists estimated that wind given off by cattle makes up 11% of Australia's carbon footprint. Methane is also emitted during the production and transport of coal, natural gas and oil. The decomposition of organic wastes (such as uneaten food) in landfill sites also contributes methane.
Nitrous oxide	Nitrous oxide is emitted during agricultural and industrial activities, as well as during the combustion of solid waste and fossil fuels.

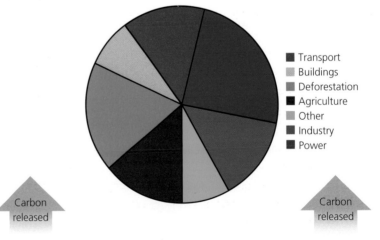

■ Transport
■ Buildings
■ Deforestation
■ Agriculture
■ Other
■ Industry
■ Power

Figure 5: Activities that produce greenhouse gases

The growth of greenhouse gases over time

Since pre-industrial times, atmospheric concentrations of many greenhouse gases have grown significantly. This is due to economic development processes spreading around the world. As countries are industrialised, their citizens become:

- Consumers of energy and goods

- Producers of air pollution (through burning fossil fuels).

Large economically advanced countries, like the USA, produce enormous amounts of new CO_2 each year. The world current level of CO_2 is the highest reached for 650,000 years (some think it is the highest for 20 million years). Even worse, it is increasing now at a rate 200 times faster than at any time in the last million years. Just since the 1800s, carbon dioxide concentration has increased beyond its natural amount by about one-third.

The growth rates of methane and nitrous oxide also fluctuate from year to year, but their long-term trend is again upwards. Methane concentration has more than doubled since the 1800s. It is well over 100% higher than at any time in the past 900,000 years. Nitrous oxide concentration has grown less, but has still risen by 16%. The table below shows one estimate of the rise for all three greenhouse gases.

An estimate of the rise of the three main greenhouse gases

Year	1800	1820	1840	1860	1880	1900	1920	1940	1960	1980	2008
Carbon dioxide (parts per million)	280	282	283	287	291	295	299	310	323	347	386
Methane (parts per billion)	700	713	727	756	796	835	871	980	1430	1656	1940
Nitrous oxide (parts per million)	270	271	275	276	283	286	290	294	304	311	318

Which places produce the most greenhouse gases?

We have already looked at the activities that produce greenhouse gases, but which nations are the biggest polluters? You can probably guess the names of some of the main culprits straight away – the United States and Europe. Both regions experienced an industrial revolution back in the 1800s and have therefore been polluting for a long time, along with Japan.

The emerging superpowers of India and China have recently caught up with these older polluters. In the 1990s, both countries began to 'take off' at breakneck speed. Their economies have been growing at around 10% per year since then – and so have their greenhouse gas emissions. China is now the world's largest single polluter.

Skills Builder 3

Look at the table of greenhouse gases.

(a) Draw a graph to illustrate changes in one of the gases shown.

(b) Identify the decade when major changes first started to occur.

(c) Suggest reasons why major changes began around this time.

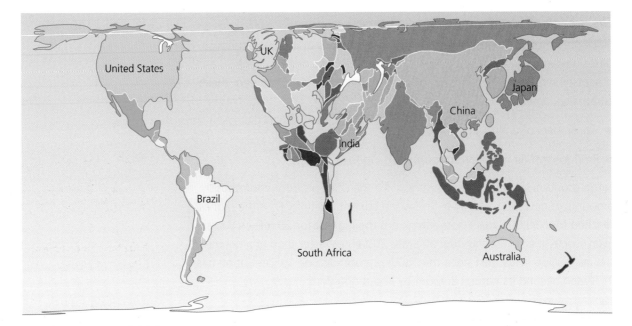

Figure 6: A map showing the size of countries in proportion to how much CO_2 they emit

Skills Builder 4

Study Figure 6.

(a) Who are the biggest polluters?

(b) Where are the smallest polluters found?

(c) What are the strengths and weaknesses of mapping data in the way shown here?

Exam Tip

⚠ Students who know that the greenhouse effect is a *natural* phenomenon will score well. Without greenhouse gases, the atmosphere would not trap the Sun's heat, leaving the Earth too cold for human life. Recently, however, human activity has been boosting the process, trapping *too much* heat.

At first, China and India were not put under much pressure to deal with their pollution. The view was that both places needed to industrialise, in order for health and welfare to improve. With a combined population of 2.5 billion people, many of whom still live in poverty, it is hard to imagine progress *not* producing a lot of greenhouse gas emissions. However, the global situation has become so urgent that both India and China now acknowledge that they need to do more to tackle their emissions.

What do greenhouse gases do?

Don't forget that the greenhouse effect is, in its natural state, a good thing for life on Earth. Without it, conditions would be up to 30 °C cooler – too cold for humans to exist. The way the greenhouse effect works is not especially complicated. Sunlight arrives at the Earth's surface and heats up the ground. Some of this warmth then escapes back into space, just like heat leaving a room when the window is left open. However, greenhouse gases act like the glass in a greenhouse and trap the escaping heat energy (Figure 7). They keep the Earth warm.

Another way of explaining it is to think of the natural amount of greenhouse gases as being like a blanket that helps you sleep at a constant, comfortable temperature. You wouldn't want to take the blanket off (and, similarly, we wouldn't want to lose the greenhouse gases in the Earth's atmosphere). But neither would you want to put an extra blanket on the bed – because it would make you too hot. The human-produced greenhouse gases are like this extra blanket, resulting in what is called the **enhanced greenhouse effect**. And if more people around the world do not change their lifestyles, the greenhouse effect will be enhanced further, with more 'blankets' being added to the atmosphere – a sure recipe for sleepless nights!

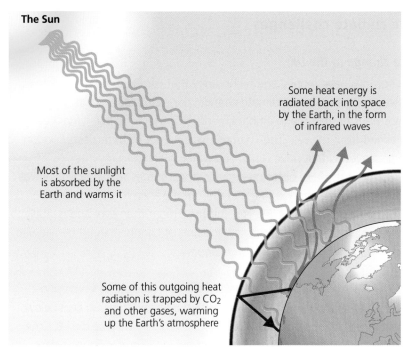

Figure 7: The greenhouse effect

The Sun

Most of the sunlight is absorbed by the Earth and warms it

Some heat energy is radiated back into space by the Earth, in the form of infrared waves

Some of this outgoing heat radiation is trapped by CO_2 and other gases, warming up the Earth's atmosphere

Projections for the future

In 2008, world carbon dioxide concentration passed 380 ppm (parts per million). In 1800, it was just 280 ppm. The figure is currently growing by 2 ppm every year, despite the best efforts made by people and governments to reduce energy use, increase efficiency and introduce 'green solutions' like solar or wind power. Worldwide fossil fuel combustion and rates of deforestation have not slowed down enough yet to make a big dent in total carbon emissions.

Methane production is also on the rise again, with increased wealth in Asia playing a big role. More people there want a meat and dairy diet. This means more cattle being raised – resulting in more methane from bovine flatulence (windy cows).

Many of the world's most knowledgeable climate change scientists belong to a grouping called the Intergovernmental Panel on Climate Change (IPCC). The IPCC believes that greenhouse gas emissions need to level out below 550 ppm. They see this limit as a 'tipping point' for Earth, beyond which events could spiral out of control:

<table>
<tr><td>ResultsPlus
Build Better Answers</td></tr>
</table>

What is the enhanced greenhouse effect? (3 marks)

■ **Basic answers** (0–1 marks)
Offer only a simple (and possibly inaccurate) description of the Earth getting warmer over time.

● **Good answers** (2 marks)
Accurately describe an increase in global temperatures due to the human production of greenhouse gases.

▲ **Excellent answers** (3 marks)
Also explain that human activities are increasing greenhouse gases beyond a natural (and desirable) level, thus making the warming effect of greenhouse gases even more powerful than it used to be.

Below 550 ppm	If greenhouse gases rise no higher than 550 ppm, then the global temperature rise should not exceed 2 °C. However, this could still be enough to cause widespread melting of glacier ice and a world sea level rise of nearly one metre, submerging low-lying areas. There would also be more storms and hurricanes due to warmer sea temperatures. Globally, many species might become extinct, although warmer conditions could encourage tree growth and greater biodiversity at high latitudes (Chapter 3). So there could be some winners as well as losers.
Above 550 ppm	If greenhouse gas emissions rise above this critical level, then conditions will rapidly worsen. Vicious circles will develop. For instance, as the ice caps melt, their bright white surface is lost. This ice surface usually reflects some sunlight back into space. Without it, more energy will be absorbed by the Earth – and temperature will rise even faster. The worst predictions suggest that the Earth could one day be ice-free if nothing is done to stop a global temperature rise of 6 °C or more. Billions of humans will lose their homes (because of sea level rises) or their fresh water supplies if this happens.

ResultsPlus
Exam Question Report

Explain one possible good effect of global warming and one possible bad effect of global warming. (4 marks, June 2006)

How students answered

Most students answered this question very poorly. The question requires two developed answers that explain why the effect is seen as *good* or *bad*. The majority of students could only write statements like 'warmer temperatures' or 'ice caps melt' to gain no more than 2 marks.

████████ 80% (0–1 marks)

A few students gave two effects, and said *why* one of these effects was good or bad.

█ 10% (2 marks)

A few students answered this question really well. They gave two effects and in both cases showed what the positive or negative impact on people or places would be. 'Melting of ice caps causes coastal flooding, damaging cities' would have been a very clear *negative* effect, for example. While 'warmer temperatures allow more crops to be grown' scores two marks as a *positive* effect.

█ 10% (3–4 marks)

Future climate challenges

Climate change in the UK

Richer countries have money and technology that can help them adapt to higher sea-levels brought by climate change. For example, Holland has long used enormous human-made embankments to protect the low-lying areas it has reclaimed from the sea. The Thames Flood Barrier (TFB) protects London from high sea levels that drive North Sea water inland into the Thames estuary. Without the TFB, occasional major storm surges would flood central London. With climate change, these surges are forecast to become more frequent and dangerous. The TFB is being used more than ever and is vitally important for London to be kept safe. Flooding of London cannot be allowed as it would paralyse the whole British economy.

The entire UK faces change and challenges if global temperatures keep rising (Figure 8 below). In a low-emission future (with the temperature rise kept below 2 °C) there may be winners as well as losers. One winner would be the UK wine industry, which is currently experiencing record growth in profits, especially for sales of fizzy champagne-style wines. Coastal tourism might also prosper. However, there will be many losses. In a high-emissions future, costs could be high:

- A complete loss of winter sports, as snow disappears from highland areas

- More cases of tropical diseases like malaria

- More severe storms and longer summer droughts, making our climate more extreme (Chapter 4)

- The economic cost of helping climate change refugees who migrate from poor countries to the UK

- Major changes for fishing industries, especially if ocean currents are disrupted by melting ice in Greenland, making British waters turn colder (though this would not happen before 2100).

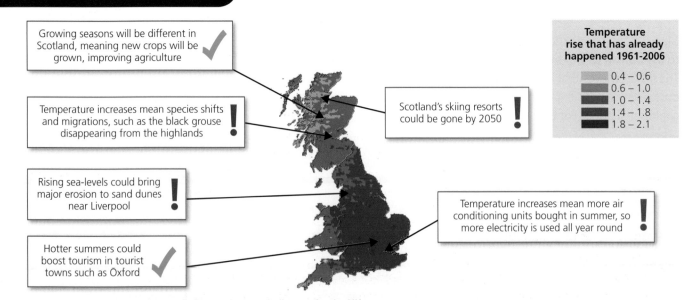

Figure 8: Rising temperatures and climate change challenges for the UK

Climate change and sea-level rise in poorer nations

Many nations are at serious risk of sea-level rise, and the poorest of them lack the resources needed to adapt to climate change. In a high-emissions future, some places could be abandoned entirely:

The Maldives	Most of the tiny islands that make up the Maldives are less than two metres above sea level. The 300,000 people that live there may soon become 'climate change refugees'. In 2008, the Maldives President asked neighbours India and Sri Lanka if he could buy some of their land!
Tuvalu	Half of the 10,000 residents of the Pacific island state of Tuvalu live within three metres of today's sea level. Many islanders think they may soon need to migrate elsewhere.

It is not just rising sea-level that brings new risks. If climate change permanently melts glaciers in the Himalayas, millions of people in Asia will lose their water supply (Chapter 4). Drought and famine would follow. Saharan Africa is also highly threatened by future water shortages. Projections show that naturally low levels of rainfall could fall further still.

Bangladesh

Bangladesh is the world's seventh most populous country, with 150 million people. It is also one of the poorest nations on Earth, with an average income of just over one pound a day. Bangladesh is also a vulnerable nation, with much of its land at or near sea-level on the delta of the River Ganges – and this land is naturally sinking and subsiding. Flooding is already frequent and damaging.

A small rise in sea-level would leave large areas of Bangladesh permanently under water (Figure 9). Climate change could also help drive tropical storms further inland, causing more farmers to lose crops (because when storms temporarily raise the water level, sea salt deposits are left behind, killing plants).

One of the many problems that flooding brings is long-term interrupted schooling when schools are flooded. If the country is to be lifted out of poverty, then safeguarding education is vital. Bangladesh needs well-educated citizens if it is to move further forwards.

Britain recently gave £75 million to help Bangladesh tackle the worst impacts of climate change. The money will be spent on special adaptation measures to protect schools, rebuilding them on raised stilts and platforms. Flood waters may then pass safely beneath the buildings. Farmers will also be helped with the introduction of crops that are more tolerant to salt.

Skills Builder 5

Study Figures 8 and 9.

(a) Describe how different parts of the UK could be affected by climate change.

(b) Describe how much of Bangladesh would be lost to (i) a two-metre rise, or (ii) a five-metre rise.

(c) Suggest how a sea-level rise could affect the economy of a poor nation like Bangladesh.

Decision-making skills

Look at the bulleted list on page 38. Make a table to list the environmental and economic costs. Rank them in what you think is the order of severity of impact.

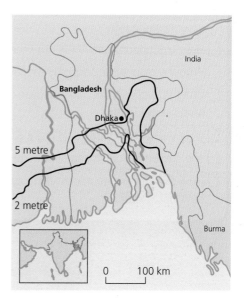

Figure 9: Land that could be lost to potential sea-level rises in Bangladesh

Quick notes (Bangladesh):
- Poorer countries are more vulnerable to natural hazards than rich countries.
- Natural hazards, like flooding, can interfere with a country's long-term economic development.
- Climate change means that the risk of flooding and storm damage is increased for Bangladesh.

examzone

Know Zone
Climate and change

The Earth's climate may seem stable, but in fact it has changed in the past and is likely to change in the future. Climate can, and does, change on a timescale of a few years and many thousands of years. In the past climate changed quite naturally, but today there is increasing evidence that humans are changing the climate.

You should know...

- ☐ How climate has changed in the past
- ☐ The timescales of climate change
- ☐ How scientists know about past climate changes
- ☐ The natural causes of climate change
- ☐ How climate change affected people and the environment during the Little Ice Age
- ☐ How large, long-term climate change contributed to extinction at the end of the last ice age
- ☐ How the greenhouse effect works
- ☐ The types of human activities increasing greenhouse gases
- ☐ How greenhouse gas levels have changed over time
- ☐ Who the main producers of greenhouse gases are
- ☐ What scientists think might happen to climate and sea-levels in the future
- ☐ How the UK's climate could change in the future
- ☐ The challenges our changing climate might bring
- ☐ How climate change might affect people in the developing world.

Key terms

Climate change	Greenhouse gases
Deforestation	Ice age
Ecosystems	Little Ice Age
Enhanced greenhouse effect	Megafauna
	Natural causes
Extinction	Orbital changes
Geological climate events	Quaternary period
	Solar output
Global warming	Volcanic activity

Which key terms match the following definitions?

A A trend whereby global temperatures rise over time, linked in modern times with the human production of greenhouse gases

B Those gases in the atmosphere that absorb outgoing radiation, hence increasing the temperature of the atmosphere

C A community of plants and animals that interact with each other and their physical environment

D A period in the Earth's past when the polar ice caps were much larger than today

E The most recent major geological period of Earth's history, consisting of the Pleistocene and the Holocene

F Long-term changes in global atmospheric conditions

G The chopping down and removal of trees to clear an area of forest

H Very large mammals, such as those that lived during the last ice age

To check your answers, look at the glossary on page 313.

ResultsPlus
Maximise your marks

Foundation Question: Describe two human activities which are increasing the amount of greenhouse gases in the atmosphere. (4 marks)

Student answer ● (achieving 2 marks)	Examiner comments	Build a better answer △ (achieving 4 marks)
Fossil fuel. Methane is increasing. This is because there are more cows grown for their meat and they give it off.	• *Fossil fuel* is on the right lines, but this is not an activity so it does not score any marks. • *Methane is increasing* is correct and scores 1 mark. • *This is because...* scores 1 mark. This part of the answer describes an activity and links well to the previous point.	Fossil fuels are burnt in power stations and this gives off carbon dioxide. Methane is increasing. This is because more cows are bred for their meat and they produce methane.

Overall comment: The student identified fossil fuels but did not develop this idea to describe how they are linked to greenhouse gases.

Higher Question: Describe two challenges the UK might face in the future due to global warming. (4 marks)

Student answer ● (achieving 2 marks)	Examiner comments	Build a better answer △ (achieving 4 marks)
In the future there could be more rainfall, leading to floods. There were severe floods in 2007 in Gloucestershire. The floods were very costly and many people had to be rescued. Flooding is a major challenge.	• *In the future...* scores 1 mark. In this part of the answer a challenge, flooding, is identified. • *There were severe...* This is a specific example which supports the challenge of flooding. This part of the answer scores 1 mark. • *The floods were...* Although this is an extension point, it is still about floods so does not gain any marks. • *Flooding is a...* This repeats the point about flooding and does not identify a second challenge. As such, it does not score any marks.	In the future there could be more rainfall, leading to floods. There were severe floods in 2007 in Gloucestershire, which were very costly and many people had to be rescued. Higher temperatures could mean much hotter summers. These could bring more forest fires and cause crops to die.

Overall comment: Although there is some good information and examples included, the answer covered only one challenge – flooding. Look carefully at questions and spot how many different points are being asked for.

Chapter 3 Battle for the biosphere

Objectives

- Learn what biomes are and why they matter.

- Explain how precipitation and temperature affect the global distribution of biomes.

- Understand the biosphere, the dynamic planet and its people are all interdependent on one another.

What is the value of the biosphere?

The world biome map

The Earth is home to large plant and animal communities called **biomes**. Together, these biomes make up our **biosphere**. You may be familiar with some of their names, such as tropical rainforest or savannah. Figure 1 shows that in many cases biome distribution roughly follows lines of latitude.

Latitude is a very important influence on air temperature because of the shape of the Earth. Sunlight arriving in the Tropics is highly concentrated whereas at the poles it is spread more thinly. The reduced concentration of solar energy nearer the poles means less energy is available for photosynthesis. As a result, the rate of vegetation growth at higher latitudes is less than in the Tropics.

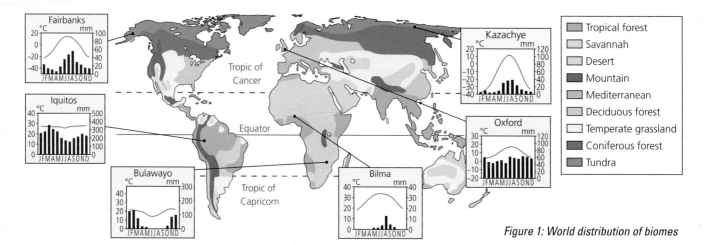

Figure 1: World distribution of biomes

Tropical rainforest

Tropical rainforest lies mostly in a band either side of the Equator. Here, the Sun's rays are concentrated, heating moist air and causing it to rise. Heavy rainfall is the result – perfect conditions for evergreen rainforest.

Deserts

Deserts are found close to the Tropic of Cancer in the northern hemisphere and close to the Tropic of Capricorn in the southern hemisphere. The air that rises over the Equator travels polewards after losing its moisture but sinks back down at the Tropics (due to complex air patterns caused by the Earth's rotation). The Sun's rays are still highly concentrated at this latitude and because the air is dry it brings desert conditions to places like the Sahara.

Deciduous

Deciduous forest grows in higher latitudes. It is found in the UK and other places along the coast of continents where rainfall is high. The Sun's rays are less strong at this latitude, and cooler winter temperatures encourage trees to shed leaves at the end of autumn (this is what 'deciduous' means).

Coniferous forest

Coniferous forest dominates by 60° north. Temperatures are so cold that trees have evolved with needle leaves that reduce moisture and heat loss. Snow slides easily off their sloping branches.

Tundra

Tundra (cold desert) is found at the Arctic Circle. The Sun's rays have little strength here, and temperatures are below freezing for most of the year. Only tough short grasses can survive.

Precipitation is the other significant climatic influence on biome distribution. World patterns are very complex, but rainfall generally tends to be high in coastal and highland regions.

Local factors affecting biomes

The world map of biomes (Figure 1) does not show the variations occurring at a local level. For example, the UK is shown as entirely covered with deciduous forest – which clearly is not true. Physical factors – especially drainage – affect local conditions. In parts of Scotland, where soil conditions are especially wet there are peat bogs rather than forests. But it is human factors – deliberate clearance of the original forests – that have caused the most change, wiping out most of the natural biome.

Altitude and distance inland are important influences too. Temperatures fall by about half a degree for every 100 metres increase in altitude. As a result, tough grasses quickly replace trees on higher mountain slopes (Figure 2). In the USA and in Asia, inland areas isolated from the sea suffer from low rainfall because winds blowing off the oceans quickly lose moisture – especially if the air passes over high mountains. The drier lands found further inland are said to be in a 'rain shadow'. On a smaller scale, the same effect is found in Great Britain.

Finally, geology has an influence. Limestone bedrock creates dry soil conditions because percolating rainwater passes through it relatively easily. In the UK, trees are rarely found in limestone areas. In the tropical rainforest, deciduous trees may replace evergreens where limestone is found.

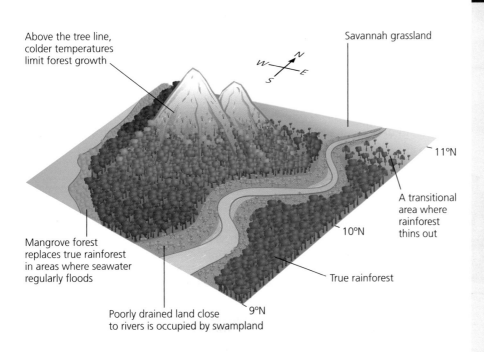

Figure 2: Local factors affecting where tropical rainforest can grow

Skills Builder 1

Study Figure 1.

(a) Describe the distribution of: (i) tropical rainforest (ii) coniferous forest.

(b) The main places where these two types of forest are found experience very different temperatures. Explain why this is the case.

ResultsPlus
Build Better Answers

Study Figure 2. Describe how local factors affect the growth of rainforest. (4 marks)

■ **Basic answers** (0–1 marks)
Are incomplete and only describe one or two of the factors shown, such as altitude.

● **Good answers** (2 marks)
Use the full range of factors and identify four places where rainforest is prevented from growing by local factors.

▲ **Excellent answers** (3–4 marks)
Distinguish between places where rainforest cannot grow at all (such as mountain tops) and places where it is present but in a changed form (such as the edges of the savannah).

The Earth's biosphere provides vital goods and services for people. Describe how the tree in Figure 4 could be used. (2 marks)

■ **Basic answers** (0 marks)
Describe a tree, but not how it is used.

● **Good answers** (1 mark)
Correctly state that the enormous tree shown in Figure 4 gives water.

▲ **Excellent answers** (2 marks)
Recognise that the baobab tree is used by people as a source of water for vital services like drinking or washing.

Figure 4: A baobab tree with its water-storing trunk

The biosphere life-support system

The biosphere acts as a life-support system for the planet, helping to regulate the composition of the atmosphere, maintaining soil health and regulating the **hydrological** (water) **cycle** (Figure 3).

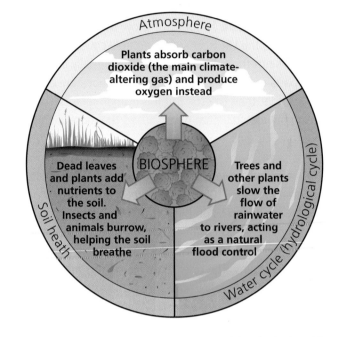

Figure 3: The biosphere's interaction with other parts of the physical world

As we learned in Chapter 2's look at the changing atmosphere, the biosphere is an important carbon store. Plants take in and store carbon dioxide while producing oxygen. Tropical rainforest is sometimes described as 'the world's lungs'. This is why **deforestation** is a major cause of global warming.

Healthy soils require constant inputs of new nutrients from falling leaves and rotting plant remains. In the absence of human interference, nutrients such as nitrogen and potassium are continually recycled and exchanged between plants and the soil beneath them. Insects and animals that live on the forest floor also play a vital role by digging and burrowing in the soil. This is very important for soil health because it allows fresh air to circulate. When forest cover is removed, soil health can quickly deteriorate.

The biosphere also interacts with the hydrological cycle. Trees and plants catch rainwater and slow down the speed at which it travels to the nearest river. You will learn more about this in Chapter 4.

The biosphere and people

The biosphere naturally interacts in good ways with the Earth's climate, soil and water. It also supports people. Plants and animals play a vital role in maintaining human life and aiding economic development.

The biosphere provides us with **goods** and **services** (Figure 5), including:

Food	When population numbers are low, natural ecosystems can be sustainably harvested, for example through berry and fruit picking. With larger populations, there is more pressure to replace natural vegetation with commercial crops. A good example of this is cereal production in the central parts of the USA. 'America's bread basket' is a region where naturally occurring grasses have been replaced by wheat and corn – increasingly used for the production of **biofuels** as well as food.
Medicines	Many naturally occurring substances act as medicines and remedies. For example, quinine is a plant extract found in the tropical rainforest. Native tribes paralyse animals with darts dipped in quinine. We call this natural store of medicines the **gene pool**. Genes are the building blocks of life found in the cells of plants and animals.
Raw materials	The biosphere provides raw materials for industrial activities. The most important of these is wood for the construction of houses and boats. Roofing material for thatched roofs also came from plants. In the dry Savannah grasslands of countries like Kenya, trees have evolved to store water so they can survive long periods of drought. The baobab tree (Figure 4) has a trunk that stores water like a barrel. It is a vital source of water for people living in rural areas where rainfall is rare.
Services	The biosphere helps humans conduct business in varying ways. For example, it provides the natural resources needed for ecotourism and many outdoor leisure activities. Along South Africa's Maputaland coast, for example, local people show tourists the sea turtle beaches during the nesting season. The biosphere can also help protect people from natural hazards such as storms and floods.

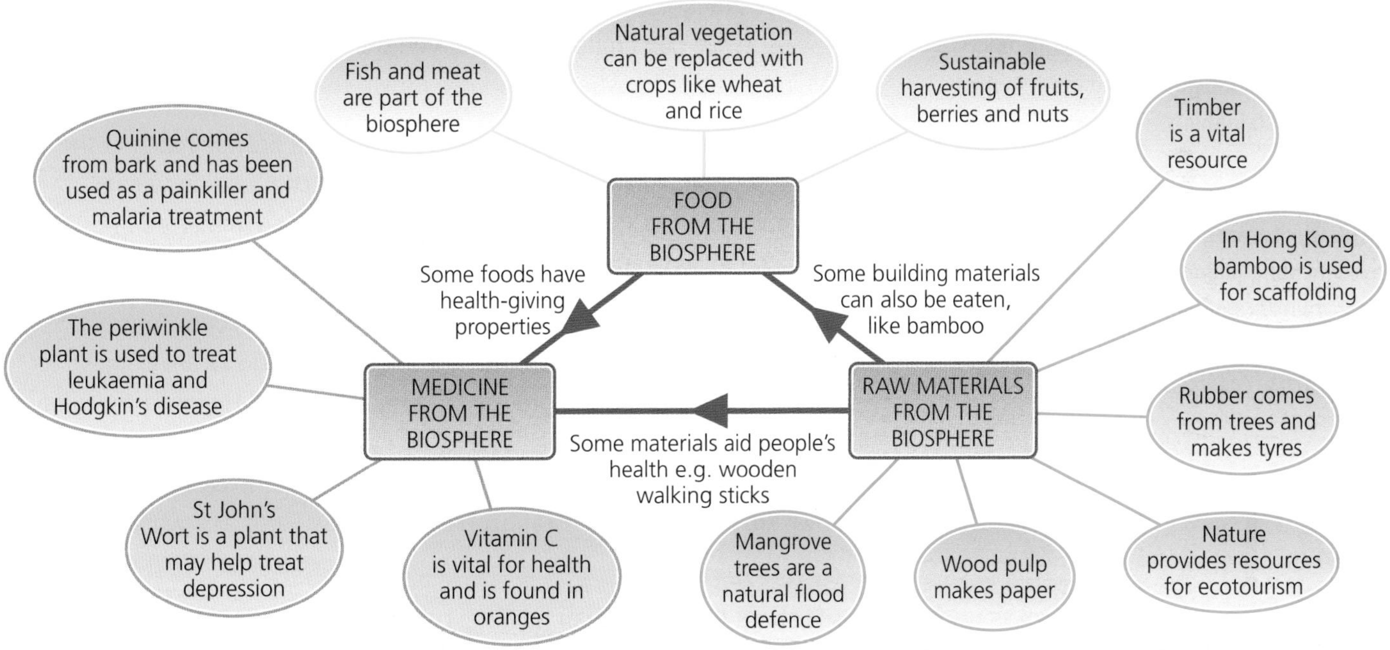

Figure 5: Biosphere goods and services

Biosphere protection

The biosphere helps protect people's health, thanks to its gene pool. Vegetation also saves people's lives in other ways. For example, it has long been recognised that India's tropical rainforest provides an important service by stopping the river Ganges from flooding. In Hindu mythology, the Lord Shiva protects people from the power of the river goddess Ganga. He ties her to the land with his hair. In this legend, Shiva's hair represents rainforest trees which intercept rainfall and stop the river Ganga (the modern Ganges) from flooding the surrounding plains.

In the UK today, natural salt marsh vegetation is sometimes seen as a better defence against the sea than artificial sea walls. Faced with rising sea-levels, the UK government has abandoned some low-lying agricultural land to the sea. The idea is to create new salt marshes. At Wallasea Island in Essex, protective ditches and dykes have been deliberately breached, allowing sea-water to spill on to agricultural land behind where it deposits silt and builds new salt marsh. This 'managed retreat' from the sea is a relatively cheap and effective way of protecting more important areas further inland. Wave energy is quickly reduced by friction when high tides move across the new salt marsh. This is a good way of using the biosphere to deliver an important service. It is also a sensible local adaptation to climate change.

The importance of forests to human life

- In Britain, wood from the deciduous forests helped make and fuel the factories that drove the Industrial Revolution. Centuries earlier, the sustainable practice of coppicing produced flexible strips of wood for longbows, while oak timber ships defeated the Spanish Armada. Today, remaining areas of forest are a resource for UK tourism.

- Coniferous forests in Scandinavia and North America provide much of the world's softwood. Pine is an especially popular material for lightweight furniture. In paper mills, the softwood is ground up and immersed in caustic soda to separate its fibres and create the pulp used to make paper.

- Tropical rainforest is a vital resource for poor countries in South America, Asia and Africa. Wood is used in many ways to help build local incomes. Forest can also be cleared to provide land for other uses. However, many of these actions are unsustainable, as they do not leave behind forested areas for future generations. The environmental impacts of tropical rainforest removal are costly, both to local people and for the Earth as a whole.

Ten facts about the Amazonian rainforest and the River Amazon can be seen on the next page.

Activity 1

In 2005, environmental campaigner Dorothy Stang was murdered after criticising powerful loggers in Amazonia. This shocking incident shows that very real battles are fought between the people who want to save the biosphere and others who use it in unsustainable ways.

Find out more about the people who try to save the rainforest at: http://ngm.nationalgeographic.com/2007/01/amazon-rain-forest/wallace-text

1 Many tropical plants are used as industrial raw materials. Plant oils, gums, resins, tannin, rubber and dyes all come from the rainforest.

2 Hardwood timber is a valuable resource that can be exported to help feed the growing population of rainforest countries like Brazil.

3 The rainforest tree canopy protects the soil. If it is removed, there is a risk of soil being washed away by the convection rain that falls daily in Amazonia.

10 Trees intercept and slow down rainwater as it moves through a river's drainage basin. This helps lower flood risk. Cutting down trees can result in greater flooding.

4 If soil is removed by rain following deforestation, it gets washed into the River Amazon. Water temperatures rise as a result of this sediment being added, making survival hard for some fish.

10 facts about the Amazonian rainforest

9 The population of South America is expected to double in the next 40 years. South American people want to clear their forest to gain living space.

5 The Amazon forest has been described as 'the world's lungs'. Its trees produce much of the world's oxygen, which humans breathe, while also soaking up carbon dioxide.

8 The tropical rainforest is a gene pool that the health of the entire planet is dependent upon. Millions of plant, insect and animal species live in Amazonia. They contain genetic material that can provide vital medical resources for the fight against diseases like leukaemia.

7 Soils are protected from the drying effect of direct sunlight by the shade that trees provide. Soils suffer from the formation of a hard crust called laterite if the forest canopy is removed.

6 The rainforest is a vital store of carbon. If it were all to disappear, global carbon dioxide levels would rise considerably – causing runaway climate change and faster temperature and sea level rises.

Windsor Forest

The UK was once covered with deciduous forest. However, over-use meant that very little was left, especially after 1750, when more farmland was required to feed the rapidly growing population. Timber was also needed for housing, shipbuilding and to prop up mine shafts. By 1919, only 3% of England's natural forest remained. Since then, efforts have been made to protect what is left and to restore the forest in some places.

Windsor Forest is one small sustainably managed area of ancient woodland that has survived centuries of change in Britain. Once a medieval Royal hunting forest, the 3,100 hectares of woodland holds more than 900 oak and beech trees over 500 years old. More than 2,000 species of invertebrate and 1,000 species of fungi have been found here, displaying the rich **biodiversity** of the native deciduous forest.

The government agency Natural England now manages Windsor as a 'post-industrial forest', meaning that it is used for recreation and **conservation** rather than timber production.

Figure 6: Windsor Forest

Quick notes (Windsor Forest):
- The biosphere provides important goods, such as timber.
- The biosphere can also be used to provide services, such as tourism and leisure.

Objectives

- Recognise that Amazonia is being degraded by human actions.

- Explain the range of indirect effects that pollution and climate change have on the biosphere.

- Understand that management is needed at a variety of scales if human use of the biosphere is to become more sustainable.

How have humans affected the biosphere and how might it be conserved?

Few places on Earth remain free from human interference. Even wilderness regions, far from settlement, are vulnerable to climate change and the sea-level rise that it threatens. Air and water pollution are problems that have 'gone global' over the past hundred years. You can be standing on a remote wilderness beach in Alaska and still find rubbish washed up along its shore. Over the next few pages, we will look at the effects of pollution and climate change on the Earth's biosphere – after first taking an in-depth look at the destruction of Amazonia.

The destruction of Amazonia

Tropical rainforest is found between 10° north and 10° south of the Equator. Along with constant high temperatures of around 28 °C, there are daily falls of convection rain (the result of frequent intense thunderstorms). This ever-wet climate provides optimum conditions for plant growth, and the biodiversity of trees, vines, shrubs and plants reaches almost unbelievable levels. One study of the Napo region of Peru found 283 species of trees in an area of forest no larger than a football pitch.

Figure 7: The destruction of Amazonia

But Amazonia is under threat. The rainforest is suffering from **degradation** – its character and quality are being constantly lowered – caused by human actions. The rate of rainforest clearance recently reached record levels. Now that deforestation is known to be a major cause of climate change, what drives countries like Brazil to cut down 100,000 square kilometres of precious forest each year?

Commercial clearances for timber harvesting (Figure 7) – especially mahogany and teak – have been occurring for centuries, fuelled by demand among the world's rich for tropical hardwood furniture and flooring. This trade continues today, despite attempts to restrict imports by countries like the UK. But there are other large-scale land use changes that have also resulted in rainforest removal in Amazonia:

The Grande Carajas development programme	The Grande Carajas development programme brought iron mines and aluminium plants to places where virgin forest once stood in Brazil. Use of wood as fuel led to further clearances.
Hydroelectric dams	The building of hydroelectric dams along the Amazon's tributaries flooded forest valleys.
New roads	The construction of new roads, such as the Trans-Amazon Highway, leads to forest loss.
Migration of farmers	Landless farmers migrate into Amazonia along the new roads. They then cut down forest for firewood or clear land to grow crops. As a result, the pattern of deforestation often follows the road network (you can look for evidence of this in Figure 8).
Forest clearance	Huge areas of forest have been cleared for commercial agriculture. Major crops such as soya beans are grown on old rainforest soils in Brazil. In neighbouring Costa Rica, around one-third of all cleared rainforest land is used for cattle ranching.

How have humans affected the biosphere and how might it be conserved?

49

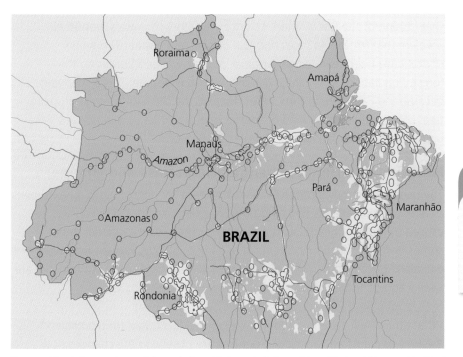

Legend:
Rivers
Roads
Areas of deforestation
ooo Urban zones

0 — 150 km

Figure 8: The deforestation and degradation of Amazonia (adapted from National Geographic Map)

The soya craze

During the past decade, removal of Brazil's forest can be linked with a new phenomenon: the rise of soya bean production. Half of all forest clearance in 2005 took place in the state of Mato Grosso – whose governor Blairo Maggi also runs the world's largest soya bean production company. Maggi is known locally as 'O Rei da Soja' (King of Soya). During 2003, his first year as governor, the rate of deforestation in Brazil rose by 40%. In 2006, environmental campaigners Greenpeace awarded Maggi their 'Golden Chainsaw Award' for being the Brazilian who most contributed to the destruction of the Amazon rainforest!

Growth in soya cultivation is a response to increased global demand. Cattle ranchers around the world, especially in Europe, are switching to soya as a safe and healthy food source for their animals.

More people now want a diet that is meat-rich. Demand for meat is soaring in emerging **superpower countries** like China, whose citizens have become richer in recent years. This means that more crops like soya need to be grown to feed the ever-growing number of cattle.

People pressure

Future population growth in Brazil will put even more pressure on Amazonia. Already numbers have increased from 60 million in the 1960s to nearly 200 million today. With a population growth rate of 1.5% per annum, more land is needed each year to provide room for farming and housing. In recent years, as many as 20 million extra people have also migrated to live in settlements along the Amazon and its tributaries. On the fringes of settlements such as Manaus, virgin forest is cleared every day to make way for more shanty dwellings.

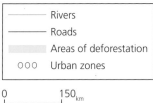

Decision-making skills

Decide which of the factors in the table on page 48 is likely to have the highest environmental impact on the tropical rainforest. Explain why.

ResultsPlus
Build Better Answers

Study Figure 8. Describe the impacts that human activities are having on Amazonia. (4 marks)

■ **Basic answers** (0–1 marks)
Neglect to mention human activity at all.

● **Good answers** (2 marks)
Correctly identify that the yellow areas are places where deforestation is taking place and emphasise that the area lost is now very large (using the scale to suggest the number of square kilometres that have been removed).

▲ **Excellent answers** (3–4 marks)
Also mention other impacts, such as urbanisation, or suggest that there is a pattern to the deforestation (you can see it follows the road and river routes through the region).

Over-fishing of the River Amazon's once well-stocked waters is another problem that Amazonia faces because of people pressure. The worst effects have been seen in the Pantanal, a giant wetland bordering the forest. Populations of the Pacu fish are in serious decline. This is worrying because the Pacu fish, in turn, helps disperse the seeds of some rainforest trees. Knowledge of this kind of species interdependency is important if the biosphere is to be managed and protected properly.

Pollution and climate change bring stress and change

The deliberate removal of forest to create space for agriculture is an example of direct human actions damaging the biosphere. Degradation can also be caused by indirect means, such as pollution. The greatest indirect threat comes from air pollution in the form of carbon dioxide. Human-produced emissions of carbon dioxide (and other equivalent gases such as methane) are widely agreed to be the cause of global warming.

What effects will a warmer world have on the distribution of biomes and the lives of plants and animals? Scientists are observing the changes now taking place in environments all over the world. Studies of changing species habitats show that nature is already on the move. Rainforest frogs – sensitive to environmental change – have disappeared in large numbers. Polar bears are extending their hunting range inland as Arctic ice melts. In places that already suffer from water shortages, the consequences of climate change may be severe for the biosphere. Areas that currently enjoy a Mediterranean climate could experience a shift towards even drier conditions. This would result in frequent wildfires which could devastate ecosystems (Figure 9).

Figure 9: Wildfires destroy a forest during a drought

Changes for the UK

The global map of biomes (Figure 1 on page 42) shows the British Isles as originally an area of deciduous forest. However, future changes in temperature and precipitation patterns for the south of England could make it more likely that grass will become the natural vegetation, rather than trees. Unusual and severe heatwaves have already affected the UK, most recently in 2006. There are also fewer frosts and winter cold spells now than in the 1960s. If these trends continue, then trees could become stressed wherever water supplies are limited (for instance in very well-drained areas with highly permeable soils).

If temperature rises between 2 °C and 4 °C, there will be major changes in the geographical distribution of British forests and other plant communities. Bird and animal species that live here would be affected by change too. Half of the UK's migratory birds have already experienced a severe decline in numbers in recent decades (Figure 10). Scientists believe that this may reflect difficulties these birds face in adapting to changing conditions, both in the UK and in Africa (where species such as the swallow spend winter).

However, while global warming will certainly bring changes to the biosphere in Britain, it may not lower our total biodiversity, because there could be winners as well as losers.

Activity 2

Study Figure 1 on page 42.

1. Consider what the effects might be of the world warming up by 1 or 2 degrees.

2. Consider what the effects of an even larger rise of 3 or 4 degrees might be.

3. What would happen to the distribution pattern? Remember that precipitation trends could change too. A warmer world could also be wetter in some places, due to increased evaporation over oceans and more storms.

How have humans affected the biosphere and how might it be conserved?

51

Changed growing season	The growing season for plants in central England has already lengthened by about one month since 1900. Exotic species such as yucca and olive now thrive in the warmer and sometimes drier conditions experienced by southern England.
New marine wildlife	There are more sightings of whales, sharks and sea turtles in British waters. These changes in marine biodiversity are thought to be a result of warming surface water temperatures.
Moving tree-line	Some scientists think that the tree-line is moving upwards into higher areas in the Scottish Highlands. This should bring greater biodiversity to some upland areas.

Climate change and the protected areas problem

In the UK, rare animal and plant species are sometimes found in particular coastal environments, such as an area of sand dunes or salt marsh. Such areas may be granted protected status by the government as sites of special scientific interest (SSSIs). Examples include the Mersey estuary (a wildfowl site) and Ainsdale sand dunes (home of the Natterjack toad).

Britain's coastal SSSIs are now greatly threatened by climate change. Rare species may start trying to move inland if their original habitats become threatened by a sea-level rise. However, they could be unable to find safety if farmland or urban settlement surrounds their homes. In the long term, Britain's SSSIs may need to be enlarged and linked together to form longer 'biodiversity corridors'.

Other pollution issues

The word 'pollution' describes an artificial over-supply of substances such as chemicals to the environment (or to a single organism). Pollution is an example of the indirect impacts that humans have on plant and animal life – it is rarely a deliberate and direct action. Air, land and water pollution all damage the biosphere in a number of ways:

- Loss of biological functions, including reproduction

- Direct mortality (death) and species decline

- Tissue and vital organ damage, due to toxins (poisons).

Non-lethal doses can be carried all through the food chain, but when they accumulate in the bodies of the top carnivores, they can become so concentrated that the animals die. This can make it difficult to get laws right governing what can be labelled a 'safe' level of pollution.

Garden warbler

Yellow wagtail

Turtle dove

Willow warbler

Figure 10: Is climate change to blame for Britain's vanishing birds?

Skills Builder 2

Study Figure 10.

(a) Which bird has been worst affected?

(b) Describe the changes observed among the UK's migrant bird population.

(c) Explain ways in which global warming could be causing these changes.

Decision-making skills

Which of the conservation frameworks in the table is the most successful? Think about the extent of their coverage (scale and focus), cost and management.

Biosphere conservation at the global scale

With nearly 200 countries in the world needing to sign agreements, global-scale attempts at conservation are quite a challenge – especially when people in poorer countries still rely heavily on farming, fishing, hunting and forestry to make a living. Despite these difficulties, important progress has been made.

Notable conservation framwork successes include:

CITES (Convention on International Trade in Endangered Species)	Throughout the world, harsh penalties now exist for poachers caught shooting endangered species, such as the rhino and tiger.
Rainforest conservation	The global importance of rainforest has led to the introduction of international 'debt-for-nature' swap agreements. The US reduced the size of Guatemala's foreign debt repayments in exchange for promises that less rainforest will be cut down.
National Parks	Covering 13% of the land surface of the Earth, National Parks are a conservation success story. Globally, they are overseen by agencies including the United Nations.
World Heritage Sites	Over 800 important sites have been awarded official recognition by the UN since 1972, many on account of their biosphere credentials.
Wetland management	The Ramsar Convention has achieved real success, protecting wetlands everywhere from unsustainable over-exploitation.

Wetlands are areas where water is the primary factor controlling the environment and the associated plant and animal life. They occur where the **water table** lies at or near ground level, or where the land is covered by shallow water. Five major wetland types are generally recognised:

- Coastal wetlands, including coastal lagoons and coral reefs

- Estuaries, including river deltas and mangrove swamps

- Lakes and their edges

- Wetlands along rivers and streams

- Marsh, swamp and bogs.

In addition, there are human-made wetlands such as fish and shrimp ponds, farm ponds, irrigated agricultural land, salt pans, reservoirs, gravel pits, sewage farms and canals.

The **Ramsar** Convention on Wetlands is a global treaty established during 1971 at a meeting in the Iranian city of Ramsar. It was the first modern global treaty on the conservation and sustainable use of natural resources. 153 states from all parts of the world have signed the treaty and 1,700 wetlands now have special protection. Ramsar sites cover 1.5 million square kilometres – a larger area than France, Germany, Spain and Switzerland combined.

ResultsPlus
Build Better Answers

Choose a local example of biosphere management. Explain the methods used to make it more sustainable. (4 marks)

■ **Basic answers** (0–1 marks)
Do little more than name a place (such as Scotland) where improvements have been made, but do not give any details.

● **Good answers** (2 marks)
Provide specific details (e.g. of wild boar coming back to the Caledonian Forest) and identify that this was important.

▲ **Excellent answers** (3–4 marks)
Also show proper understanding of sustainability and identify that all the plant and animal species that make up the biosphere should be preserved and passed on for future people to use and enjoy.

How have humans affected the biosphere and how might it be conserved?

53

Why conserve wetlands?

Wetlands are among the world's most ecologically productive environments. They are places of extremely rich biodiversity, supporting high concentrations of birds, mammals, reptiles, amphibians and fish (Figure 11). A wetland is considered internationally important if it:

● Contains a representative, rare, or unique example of a natural wetland type found locally

● Supports vulnerable, endangered, or critically endangered species

● Regularly supports 20,000 or more waterbirds

● Supports a significant proportion of native fish species.

Wetlands are also important sources of water for people, but they are threatened by population growth and the ever-rising demand for irrigation. Climate change worsens the situation by threatening higher temperatures and water shortages in some vulnerable regions, such as south-east Asia.

Sustainable biosphere management at the local level

The principle of **sustainability** is an important factor influencing the management of the Ramsar sites. It is hoped that humans will use each protected Ramsar wetland so it can benefit present generations and meet the needs of future generations. Another good example of sustainable management plan can be seen in Scotland.

Figure 11: Wetlands have rich biodiversity

The Caledonian Forest

The remaining fragments of Scotland's Caledonian Forest are an important biological resource. They are all that remains of the original forest that first colonised Scotland at the end of the last Ice Age (10,000 years ago). The forest is an important 'environmental inheritance' for people living in the Scottish Highlands. The European Union has provided funding for the restoration of this internationally important habitat for the benefit of future generations. One of the objectives is to bring back lost animal habitats and populations within a part of the forest known as Glen Affric.

The aim is to restore biodiversity to its natural level by adopting a countryside management strategy known as environmental stewardship. One of the first decisions taken was to reintroduce wild boar (Figure 12) to Glen Affric, a species that has been hunted to extinction over the years. A close relative of pigs, wild boars feed on all sorts of forest floor organisms (such as fungi and insects). They disturb forest floor vegetation in ways that can aid the growth of trees. Forest managers describe the boars as a very useful 'ground disturbance force'.

There are similar proposals to reintroduce other lost animals in some Scottish forest, including wolves, lynx and brown bears! These are all regarded as dangerous animals by the public, which makes the idea highly controversial.

Figure 12: Wild boar reintroduced to British forest

**Quick notes
(The Caledonian Forest):**
● Sustainable living involves protecting the natural environment so that future people may enjoy it too.
● Where damage has been done in the past, it is still possible to make repairs.
● Controversial decisions are sometimes taken by the people responsible for protecting the natural environment.

examzone
Know Zone
Battle for the biosphere

The biosphere contains a huge diversity of plant and animal species. This biodiversity is an important resource, as it provides humans with important goods and services. These vital resources are increasingly threatened and degraded by human activity, and urgent action is needed to conserve them for future generations.

You should know...

- ☐ How to describe the distribution of biomes across the Earth's surface
- ☐ How climate (temperature and precipitation) influences the distribution and types of biomes
- ☐ How local factors, such as altitude, also influences biome distribution
- ☐ How the biosphere provides important services to humans
- ☐ How the biosphere provides goods (resources) that humans use
- ☐ How goods and services support human life on Earth
- ☐ How humans are directly degrading biomes by their actions, such as deforestation
- ☐ A named example of biome degradation
- ☐ How global climate change is affecting biomes
- ☐ Why humans need to conserve biomes and biodiversity
- ☐ How to define sustainable and unsustainable
- ☐ Why humans need to use biomes more sustainably in the future
- ☐ How global actions and agreements could help make this possible
- ☐ Examples of global actions and agreements
- ☐ How local and national management could conserve biomes
- ☐ Examples of local and national management

Key terms

Biodiversity
Biofuels
Biome
Biosphere
CITES
Conservation
Deforestation

Degradation
Gene pool
Goods
Hydrological cycle
Ramsar
Services

Superpower countries
Sustainability
Unsustainable
Water table
Wilderness

Which key terms match the following definitions?

A The chopping down and removal of trees to clear an area of forest

B A naturally occurring process or event which has the potential to cause loss of life or property

C Uncultivated, uninhabited and inhospitable regions

D The world's most powerful and influential nations – the USA and, increasingly, China and India

E The number and variety of living species found in a specific area

F The ability to keep something (such as the quality of life) going at the same rate or level

G A plant and animal community covering a large area of the Earth's surface

H The level in the soil or bedrock below which water is usually present

To check your answers, look at the glossary on page 313

ResultsPlus
Maximise your marks

Foundation Question: Describe some of the goods and services the biosphere provides humans with. (4 marks)

Student answer ● (achieving 2 marks)	Examiner comments	Build a better answer △ (achieving 4 marks)
The biosphere gives humans goods and services. Plants add oxygen into the air. Humans might use the biosphere for food. It could be to do with floods.	• *The biosphere gives...* This part of the answer just repeats the question so does not score any marks. • *Plants add oxygen...* is correct and scores 1 mark. • *Humans might use...* scores 1 mark but food is not very specific. • *It could be...* does not score any marks because it is not linked to goods or services.	People can cut trees down for timber, which are goods. Plants add oxygen into the air. Humans can hunt animals for food. Forests can stop flooding from happening.

Overall comment: The student would have scored more marks if they had focused more carefully on goods and services. Also, remember that there is no need to repeat the question in your answer.

- -

Higher Question: Describe **two** services the biosphere provides and explain why they are important. (4 marks)

Student answer △ (achieving 3 marks)	Examiner comments	Build a better answer △ (achieving 4 marks)
The biosphere regulates water movement around the hydrological cycle. This can prevent flooding from destroying homes. It also makes the atmosphere have the right amount of gases. This is really important.	• *The biosphere regulates...* This is a good start to the answer as the student describes a service and uses some good geographical terminology. This part of the answer scores 1 mark. • *This can prevent...* This also scores 1 mark because it links to a service already described and explains why flooding is important. • *It also makes...* This describes another service and scores 1 mark, but it is only partly correct. • *This is really important.* No explanation is given so this part of the answer does not score any marks.	The biosphere regulates water movement around the hydrological cycle. This can prevent flooding from destroying homes. It also regulates the amount of carbon dioxide and oxygen in the air. This means the air is breathable for humans and animals.

Overall comment: Although there is some good information and examples included, the answer covered only one challenge – flooding. Look carefully at questions and spot how many different points are being asked for.

Chapter 4 Water world

Objectives

- Recognise the main flows and stores of the hydrological cycle.

- ◉ Explain how changes in water supplies can impact on people and ecosystems.

- ◉ Consider how hydrological systems might respond to climate change.

Key:

361	Water flow. The number of cubic kilometres of water that falls or rises each year.
13	Water store. The number of cubic kilometres of water that is stored.

Why is water important to the health of the planet?

The hydrological cycle

The Earth's water is continuously on the move. It is constantly recycled, moving from place to place via key processes called **water flows**. Sometimes it remains held for a period of time in a **water store**, such as glacier ice, before continuing its journey again. The complete picture of what happens is called the global **hydrological cycle** (Figure 1). The key flows are detailed in the table below.

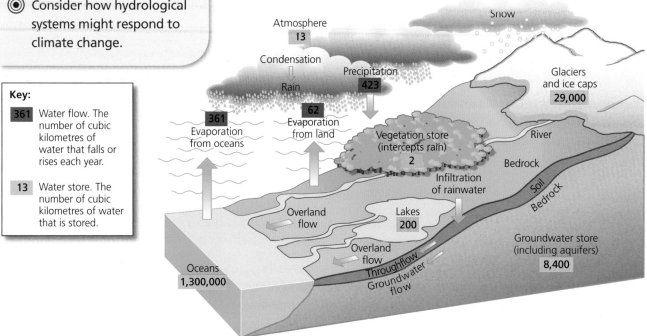

Figure 1: The global hydrological cycle

Evaporation	Water will return to the atmosphere in vapour form, once it has been warmed by sunlight.
Condensation	When water vapour cools down again, tiny water droplets form, often becoming visible as cloud.
Precipitation	Tiny water droplets collide and grow, building bigger droplets that then – under gravity – fall to the ground as rain or snow.
Overland flow	Some rainfall flows quickly over the surface of the ground to reach rivers and lakes. Overland flow is often responsible for causing flash floods.
Throughflow	Sometimes rainwater has time to soak into the soil (this is called **infiltration**). The water then flows slowly through the soil until it reaches a river.
Groundwater flow	Once it has entered the soil, some rainwater soaks into the rock below. It moves very slowly through tiny air spaces in the rock, called **pores**, or it flows along cracks, called **joints**.

Water stores

While moving through the stages of the hydrological cycle, water sometimes stops flowing for periods of time. Instead, it becomes temporarily stored in lakes, rivers, ice caps or the oceans. 99% of the Earth's water is kept in the last two of these – the ice caps and the oceans. Movement through the cycle begins again when the Sun's energy triggers evaporation or ice melting.

The Earth's biosphere also captures and stores water. Vegetation leaves are a temporary storage space where water sits for a time, until it evaporates or drips to the ground. In vegetated drainage basins, therefore, the movement of rainwater becomes a **regulated flow** – because of the plants. This means that river levels do not rise and fall too quickly, even when sudden bursts of heavy rainfall arrive. For this reason, forest is an excellent flood defence. It intercepts rainfall that might otherwise reach the river too quickly and cause a flash flood.

The Earth's lithosphere also acts as a water store. Soils and rocks can hold water for a period of time. The actual volume that can be stored is determined by how **permeable** substances are, along with other important geological factors such as slope angle and jointing. Sometimes a layer of permeable chalk lies on top of a different rock type that is **impermeable** – it will not let water pass through. Rainwater can soak down into the tiny pores (air spaces) found in the chalk, but the impermeable layer below acts as a barrier, and a water store – called an **aquifer** – develops.

Skills Builder 1

Study Figure 1.

(a) Name the two largest water stores.

(b) Describe the fastest route that rain falling over land can take to reach the ocean.

(c) Explain why tree planting slows water movement through the hydrological cycle.

(d) Explain why some clouds do not produce precipitation.

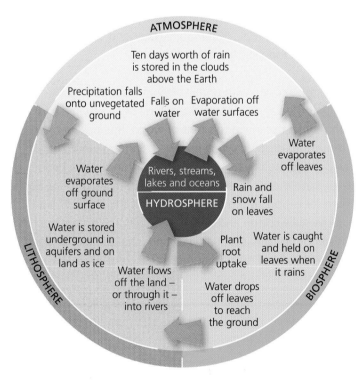

Figure 2: The links between the hydrosphere, atmosphere, biosphere and lithosphere

Unreliable water supplies

The flows of the hydrological cycle vary over different time scales, especially in places that naturally experience a fluctuating or seasonally changing climate:

Seasonal variations	Some climates have distinct wet and dry seasons.
Longer natural cycles	Natural weather cycles bring clusters of drier or wetter years in some regions of the world, for reasons not always fully understood.
Climate change	Year-on-year temperature rises are being recorded across the world. These data suggest global warming is taking place, affecting all areas.

The Sahel region (see page 119), where rainfall is notoriously unreliable, suffers from all three problems (Figure 3). The climate here naturally has a marked wet and dry season. The wet season is short, and when the rain does fall, it comes in high-intensity bursts. Torrential rains in August and September 2006 affected thousands of people throughout the region and 50,000 people were left homeless in Niger. However, the rainwater ran over the land so quickly that little was captured or stored, so water availability did not actually increase for many local people.

Since the 1970s, the region has also been in the grip of a longer-term drought. Figures suggest rainfall in some parts of the Sahel has fallen by 30% over this time. During most years of the past three decades, the Sahel countries – Burkina Faso, Chad, Mali, Mauritania, Niger and Senegal – have suffered chronic water and food shortages.

In the past, these countries coped with unreliable water supplies. Long-term records tell us that the Sahel's climate has always been variable and that cyclical drought is a fact of life in this part of world. But local people used to be more resilient to drought when it arrived. Farmers were often nomadic and moved to find new water if local supplies ran short. Changes in agriculture, however, have meant that people cannot do this any more. Food is increasingly grown for export by big businesses – using all the best land. The expansion of large-scale cotton and peanut production has permanently pushed many subsistence farmers into marginal lands that are too dry for most farming. Hunger, thirst and malnutrition follow.

Activity 1

Local people can experience water shortages in places like the Sahel, even when rain has fallen recently.

Think about how each of the following can impact on water availability for local people:

- lack of technology
- civil war
- cash crops
- population increases.

Is it possible that water shortages actually owe more to human rather than physical factors?

Figure 3: The Sahel suffers from seasonal variations, longer natural cycles and climate change

Case study: Australia's unreliable water

Australia is a naturally arid continent (Figure 4). However, water supplies have recently become even more unreliable than usual. Since the 1990s, Australia has experienced a decade-long dry spell. Rainfall has fallen to 25% of the long-term average. In 2007, the Murray River reached its lowest level in over 100 years of records. Some Australian geographers say that this is the worst drought for 1,000 years. In 2009, the country experienced terrible bush fires.

Several reasons exist why water shortages are even greater than usual. El Niño is a complex global weather pattern that periodically reduces Australian rainfall. It has caused further drying-up of this already arid continent. Over-use of water by agriculture and industry has made the situation even worse. It may also be that we are seeing the first major effects of climate change and global warming in Australia. Impacts of increased aridity include:

- Wheat crop yields have fallen by around one half and people face big food price rises.
- Hundreds of winemakers could be forced out of their AU$2.6 billion business.
- The government says it will pay severely affected farmers AU$150,000 each to walk away from their land. This 'exit with dignity' strategy will help farmers retrain or relocate.
- Local **ecosystems** are suffering. Wells dug by sheep farmers are drying up, and local native animals like kangaroos lose their use of them too.

A new technology called electromagnetic imaging is helping some farmers survive by finding new hidden water stores beneath their fields. However, if current trends continue, Australia also needs to seriously change its water consumption habits.

Figure 4: The Australian desert, where rain may be starting to fall even less frequently

Case study quick notes:
- Unreliable water supplies are a natural feature of life in some places.
- Unreliable water supplies threaten a range of agricultural and industrial activities and seem to be getting worse.
- Farmers must try adapting to the increased aridity or else 'retreat' and change their occupation.

Describe two ways in which climate change could impact on water supplies. (4 marks)

■ **Basic answers** (0–1 marks)
State that climate change would make the sea level rise. People do not take their drinking water from the sea, so this is not a good answer about changing water supplies.

● **Good answers** (2 marks)
Describe two climate change impacts, one of which is well connected with the idea of changing water supplies. For example, the disappearance of glaciers might deprive some places of their regular drinking supplies from annual meltwater.

▲ **Excellent answers** (3–4 marks)
Provide two developed impacts. For example, if more storms start passing over the UK, it could add water to reservoirs (meaning that climate change could actually help water supplies in some cases).

Climate change and the world's water supplies

How might climate change and global warming impact on the world's hydrological systems? Some places where water is naturally scarce may become even more drought-stressed in the future, as we have already learned. However, a warmer world will also be one where more evaporation takes place over the oceans – and what goes up must come down! Climate change scientists therefore believe that some already wet places will become even wetter.

There are reports of more rain-bearing storms forming over the Atlantic, close to the UK. Heavy inputs of water into Britain's drainage basins were seen in 2004 (the flash flooding of Boscastle), 2007 (widespread urban flooding) and 2008 (flooding in Wales and north-east England). Many scientists see these events as symptoms of a changing hydrological cycle. They think that in future, the UK will be faced with:

● Longer, hotter summers, especially in London and the south-east (like the extraordinarily dry summers of 1976 and 2006 that stretched UK water supplies to their limit)

● Warmer, wetter winters, with more rain-bearing storms predicted to pass over all parts of the UK, and not just northern regions.

Climate change and water supplies in America

Climate change may bring reduced rainfall inputs to places where water is already naturally scarce for all or part of the year. People and ecosystems in such areas could become seriously threatened. One such imperilled region is the south-west of the United States. Already far too dependent on irrigation from the Colorado River (see pages 64–65), this region is naturally very dry. Yet despite this arid climate, people like the hot weather there, and the state of Arizona saw a 40% rise in population in the 1990s. Just across the border, in Nevada, the growing city of Las Vegas doubled its consumption of water between 1985 and 2000. However, severe drought conditions have recently begun to affect the entire region. For Las Vegas, 2002, 2004 and 2007 were some of the very driest years on record.

The city's authorities have responded by removing the dried-up remains of some sports pitches and re-covering the ground with plastic turf. Drought-tolerant 'desert landscaping' is now encouraged instead of planting grass lawns in front of new buildings. This could be bad news for Las Vegas wildlife such as snakes and grey squirrels, who have taken up residence in the artificially irrigated areas.

Figure 5: Fountains in Las Vegas. Can this kind of water use continue in the Nevada desert?

Climate change and water supplies in Asia

Threatened water supplies in Asia give climate change experts their greatest cause for concern – because half of the world's population live in this region. Major rivers in China, India and Vietnam are fed by seasonal melting of Himalayan glaciers, especially those on the Tibetan Plateau. Every summer, icy water pours off the Tibetan plateau, and flows from this massive hydrological store feed the Yellow River, the Mekong and other major rivers. As long as new glacier ice accumulates each winter, cities and ecosystems in the region are guaranteed a sustainable water supply. However, a much warmer climate would lead to permanent melting – and ultimately to the disappearance of Himalayan glaciers. Billions of people would experience severely reduced water supplies.

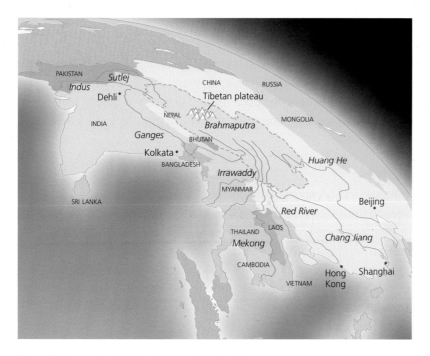

Figure 6: Seasonal glacial meltwater flows in Asia

Skills Builder 2

Study Figure 6.

(a) Name one city that receives seasonal flows of meltwater from the Himalayas.

(b) Describe possible benefits that seasonal ice melting brings to people in this region.

(c) Explain why the permanent melting of Himalayan ice would bring disaster.

Objectives

- Recognise ways in which people interfere with water quality and water supplies.

- Explain the costs and benefits brought by management of the Colorado River.

- Appreciate how innovative thinking can help solve local water problems.

ResultsPlus
Exam Question Report

Describe the causes of river pollution and explain how people dealt with this pollution. (5 marks, June 2006)

How students answered

Many answers consisted of simple statements or words that did not really describe a problem (a few just wrote the word 'chemicals').

50% (0–1 marks)

Some students described a couple of problems (industry and farm pollution) but without giving specific details.

40% (2–3 marks)

Few students gained 4 or 5 marks by providing some detail about two or more causes of pollution and attempts to tackle this problem.

10% (4–5 marks)

Figure 7: Pollution in the Kabul River, Afghanistan

How can water resources be sustainably managed?

Unhealthy hydrology – problems with water quality

The human pressure on rivers is enormous – they are used for transport, for industrial processes, for drinking and for sewage disposal. Rivers in industrialising nations like India and China experience especially damaging impacts because, as yet, they have few laws in place to protect their water and ecosystems. **River pollution** in rivers outside Europe and North America is often appalling (Figure 7). Three common types of damaging pollution are shown in the table.

Human excrement	Cholera bacteria spread quickly in river water when contaminated human waste is added. As long as toilet waste goes untreated, the cycle of cholera infection will continue.
Toxic chemicals	Liverpool's River Mersey used to be hugely polluted. Alkali factories used the polluting Leblanc process to make vital chemicals for the manufacture of textiles. Vast amounts of calcium sulphide by-product were leached into local streams, making conditions poisonous for wildlife. Fortunately, this is no longer the case today. The factories have mostly disappeared, while the UK Environment Agency can issue heavy fines to any remaining polluters.
Plastic bags	In China and Bangladesh, use of thin plastic bags (less than 0.025 mm thick) is prohibited because of pollution concerns. The bags are blamed for blocking waterways and sewers. This form of pollution worsens flooding problems, especially in the monsoon season. Plastic bags also contribute to health problems in many countries. They get trapped by vegetation along river banks, providing pools of still warm water where malaria-bearing mosquitoes will breed.

Disrupted water supplies

It is not just water quality that can be harmed by human actions. Water supplies are also interfered with in damaging ways. The flows of the hydrological cycle can be speeded up or slowed down, and natural water stores such as lakes may shrink because of over-use by people – the Aral Sea is a good example of this.

Changing flows

Overland flow is a very important hydrological flow, as Figure 1 showed. Too much overland flow may result in an over-supply of water to rivers, possibly leading to flooding. One common cause for accelerated overland flow is **deforestation**. Without trees to intercept and store rainfall, overland flow is more likely to occur. Another cause is urbanisation. In urban areas, the soil is covered by tarmac and concrete. As a result, rainwater travels much more quickly towards the nearest river via overland flow. Drainpipes also speed up urban water flows.

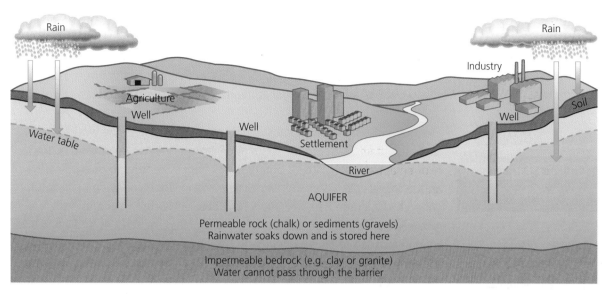

Figure 8: The features of an aquifer

Changing stores

Water stores are essential for human societies. However, they need careful management if they are to remain a sustainable resource. Water stores are naturally recharged whenever it rains. However, if too much water is taken from them too quickly then they will dry out before they have a chance to recharge. Many major cities like Mexico City are sited above **groundwater** aquifers. However, **over abstraction** (over-pumping) sometimes takes place. When water is removed from the pore spaces in permeable rock, the rock loses strength. If too much weight has been added overhead by buildings, empty pores close up (imagine pressing down on a Swiss cheese full of holes. Ground beneath the city becomes compacted, leading to a permanent loss of water storage capacity.

Coca-Cola and the Plachimada aquifer

In the drought-prone Indian state of Kerala, an aquifer lies close to the village of Plachimada. In 2000, this water supply attracted the attention of the transnational corporation Coca-Cola. The company was keen to increase sales of its famous drink in India, but the most important ingredient in its production – clean water – was not always easy to find locally. Groundwater supplies were needed.

First, Coca-Cola set up a subsidiary firm called Hindustan Coca-Cola Beverages. Next, they established a bottling plant near Plachimada. Finally, six wells were dug, tapping into the precious groundwater store. Very soon afterwards, water shortages began to be reported. By 2004, news from the region described a desperate situation. Wells used by the Plachimada villagers had dried up.

It is likely that drought in the region would have brought water shortages anyway, even without Coca-Cola's arrival. However, it was probably not a smart move to make industrial use of this aquifer, given the region's known drought risk. Coca-Cola has since helped the people of Plachimada survive the drought by driving tankers full of fresh water to the village.

Skills Builder 3

Study Figure 8.

(a) Name one impermeable rock type.

(b) Explain how the presence of impermeable rock helps create a water store.

(c) Suggest how aquifers impact on population distribution in arid regions.

Quick notes (Coca-Cola):
- Economic development is more likely to be located near hydrological stores.
- Water stores in dry areas must be used carefully if they are to remain a sustainable resource for local people.

Large-scale water management: The Colorado River

Some of the world's biggest construction projects have been involved with **water management** on an immense scale. China's Three Gorges Dam project, which is nearing completion after 15 years work is well worth investigating. Also worthy of an in-depth look is the management of the Colorado River.

The Colorado is one of the most highly managed rivers in the world. Its water feeds the needs of seven US states and Mexico. The natural regime of the river was completely altered during the twentieth century to give US cities and farms reliable water supplies all year round. This is because in its natural state, the Colorado had a very low flow between September and April, whereas during the summer months, snowmelt in the Rockies and Wind River Mountains used to cause a big rise in river levels. In the most extreme years of the early twentieth century, the Colorado's discharge was 13 times higher in mid summer than in winter.

Since then, water management projects have radically changed the regime of the Colorado. Flood peaks have been smoothed out by dam building and reservoir construction. The Hoover Dam was built in 1935. Regarded as a technological marvel at the time, the dam stores the equivalent of two years' river flow in Lake Mead.

The Glen Canyon Dam followed in 1963. It also holds an enormous amount of water. The two lakes build up reliable supplies for thirsty desert cities like Las Vegas. Water is piped along man-made constructions called aqueducts.

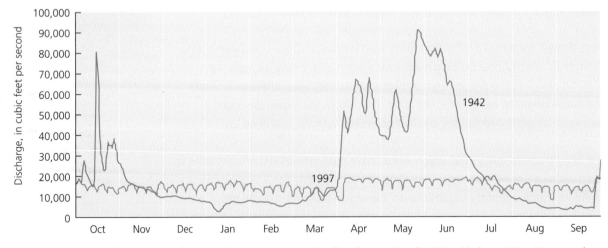

Figure 9: The Colorado River regime before and after construction of the Glen Canyon Dam (10,000 cubic feet = 283 cubic metres)

However, these direct changes to the annual pattern of river flow have had negative knock-on impacts for river processes, river valley landforms and river ecosystems:

● Silts and sands get trapped behind the dams. Rates of transportation of sediments have fallen nearly to zero since the dams were built. This makes the water colder (silt heats up in sunlight, warming water around it). Four of the Colorado's fish species, including the squawfish, have been lost. Only trout, which prefer clear cool water, have risen in numbers as a result of the changes.

● Sandbanks along the sides of the river in its lower course have been starved of sediment and are reduced in size. Plants and animals that live on the sandbanks – such as the Kanab snail and the willow flycatcher – have also declined. The sandbanks were once used for fishing and rafting but they cannot be used any longer.

US government agencies introduced new management measures in 1996 to tackle these problems (Figure 11). Four giant steel pipes have recently been fitted to Glen Canyon Dam. Every few years they are opened, releasing massive jets of water from the reservoir. This escaping water has high energy levels. It can flush out large amounts of sediment that have built up behind the dam. As the water heads downstream it begins to lose energy and sediments are deposited, helping rebuild the lost sandbanks. In time, scientists believe they can restore river and ecosystem conditions to a more natural state.

Figure 10: View of the Colorado River and the surrounding area

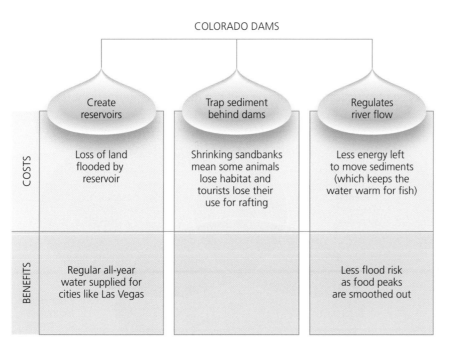

COLORADO DAMS

	Create reservoirs	Trap sediment behind dams	Regulates river flow
COSTS	Loss of land flooded by reservoir	Shrinking sandbanks mean some animals lose habitat and tourists lose their use for rafting	Less energy left to move sediments (which keeps the water warm for fish)
BENEFITS	Regular all-year water supplied for cities like Las Vegas		Less flood risk as food peaks are smoothed out

Figure 11: Costs and benefits for the Colorado River

Decision-making skills

Why are the following four people likely to have conflicting views about the use of the Colorado River?:

- Californian farmers who need the irrigation water in huge quantities
- Native Americans on whose land the giant dams are built
- Californian urban dwellers who need huge quantities of water for golf courses and gardens
- Mexicans near the mouth of the Colorado who would also like some water for their farming

ResultsPlus
Build Better Answers

Using examples, explain how human activity can interfere with the size of water supplies. (4 marks)

■ **Basic answers** (0–1 marks)
Ignore the word 'size', and instead deal with pollution and water quality.

● **Good answers** (2 marks)
Clearly describe how water supplies are improved by dam building (e.g. the Colorado) or made worse by over-using underground supplies.

▲ **Excellent answers** (3–4 marks)
Give good details of more than one example, or show sound understanding of physical processes (e.g. mentioning the Plachimada aquifer, or explaining that aquifer use becomes unsustainable if all the water is extracted before new rain refills it).

Decision-making skills

Draw a diagram to show how small-scale water projects can be environmentally, socially and economically sustainable.

Small-scale water management

In water-stressed areas, small-scale solutions can be introduced that keep water supplies more constant throughout the year. Poor countries especially can benefit from the use of **intermediate technology** – appropriate, small-scale, practical solutions that local people can apply and maintain themselves. They include the use of:

Lined wells	The traditional method of obtaining groundwater in rural areas of the developing world, and still the most common, is by means of hand-dug wells. To avoid sewage water contaminating the well water, the sides need to be lined with concrete rings, and a concrete cover slab should be placed over the well when not in use.
Handpumps	Raising water from the ground using a pump is preferable to using a bucket and rope. It is a more efficient method, and there is less chance of the water becoming contaminated by mud and dirty hands.
Rain barrels	These allow **water harvesting** to take place. Roof gutters can be arranged so that rainwater flows into storage barrels. The best barrels contain a purification system. The rainwater soaks through filters and layers of sand, gravel or charcoal which help remove solids and various impurities.

In addition to increased supplies, improved water quality is another common aim of intermediate technology projects (Figure 12). Without improved water quality, cholera or diarrhoea often remain persistent problems for communities. Cholera bacteria find their way into the human small intestine after being swallowed. There they thrive, drawing water into the bowel by osmosis. As a result, the sufferer experiences violent diarrhoea. New generations of thriving bacteria rapidly exit the unhappy victim. Soon they are back in the river or they have leached into unlined wells, once again polluting potential drinking water. This cycle of infection will continue without help from intermediate technology.

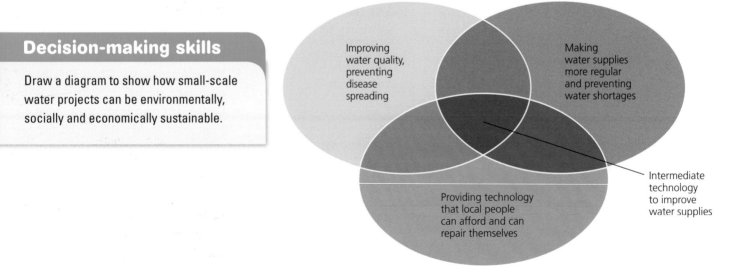

Figure 12: Intermediate technology for improved water supplies

Introducing the Afridev handpump to Tanzania

Tanzania is a desperately poor country in east Africa, where life expectancy is just 46. 70% of Tanzania's rural population and 30% of urban dwellers have no access to safe water. Diarrhoea accounts for at least 20% of infant deaths. It is a common problem in places where water supplies are contaminated with sewage, and can prove fatal, particularly to children.

International agency WaterAid brings intermediate technology to communities needing sewage-free water. Working in partnership with Tanzania's Anglican Church, for example, WaterAid helped the community of Chessa village construct a new well, 24 metres deep, fitted with an 'Afridev' handpump (Figure 13). Fifteen families can now drink safe underground water.

A key principle of WaterAid projects is that communities take ownership of the technology and are responsible for its upkeep. For this reason the technology must be simple, so that ordinary people can fix it when it breaks. The Afridev handpump provided in Chessa is not very sophisticated. Better handpumps exist with a lower rate of breakdown. However, when more advanced machines break down, a specialist engineer needs to be called in – which can leave local people without water while they wait for repairs.

Figure 13: An Afridev handpump

An Afridev handpump may break down fairly often but it can quickly be repaired by a local caretaker. In Chessa, brothers Vincent and Lazaro William volunteered to be the Afridev handpump's caretakers. After one week's training, they were given a toolkit consisting of a spanner and wrench. They take the pump apart once a year to check the whole system, and fix whatever problems arise on a day-to-day basis.

Also helping people in Tanzania is local radio soap show Pilika Pilika (Busy, Busy). WaterAid supports the show and uses it to spread messages on hygiene, sanitation and water management. In one recent episode, a major character had an unpleasant experience when he tripped and fell into a well that had been heavily polluted with toilet water. The audience were reminded that they need to keep their water supplies free of sewage!

Quick notes (Afridev handpump):
- Poor water supplies are still a major cause of death and illness for low-income countries like Tanzania.
- Charities like WaterAid do vital work providing aid for improved water supplies in poor countries.
- Intermediate technology provides clean water in a way that allows local people to take ownership of the project.

Activity 2

WaterAid's website (www.wateraid.org/) contains a wealth of up-to-date case studies (from 17 countries) and technology details appropriate for GCSE students.

You could work in groups and prepare short presentations to share with your classmates, perhaps using PowerPoint.

examzone

Know Zone
Water world

Water is vital for life on our planet. Water moves around the Earth through an amazing system called the hydrological cycle. This ensures our water supply by regulating the flow of water and keeping water clean. If humans interfere with the hydrological cycle it can threaten our water supply and quality of life.

You should know...

☐ How the hydrological cycle links to the biosphere, atmosphere and lithosphere

☐ Why the global hydrological cycle is called a system

☐ The key processes, flows and stores of the hydrological cycle, and their names

☐ Why some areas, like the Sahel, have unreliable water supply

☐ How changing water supply has affected Australia

☐ How too much or too little water can lead to problems in the UK, Asia and America

☐ How humans reduce water quality through pollution

☐ How future climate change might affect water supplies

☐ How human use of stores and flows can reduce water supply

☐ How large-scale water management can interfere with water supply

☐ How the Colorado River management has had positive and negative impacts

☐ What intermediate technology is

☐ How intermediate technology might improve water supply in the developing world.

Key terms

Aquifer	Over abstraction	Water
Deforestation	Permeable	management
Ecosystem	Pollution	schemes
Groundwater	Pores	Water store
Hydrological	Precipitation	Water table
cycle	Regulated flow	
Impermeable	River pollution	
Infiltration	Throughflow	
Intermediate	Water flow	
technology	Water	
Joints	harvesting	

Which key terms match the following definitions?

A When water is being used more quickly than it is being replaced

B Storing rainwater or used water ('grey water') for use in periods of drought

C A community of plants and animals that interact with each other and their physical environment

D A technology that the local community is able to use relatively easily and without much cost

E Small air spaces found in a rock or other material that can also be filled with water

F An underground store of water, formed when water-bearing (permeable) rocks lie on top of impermeable rocks

G When moisture falls from the atmosphere – as rain, hail, sleet or snow.

H Lines of weakness in a rock that water can pass along

To check your answers, look at the glossary on page 313

ResultsPlus
Maximise your marks

Foundation Question: Describe how deforestation could affect water-cycle processes. (3 marks)

Student answer ■ (achieving 1 mark)	Examiner comments	Build a better answer △ (achieving 3 marks)
Deforestation could increase interception. *There would be more surface run-off.* *This will soak in water.*	• *Deforestation could...* is incorrect so does not score any marks. • *There would be...* This sentence is good and includes geographical terminology. This scores 1 mark. • *This will soak...* This part of the answer is not precise and the meaning of *soak in* is not clear.	Deforestation would decrease interception. There would be more surface run-off. The amount of infiltration will fall.

Overall comment: This answer shows how important it is to learn key terms, such as infiltration and interception, and use them accurately. You should avoid using phrases like 'soak in', which are not correct geographical terms.

- -

Higher Question: Explain how human activity could change the amount of infiltration. (3 marks)

Student answer ● (achieving 2 marks)	Examiner comments	Build a better answer △ (achieving 3 marks)
If humans build impermeable surfaces like roads and roofs infiltration falls. *Around some rivers trees are planted to intercept water.* *This decreases infiltration.*	• *If humans build...* scores 1 mark. It is a good start to the answer and uses an example. • *Around some rivers...* is also correct and uses a key term, so scores 1 mark. • *This decreases infiltration* is incorrect.	If humans build impermeable surfaces like roads and roofs infiltration falls. Around some rivers trees are planted to intercept water. This increases infiltration.

Overall comment: This is a good answer, including key terms such as infiltration and interception. The student explains how humans affect processes, but made a mistake with the last point – possibly because they were rushing.

Chapter 5 Coastal change and conflict

Objectives

- Know that there are different types of rock and that they affect the coastline.

- Be able to describe some of the landforms found on coastlines with hard or soft rock.

- Understand how geology and a range of processes impact on the coastal landforms.

How are different coastlines produced by physical processes?

Coasts and geological structure

Many coasts around Britain are under attack from the waves. Britain's **geological structure** is varied and the various coasts are made up of different rocks. In some places, such as Land's End in Cornwall, the rocks are made of granite. Granite is a 'hard' rock – it is resistant to **erosion** – with relatively few points of weakness such as joints and faults (cracks in the rock). This hardness means that it is not easy for the waves to erode the rock (wear it away). So the coast here has high, steep cliffs, with very few beaches (Figure 1).

Figure 1: Granite cliffs near Land's End

The coast looks different in other places, such as the east coast of Yorkshire. South of Scarborough, the coast is made up of softer boulder clays which are more easily eroded by the waves. The result is that the rocks are eroded quickly and the coastline retreats nearly two metres every year. Figure 2 shows what this coast looks like. The table below summarises the differences between the two coasts.

Contrasts in the cliffs in hard rock and in soft rock areas

	Hard rock coast	Soft rock coast
Shape of cliffs	High, steep and rugged	May be high but are less rugged and not so steep.
On the cliff face	Cliff face is often bare, with no vegetation and little loose rock.	There may be piles of mud and clay which have slipped down the face of the cliff.
At the foot of the cliff	A few boulders and rocks which have fallen from the cliff.	Very few rocks, some sand and mud.

Figure 2: The Yorkshire coast, south of Scarborough

How do waves erode the coast?

Wave erosion involves several different processes, including:

Hydraulic action	This results from the force of the water hitting the cliffs, often forcing pockets of air into cracks and crevices in a cliff face.
Abrasion	This is caused by the waves picking up stones and hurling them at the cliffs and so wearing the cliff away.
Attrition	Any material carried by the waves will become rounder and smaller over time as it collides with other particles and all the sharp edges get knocked off.

What landforms are created at the coast by wave action?

Although headlands are made of resistant rock, they gradually become eroded. Waves approaching the coastline bend around headland and erode it from both sides. Any points of weakness, such as cracks or joints in the rock, become eroded more quickly than the surrounding rock. This forms small caves on either side. With more erosion taking place over time, the two caves may join to form an arch. More erosion, and **weathering** of the arch roof, leads to the roof collapsing. This leaves an isolated **stack**. The stack itself gradually becomes undercut by wave erosion, eventually collapsing to leave a **stump**, which may only be visible at low tide. All of these landforms are shown in Figure 4.

Concordant and discordant coasts

Coasts in some places, such as the Swanage area (Figure 5), are made up of both hard and soft rocks. The harder rocks, which here are chalk and limestone, are more difficult for the waves to erode, so they stand out as **headlands**. The softer rocks, which here are clays, are more easily eroded and so form the **bays**. In the Swanage area the alternating hard and soft rocks are at right angles to the coast. This coast is called a **discordant coast**.

In other places, such as the coast near Lulworth Cove, the alternating bands of hard and soft rock run parallel to the coast. Figure 6 on page 72 shows the resulting coastline when the rocks are eroded at different rates. This is called a **concordant coast**. The entrance to the cove is narrow where the waves have cut through the resistant limestone. Then the cove widens where the softer clays have been more easily eroded. At the back of the cove is a band of more resistant chalk, so erosion is slower here.

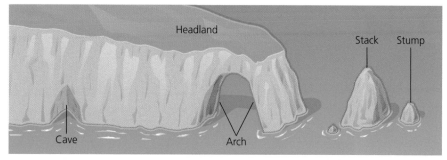

Figure 4: Landforms produced by the erosion of a headland

Figure 6 on page 72

Skills Builder 1

Figure 3: Landforms of erosion at Dyrholaey, southern Iceland

Study Figure 3.

(a) Name the landforms labelled A and B.

(b) Describe how these landforms might change over time.

(c) Explain how natural processes might cause these changes.

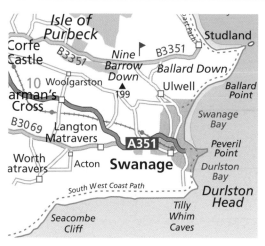

Figure 5: The Swanage coast

Skills Builder 2

Study Figure 6.

(a) Label the band of limestone rock.

(b) Label the band of clay.

(c) Label the band of chalk.

(d) Explain how the cove was formed.

Skills Builder 3

1. Find a photograph of a bay and headland coastline from the Internet.

2. Label the bays and headlands, using drawing tools.

3. Annotate to suggest how the bays and headlands have been formed, using text boxes.

Figure 6: Aerial view of Lulworth Cove

Different types of waves

There are different types of wave that move towards the coast. When waves break, water rushes up the beach, due to the energy from the wave. This is called the **swash**. When the water has lost its energy further up the beach, it runs back down again, under gravity. This is the **backwash**. Big waves break with lots of force and energy and this means they have the power to carry out erosion of beaches or rocks on the coast. These are called **destructive waves** and you can see their main features in Figure 7. As well as being tall and breaking with lots of power, they usually arrive quickly and have a high frequency – a lot of them come in a short period of time. Their backwash is greater than their swash and so sediment is taken away from the beach, back into the sea.

Constructive waves are very different. In calm conditions – without much wind – the waves are usually small, weak and with a low frequency (Figure 8). They don't break with much force and they tend to add sand and other sediment to the coastline by **deposition**. The swash is greater than the backwash and so sediment is pushed up the beach, helping to build it up into a beach.

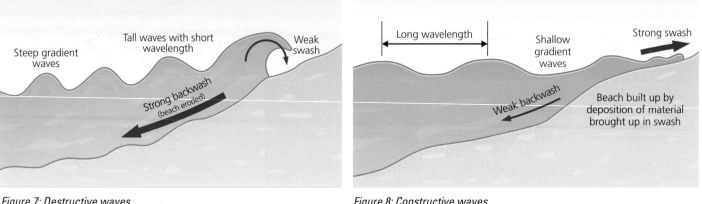

Figure 7: Destructive waves

Figure 8: Constructive waves

Retreating cliffs

The landforms that are produced by the processes working on cliffs can be seen in many coastal areas. The wave-cut platform is a fairly flat rocky area at the base of a cliff that is being eroded. Waves cause undercutting at the base of the slope, as Figure 9 shows. This forms a notch that gradually gets bigger. The rock above eventually loses its support and then it collapses. The debris is gradually washed away by the waves.

The process is repeated, and so the cliff slowly retreats backwards and becomes steeper. At Seven Sisters in Kent, the cliffs are up to 160 metres high, while beneath them the wave-cut platforms extend up to 540 metres out to sea.

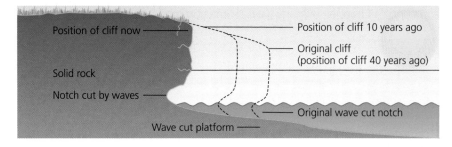

Figure 9: Retreating cliffs

Longshore drift

Where waves approach the coastline at an angle, when they break their swash pushes beach material up the beach at the same angle. The backwash then drags the material down the beach at a 90° angle, due to the force of gravity. This produces a zig-zag movement of sediment along the beach known as **longshore drift**, which you can see in Figure 10. The smallest material, such as fine sand, is easily moved and so ends up furthest along the beach. The largest materials – pebbles perhaps – are heavy and are not moved as far.

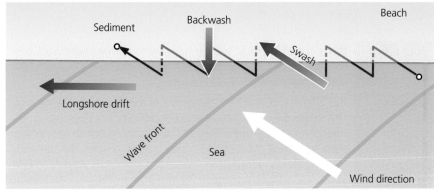

Figure 10: The process of longshore drift

The formation of beaches, spits and bars

Constructive waves add sediment to the coastline and lead to the formation of a beach. This is particularly the case on low-energy coastlines that are sheltered from strong winds and waves, such as in bays between two headlands. A good example is Swanage Bay in Dorset.

Skills Builder 4

(a) Draw a sketch of the spit in Figure 11.

(b) Label the spit, the curved end and the salt marsh.

(c) Annotate your sketch to explain how the spit has formed and developed.

Figure 11: Aerial view of the Dawlish Warren spit

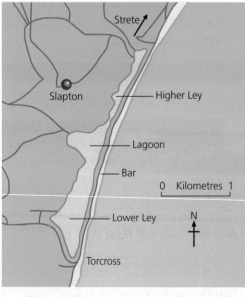

Figure 12: The bar and lagoon at Slapton in Devon

Longshore drift can carry beach sediment beyond a bend in the coastline leading to an extension of the beach into the open water, known as a **spit**. The end of the spit often becomes curved because it is exposed to strong winds and waves. Deposition can then happen in the sheltered water behind the spit, forming a salt marsh. You can see both of these features of a spit in Figure 11.

If longshore drift continues to extend the length of the spit, it may join up with the coastline on the other side of an opening such as a bay. This results in the creation of a 'bar' with a 'lagoon' behind, such as the one at Slapton, Devon, which you can see in Figure 12.

Weathering

Weathering can happen in many ways. Some common ones in coastal areas are:

Mechanical weathering	Salt crystal growth, for example, happens because sea water contains salt. When spray from waves lands on rocks, the water can be evaporated leaving the salt behind. The salt crystals grow and create stresses in the rock, causing it to break down into small fragments.
Chemical weathering	Solution, for example, happens because all rain is slightly acidic. CO_2 dissolves in water to form carbonic acid which, when rain falls on rocks, can react with weak minerals, causing them to dissolve, and the rock to decay.
Biological weathering	The roots of vegetation, for example, can grow into cracks in a rock and slit the rock apart.

Sub-aerial processes refer particularly to the disintegration of rock through processes of weathering (such as freeze-thaw and thermal expansion) and the impact of wind and rain.

Mass movement

There are several different processes of mass movement. In coastal areas, the main forms of **mass movement** are:

Rock fall	this is one of the most sudden forms of mass movement. Rock fall occurs when fragments of rock weathered from a cliff face fall under gravity and collect at the base.
Slumping	this often happens when the bottom of a cliff is eroded by waves. This makes the slope steeper and the cliff can slide downwards in a rotational manner, often triggered by saturation due to rain, which both 'lubricates' the rock and makes it much heavier. You can see how the weight added by rainwater and erosion by waves combine to cause a rotational slump in Figure 13 (page 75).

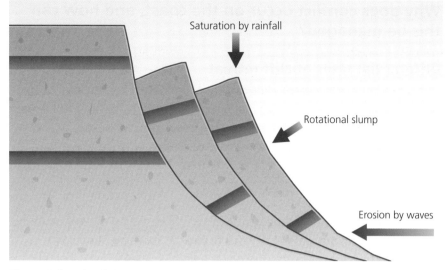

Figure 13: Rotational slump

Other factors which affect coastlines

Changing sea levels and storm activity

It is widely predicted by scientists that sea levels will rise by about 0.3 metres by the year 2100, mainly due to global climate change. This is going to put many low-lying coastal areas around the world at risk of flooding. These places include Bangladesh, the Netherlands, the Maldives and even parts of eastern England, such as the Thames Estuary. Insurance companies in the UK predict that flooding will be eight to twelve times more frequent here by 2100. These predictions have led governments to take action. In London, although the Thames Barrier is already in place (see Figure 14) to hold back very high tides, an Environment Agency project called 'Thames Estuary 2010' will install a series of new flood walls along the river. There are also plans to leave areas of open space, on to which flood water can go without damaging buildings. Both of these measures should help prevent the potentially serious effects of flooding.

Storms at sea

Strong winds and storms can increase the height of waves and tides and so cause flooding. The Environment Agency monitors sea conditions over a 24-hour period, 365 days a year. The Storm Tide Forecasting Service provides the Environment Agency with forecasts of **coastal flooding**, surge and wave activity, together with warnings when hazardous situations are seen to be developing. If people are concerned about flooding from the sea, they can contact the Environment Agency's 24-hour Flood Line, or seek advice via their website. They will advise them of what precautions they should take and what action is needed in the event of a flood emergency.

The Health Protection Agency also provides advice about how flooding can affect people's health and what precautions they should take. The following is an extract from their website:

ResultsPlus
Watch out!

75

■ Try not to confuse the words 'prediction' – suggesting what flooding is likely to happen in the future – and 'prevention', which relates to stopping the impacts of any flooding.

Figure 14: Aerial view of the Thames Barrier

Activity 1

1. Use the Internet to find another location that suffers from coastal flooding.

2. Research how this area tries to predict or prevent coastal flooding.

The main threats to health during and immediately after a flood are drowning, and injuries caused by accidents in flowing water. Walking or even driving through floodwater is risky – six inches of fast-flowing water can knock you over, and two feet of water will float your car. Manhole covers may have come off and there may be other hazards that you can't see. www.hpa.org.uk

Saturation by rainfall

Rotational slump

Erosion by waves

Objectives

- Know the factors affecting rates of cliff retreat.

- Be able to explain how people and the environment are affected by coastal retreat.

- Understand why methods of engineering used to protect the coast have both advantages and disadvantages.

ResultsPlus
Watch out!

Good examination answers on this topic will typically show awareness that rapid rates of cliff retreat are likely to be the result of two or three different causes acting together on a coastline, rather than just one.

Skills Builder 5

(a) Use an atlas to draw a sketch map of the British Isles.

(b) Add arrows to show the fetch that would directly affect the south-west of Ireland, the south coast of England and the coastline of eastern England

(c) How would these different fetches affect the coastal processes in these three locations?

Why does conflict occur on the coast, and how can this be managed?

Differential rates of cliff retreat

Erosion and retreat of coastal cliffs can happen at very different rates. Some cliffs erode at rates of nearly two metres per year, such as those at Holderness in Yorkshire, whilst others are barely eroded at all. The rate of erosion and retreat is influenced by factors such as the **fetch** (the distance of sea over which winds blow and waves move towards the coastline), the geology and coastal management strategies.

Coasts that face a major ocean, such as the south-west coast of England facing the Atlantic Ocean, have a very long fetch, and the winds are strong and persistent. This produces destructive waves with high energy that can erode cliffs at rapid rates. But the fetch is only one of the factors that affect rates of erosion and retreat. The geology of south-west England is mainly granite, which is a very resistant rock. The actual rates of erosion are, therefore, very slow – typically only a few millimetres per year. One of the reasons for the high rates of erosion and retreat at Holderness is that the geology is very weak clay.

The final factor that affects these rates is that of **coastal management**. If coastal defences such as concrete sea walls protect weak geology then rates of erosion will be much slower.

The effects of coastal retreat

Cliff retreat can have many impacts on both people and the environment. Many houses, apartments and hotels are built on cliff tops to take advantage of the wonderful sea views. Cliff retreat in many coastal areas has put these properties at risk of collapse into the sea. At Holbeck near Scarborough on the Yorkshire coast, the Holbeck Hall Hotel collapsed in June 1993 because the sea had undercut the cliff on which it stood (Figures 15 and 16).

In the UK, loss of property to cliff retreat is currently not covered by insurance policies. In the USA, about $80 million per year is paid out by a national insurance programme.

Figure 15: The Holbeck Hall Hotel, near Scarborough

Why does conflict occur on the coast, and how can this be managed?

77

Figure 16: The collapse of the Holbeck Hall Hotel

People and coastal retreat

In 1994 Sue Earle, a farmer at Cowden, near Hornsea, East Yorks had to watch while her home of 25 years was demolished because of the rapid erosion of the cliff. Miss Earle fought a losing battle against the waves of the North Sea. When Sue first moved to the farm 39 years earlier, it had been 150 metres from the sea. Unfortunately the soft clay cliffs were eroded over the years by powerful storm waves. So in 1994 – when the farm was just two metres from the cliff edge – she was forced to move out, and the farm was demolished. Sue argued that the problem accelerated when sea defences were built nearby to protect the village of Mappleton.

More recently, another family faced problems from cliff erosion at Barmston, just north of Hornsea. The Norcliffe family owned a bungalow holiday home at Barmston. By 2004 the porch was only 1.5 metres from a 25-metre drop – and the family had to move out. They were furious because this area of coastline is being left to fall into the sea. There are no defences against the waves. The Humber Estuary Coastal Shoreline Management Plan for Barmston concludes that the potential economic damage to caravan parks and isolated farms was not enough to justify the expense of building defences against the sea. The preferred option is to **'do nothing'**. The East Riding Council's coastal management team established a 'rollback' policy for caravan parks. This allows the caravan owners to retreat inland with their caravans, as the sea advances. The problem for the Norcliffe family was that the policy does not apply to houses. Neither Sue Earle nor the Norcliffe family have been able to gain any compensation for the loss of their houses.

Figure 18: The demolition of Sue Earle's farm

Coastline Year
1846
1887
1955
1978
1994

0 Kilometres 1

People, property and the coast in Holderness

The Holderness coast of East Yorkshire is a 60 km long stretch of low cliffs (20–30 metres high). The cliffs are composed of soft, easily eroded boulder clay. The cliff line is retreating at a rapid rate of nearly two metres per year, which is the fastest in Europe. Most of the erosion takes place in storms and tidal surges. The highest recorded loss was six metres of land in a storm in 1967. There is a lot of historical evidence as to the rate of retreat of the cliffs. Over 4 km of land has been lost since Roman times. Dozens of villages have been washed away by the sea, as Figure 19 shows.

Why is cliff erosion such a problem here?

- The cliffs are made of soft glacial clay.

- The coast is very exposed, and waves have a long fetch over the North Sea.

- As sea-levels rise, more of the cliffs come under attack from the waves.

- The waves are mainly destructive ones.

- Most of the material which falls into the sea as the cliffs collapse is washed out to sea. The rest is moved along the coast by longshore drift. So the beaches in Holderness are narrow and give little protection to the cliffs.

What do people think about the erosion of the Holderness coast?

People who live on the coast in Holderness have different views about what should be done about the erosion of their cliffs. There is now a range of strategies in the Shoreline Management Plan which engineers have developed in order to prevent or reduce cliff erosion. These strategies are:

1. Do nothing.

2. Advance the existing defence line by even more **hard engineering**.

3. Hold the existing defence line by maintaining or improving the standard of protection.

4. Retreat the existing defence line (this is called **strategic realignment**) and so allow the caravan parks to move inland.

So it is possible to protect the communities of people who live close to the edge of the cliffs. However, these strategies are expensive to put in place, so planners and engineers have to do a cost–benefit analysis to assess whether or not it is worth trying to protect a stretch of coastline. One important part of any assessment of what to do is how sustainable each option may be. Clearly doing nothing is sustainable, in that no additional resources will be needed. Building defences in options 2 and 3 is a lot less sustainable because it would involve building extensive concrete and steel defences against the sea.

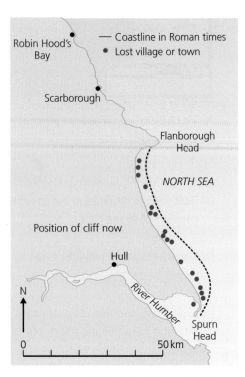

Figure 19: The lost villages of Holderness

Why does conflict occur on the coast, and how can this be managed?

79

Option 4, the strategic realignment, would involve some additional resources such as an extra field for caravan sites and holiday homes, but it would probably be less environmentally damaging than options 2 and 3. The costs in any cost–benefit analysis include things such as:

● The loss of farmland if nothing is done.

● The loss of caravan parks and holiday chalets if nothing is done.

● The costs of building defences against the sea.

● The loss of houses if nothing is done.

● The loss of roads if nothing is done.

The benefits are mainly:

● People do not lose their homes or businesses.

● Erosion of the cliffs stops.

● People do not lose their farms and land.

● The coastline is stabilised.

But there are lots of different opinions, as shown below.

Activity 4

(a) What do you think a local ice cream seller would say about the erosion of the coast?

(b) What do you think a person owning a holiday bungalow would say about the erosion of the coast?

(c) What do you think a person owning a nightclub in Scarborough would say about the erosion of the coast?

'Erosion is a natural process. You cannot stop it. We cannot afford to defend the whole coast.' *Chief Engineer of East Yorkshire*

'My whole business is threatened by the erosion of the cliffs. I cannot afford to buy more land to move the caravans.' *Local caravan park owner*

'I have lived here all my life and I do not want to move, but I will have to because my home is getting near the edge of the cliff. I am too old to move.' *Local older person*

'Not many people live along the coast, so I do not think we should try to build expensive sea defences for a few people. We should be spending our money in the towns where there are big problems of old houses that need replacing.' *Local politician*

'I have lost valuable farmland and received no compensation. It is not fair – it's not my fault the coast is eroding.' *Local farmer (female)*

'There is nothing to do here – especially at night – so I don't think it matters if the cliffs are eroded.' *Local young person*

So there is disagreement about what to do with the Holderness coast in the future. Some towns and villages have been protected against the sea, but other will not be, according to present plans. More than this, some local people (like Sue Earle) argued that building sea defences in some parts of the coast in fact only made it worse for other places further along the coast, where erosion was even stronger than before.

Decision-making skills

Read the different opinions on coastal management above.

Arrange these opinions on a scale from *strongly for* to *strongly against* to show the likely views on coastal management.

80

When you are answering questions about the advantages and disadvantages of different methods, you do not have to 'balance' them by giving an equal number of each, but you should give at least one of each.

Coastal management options – from hard engineering to holistic approaches

Coastal areas can be defended against wave erosion in a variety of different ways. Recently the focus has been more on a **holistic approach** to coastal management. This means considering all the social and economic factors (such as tourism) that affect a coastal area, as well as the physical conditions of the coast, such as the size and scale of the waves, beach and cliffs. In this way all the many factors affecting a coast will be considered in an integrated way. The purpose of all this is to make our management of coastal areas and their resources more sustainable. Many of the methods used in the holistic approach to managing coastal areas are described as being soft engineering, which is generally considered to be more environmentally friendly than more traditional hard engineering. The advantages and disadvantages of all the methods are described below.

Hard engineering methods

Sea wall – a long concrete barrier built at the base of a cliff offshore

Advantages	Disadvantages
It protects the base of the cliffs against erosion because it is made of resistant concrete. Land and buildings behind it are protected. If it is 'recurved', it can reflect wave energy.	It is expensive to build, and the cost of maintenance is high. It restricts access to the beach and it may be unsightly.

Groynes – wooden, rock or concrete 'fences' built across the beach, perpendicular to the coastline

Advantages	Disadvantages
These prevent the movement – by longshore drift – of beach material along the coast. The beach can then build up as a natural defence against erosion – and as an attraction for tourists.	They may look ugly and they do not last very long because the wood rots. Sand is prevented from moving along the coast, and places elsewhere may lose their beach and the natural defence it provides.

Rip rap – large boulders of resistant rock

Advantages	Disadvantages
These absorb wave energy and protect weak cliffs behind. They look quite natural.	They can be expensive. They still let some wave energy through. They can restrict access for the very young and the elderly.

Revetments – slatted wooden or concrete structures built at the base of a cliff

Advantages	Disadvantages
These absorb and spread wave energy through slats. They do not interfere with longshore drift.	Regular maintenance is needed and they are quite expensive.

Off-shore reefs – rock or concrete barriers built on the sea bed a short distance from the coastline

Advantages	Disadvantages
Waves break on the barrier before reaching the coast. These significantly reduce wave energy and allow a wide beach to develop.	They are very expensive to build and can interfere with boats.

Why does conflict occur on the coast, and how can this be managed?

81

Soft engineering methods

Beach replenishment – adding sand taken from somewhere else, often offshore	
Advantages	**Disadvantages**
This looks completely natural. It provides a beach for tourists. The beach absorbs wave energy and protects the land or buildings behind. Quite cheap.	The sea keeps on eroding it away – so it has to be replaced every few years.

Managed retreat – people and activities are gradually moved back from the vulnerable areas of coast	
Advantages	**Disadvantages**
People and activities are gradually moved back from the vulnerable areas of coast. Natural processes are allowed to happen.	Compensation has to be paid. There is quite a lot of disruption to people's lives and to businesses.

Cliff regrading – making the cliff face longer, so that it is less steep	
Advantages	**Disadvantages**
The angle of the cliff is reduced, making mass movement less likely. This method is relatively cheap.	Other methods need to be used at the base of the cliff to stop it being steepened again by erosion. Properties on the cliff may have to be demolished.

Integrated Coastal Zone Management

Coastal zones all over the world are changing rapidly. They are subject to the processes of erosion, but they are also subject to other pressures, such as the demand for space to build holiday homes or caravan parks. In some places the demand is to conserve the plants, animals, fish and insects of the area. So now planners everywhere are trying to find more sustainable ways of managing the coast – to reduce the damage to the environment. The main way of doing this is to treat the whole of the coastal area from the shoreline to several kilometres inland as one area for planned development. This approach is called **Integrated Coastal Zone Management (ICZM)**. In the case of the Holderness area in Yorkshire, for example, the East Riding Integrated Coastal Zone Management Plan was produced in 2002. It sets out which areas will be protected from erosion by the sea and in which areas the 'do nothing' approach will be employed. Local people are still very unhappy with this approach and argue that local economic and social needs are being ignored. So debate continues as to the best method to protect coastal areas.

Activity 5

With a partner, discuss and decide whether you think hard or soft engineering methods are more suitable to protect busy tourist resorts from coastal erosion.

Activity 6

With a partner, discuss and decide whether you think Integrated Coastal Zone Management is more sustainable than hard or soft engineering.

Figure 20: Coastal protection at Mappleton, East Yorkshire

Decision-making skills

Look at the tables on pages 80 and 81. Summarise the factors which influence the choice of coastal management methods, e.g. whether hard or soft engineering, what type of hard engineering?

Coastal management – a case study of Swanage

Swanage Bay and Durlston Bay in Dorset have both suffered from some significant coastal erosion during the last century. Rates of erosion in both places are estimated to have been about 40–50 cm per year.

In Durlston Bay, several methods were used to protect the cliffs from erosion and to safeguard the apartments and houses on the cliff top. Erosion mainly occurred at one particular point, where there was a major weakness in the rock. The methods (Figure 21) included:

- Regrading of the cliff – this extended it forward at the base, making the slope longer and therefore less steep.

- Installing drainage – this removed excess water, so that the slope was not as heavy or as well lubricated after rain.

- Placing rip rap – large granite boulders each weighing about 8 tonnes – at the base of the cliff to resist wave attack.

Figure 21: Management methods in Durlston Bay

Why does conflict occur on the coast, and how can this be managed?

83

Swanage Bay needed different methods of defence from Durlston Bay, because the erosion occurred along a considerable length of cliff rather than at one point. On the cliff top, the houses and hotels (such as the Grand Hotel) were losing their gardens and were in danger of collapsing. The methods used here included:

● Sea wall – this was built in the 1920s and provided a promenade (walkway) as well as a barrier to wave attack.

● Cliff regrading – a series of steps were made in the cliff to lower slope angles.

● Groynes – a series of mainly timber groynes were installed in the 1930s, and 18 of them have recently been replaced with new ones. These reduced longshore drift and helped make sure that the beach remained in place to absorb the energy of breaking waves.

● Beach replenishment – 90,000 m³ of sand was dredged from Studland Bay and pumped on to the beach at Swanage. This works with the groynes to ensure a good size of beach.

The cost of the recent new groynes and the beach replenishment was about £2.2 million. You can see some of the methods used at Swanage in Figure 22.

Cliff regraded

Sea wall

Beach replenishment

Figure 22: Swanage Beach and cliffs

Know Zone
Coastal change and conflict

Coasts are packed with physical geography and important processes. They are popular and often crowded places, where physical and human geography often combine to cause conflict.

You should know...

☐ How rock type (geology) and structure influence coastal landforms

☐ How landforms such as cliffs and stacks form

☐ The key terminology of coastal landforms

☐ The difference between erosion, weathering and mass movement

☐ How these processes help form coastal landforms

☐ The different types of waves, constructive and destructive

☐ How wave action moves sediment and changes the profile of beaches

☐ How sediment is transported and deposited on coasts

☐ How some coasts are threatened by rapid erosion and rising sea levels

☐ How erosion can cause conflict

☐ The range of management options for coasts

☐ The coasts and benefits of different options for Holderness and Swanage

Key terms

Backwash
Bay
Coastal flooding
Coastal management
Concordant coast
Constructive wave
Deposition
Destructive wave
Discordant coast
Do nothing
Erosion
Fetch
Geological structure
Hard engineering

Hard rock coast
Headland
Holistic approach
ICZM
Longshore drift
Mass movement
Soft rock coast
Spit
Stack
Strategic realignment
Stump
Sub-aerial processes
Swash
Weathering

Which key terms match the following definitions?

A The dropping of material that was being carried by a moving force

B A coastline created when alternating hard and soft rocks occur parallel to the coast, and are eroded at different rates

C The distance of sea over which winds blow and waves move towards the coastline

D A coastline created when alternating hard and soft rocks occur at right angles to the coast, and are eroded at different rates

E Water from a breaking wave which flows under gravity down a beach and returns to the sea

F Large, powerful waves with a high frequency that tend to take sediment away from the beach, because their backwash is greater than their swash

G The movement of material along a coast by breaking waves

H Material deposited by the sea which grows across a bay or the mouth of a river

To check your answers, look at the glossary on page 313

ResultsPlus
Maximise your marks

Foundation Question: Describe the advantages and disadvantages of different hard engineering methods used to protect coastlines. (6 marks)

Student answer ● (achieving Level 2)	Examiner comments	Build a better answer △ (achieving Level 3)
Hard engineering like wooden groynes can be used to build up beaches to protect coasts.	• *Hard engineering like...* This is a good example and is correct – groynes do build up beaches.	Hard engineering like wooden groynes can be used to build up beaches to protect coasts.
Some people think they are ugly and stop you moving along the beach easily.	• *Some people think...* This includes a disadvantage of groynes so balances the first statement.	Some people think they are ugly and stop you moving along the beach easily.
On some coasts 'do nothing' is used which just lets the coast erode.	• *On some coasts...* This is not an example of hard engineering so does not score any marks.	Offshore reefs reduce wave energy and this reduces erosion. They cannot be seen at high tide.
This can be cheaper, but you might have to compensate some people who lost land.	• *This can be...* Although this part of the answer is correct, it is not answering the question.	They are expensive and might be a hazard to boats.
Sea walls made of concrete are expensive at over £5,000 per metre.	• *Sea walls are...* This is a good example with appropriate facts on costs.	Sea walls are made of concrete and are expensive at over £5,000 per metre.
They can be ugly and might stop people getting onto the beach easily.	• *They can be...* This is correct but they are all disadvantages.	Sea walls protect the base of the cliff from erosion.

Overall comment: There is some good material in this answer, but there is not enough balance between advantages and disadvantages. In addition, the second example is not hard engineering.

- - -

Higher Question: Explain why some cliffs erode more rapidly than others. (6 marks)

Student answer ● (achieving Level 2)	Examiner comments	Build a better answer △ (achieving Level 3)
In Holderness cliffs are made of soft boulder clay, which is easily eroded.	• *In Holderness cliffs...* This contains a good example of a rock type and location.	In Holderness cliffs are made of soft boulder clay, which is easily eroded.
Harder rocks like chalk at Flamborough head are not eroded as easily.	• *Harder rocks like...* This is another good example, but using geographical terms such as abrasion as well as erosion would improve the answer	Harder rocks like chalk at Flamborough head are not eroded as easily by abrasion.
Some cliffs contain weaknesses like joints and faults.	• *Some cliffs contain...* This is another good reason.	Some cliffs contain weaknesses like joints and faults.
They get attacked by hydraulic action and lots of weaknesses means lots of erosion.	• *They get attacked...* In this sentence hydraulic action is a good term to use.	They get attacked by hydraulic action and lots of weaknesses means lots of erosion.
Big storms cause lots of erosion.	• *Big storms cause...* This point is correct but needs to be explained.	Big storms cause erosion, as some coasts are exposed to a long fetch and large waves. Human actions, e.g. groynes at Mappleton, have made erosion faster elsewhere.

Overall comment: This is a good answer with some examples, but the student needed to add further explanation and remember to use specific geographical terms such as abrasion.

Chapter 6 River processes and pressures

Objectives

- Define the key drainage basin terms.

- Be able to describe the change in characteristics of a river and its valley, from source to mouth, including landforms.

- Understand how processes of erosion, transport and deposition work with geology and slope processes to affect the long profile.

ResultsPlus
Exam Tip

⚠ Examination questions will often ask you to define one or more of these very important terms – so you must learn them.

Characteristic	Change from source to mouth
Width	increases
Depth	increases
Velocity	increases
Discharge	increases
Gradient	decreases

How do river systems develop?

Drainage basin terms

A **drainage basin** is the area of land drained by a river and its tributaries. When it rains, water eventually finds it way into rivers, either by moving across the surface or by going underground and moving through the soil or the rock beneath.

There are a number of important technical terms associated with drainage basins. These are shown in Figure 1, and their meanings are given underneath.

Watershed – the boundary of a drainage basin. It separates one drainage basin from another and is usually high land, such as hills and ridges.
Confluence – a point where two streams or rivers meet.
Tributary – a stream or small river that joins a larger stream or river.
Source – the starting point of a stream or river, often a spring or a lake.
Mouth – the point where a river leaves its drainage basin as it flows into the sea.

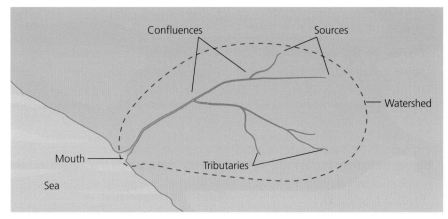

Figure 1: Drainage basin terms

Change in characteristics from source to mouth

A river changes with increasing distance downstream from its source towards its mouth, as it moves from its **upper course**, to its **mid-course** and finally into its **lower course**.

The key characteristics of a river are:
- width – the distance from one bank to the other
- depth – the distance from the surface of the water to the bed
- velocity – how fast the water is flowing
- discharge – the volume of water moving through the river channel in a given time
- gradient – the steepness of the riverbed.

The changes in these characteristics are summarised in the table on the left.

Afon Nant Peris, Wales

The river Nant Peris rises at 400 m in the Llanberis Pass in North Wales. The river appears as a trickle of water next to the A4086 road. This area receives 2500 mm of rain each year and 80% of it flows out along the river. Figure 2 shows the **long profile** of the Afon Nant Peris, that is the view of the river from its source in the LLanberis Pass to its mouth at Caernarvon. Like most rivers, the Afon Nant Peris has an uneven long profile – steep in some places and much less so in others.

As it flows down the pass, the channel of the river widens and deepens. The river erodes the valley floor, transporting the debris and, later, depositing elsewhere. In its upper course, the sides of the valley are steep and the river's course is not straight. It flows around the interlocking spurs of higher land. The river is eroding downwards rapidly, and transporting lots of rocks, sand and mud down river.

In its middle course the river flows across a broad area of flat land called a **flood plain**, which it has created. Often the river flows in large curves, called **meanders**. There is both **erosion** and **deposition** in this section of the river, but the deposition is temporary. Sooner or later the **sediments** deposited will be eroded by the river and moved. The gradient here is lower than in the upper course but velocity (speed) is greater because the river is deeper so there is less friction with the banks and bed of the river.

In its lower course the river is known as the Afon Selant and meanders across a broad flood plain. The river is close to sea level so there is little downward erosion. However velocity and discharge are now at their maximum. The river forms **ox-bow lakes** which may eventually fill with sediment and dry up.

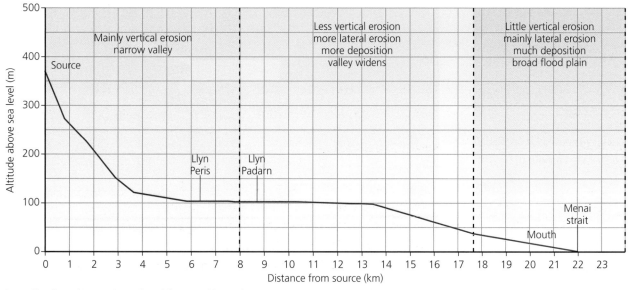

Source: Flint, Fline and Punnett 'Geography and Change' Hodder Staughton 1996 ISBN 0340 647 868 Page 71

Figure 2: The long profile of the Afon Nant Peris

River erosion

River erosion involves several different methods, including:

Hydraulic action	This results from the force of the water hitting the riverbed and banks and wearing them away. This is particularly important during high-velocity flow.
Abrasion	This is caused by the river picking up stones and rubbing them against the bed and banks of the channel in the flow. This wears the bed and banks away.
Attrition	Any material carried in the river will become rounder and smaller over time as it collides with other particles and the sharp edges get knocked off.
Corrosion	The dissolving of rocks and minerals by river water flowing over them.

Formation of landforms

Rivers produce a wide range of landforms, many of which help make river landscapes attractive places for people to visit and enjoy.

Interlocking spurs

Near their source, rivers do not have a lot of power as they are very small. They tend to flow around valley side slopes, called spurs, rather than being able to erode them. The spurs are left **interlocking**, with those from one side of the valley interlocking with the spurs from the other side (Figure 4).

Figure 4: Interlocking spurs

ResultsPlus
Exam Question Report

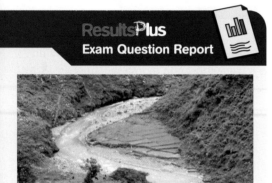

Figure 3: A meander bend

Figure 3 shows a meander bend. Explain the changes which might happen to this meander over a period of time. You may use a diagram or diagrams in your answer.
(5 marks, June 2007)

How students answered

Most students answered this question poorly. They often described the changes, stating that the bend would get bigger and perhaps become an ox-bow lake. But they did not explain why this happened.

60% (0–1 marks)

Many students gave some explanation of the changes by mentioning erosion and/or deposition and stating where they were happening.

31% (2–3 marks)

Some students answered this question well. They not only mentioned erosion and deposition, but also gave detail of specific process mechanisms such as abrasion. They also established a clear sequence in the changes happening.

9% (4–5 marks)

Waterfalls

Where a river flows over bands of rocks of differing resistance, the weaker rock is eroded more quickly and a step may develop in the river's bed. The increased velocity gained by the water as it falls over the step further increases the rate of erosion of the weaker, downstream band of rock. Abrasion and hydraulic action at the base of this step causes undercutting and the formation of a plunge pool. Eventually the overhanging, more resistant rock collapses due to a lack of support, making the **waterfall** steeper. The waterfall in Figure 4 is almost vertical. Over time, repetition of this process means that the position of the waterfall retreats in an upstream direction.

Meanders

Meanders are large bends in a river's course, found well below the source, often in its lower course as the river approaches its mouth. They are common landforms found on a river's flood plain. The flow of the river swings from side to side, directing the line of maximum velocity and the force of the water towards one of the banks. This results in erosion by undercutting on that side and an outer, steep bank is formed, called a river cliff. Deposition takes place in the slower moving water on the inside of the bend, leading to the formation of a gently sloping bank, known as a slip-off slope. The cross-section of a meander is, therefore, asymmetrical – steep on the outside, gentle on the inside.

Ox-bow lakes

As meander bends grow and develop, the necks of land between them become narrower. Eventually the river may erode right through a neck, especially during a flood. Water then flows through the new, straight channel and the old bend is abandoned by the river. The bend gradually dries up and, helped by deposition at the neck by the river, becomes sealed off as a horseshoe-shaped lake, as Figure 6 shows.

Figure 5: Waterfall on Hills Creek, West Virginia

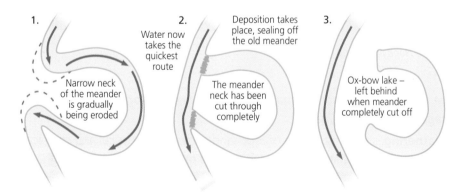

Figure 6: Formation of an ox-bow lake

ResultsPlus
Watch out!

■ Meanders are not the same as interlocking spurs. Although the river has a winding course in both cases, with interlocking spurs it is flowing around obstructions. This is not the case in meanders.

Activity 1

Draw up a table to show all of the landforms that are produced by river processes. Use three headings – 'Erosion', 'Deposition' and 'Erosion and Deposition' – putting each landform under the correct heading.

Activity 2

Consider these four river landforms – waterfall, flood plain, meander, interlocking spurs.

(a) Which two are most likely to be found in the upper course of a river near the source?

(b) Describe the appearance and position of a flood plain.

(c) Explain how a waterfall is formed.

Flood plains

A flood plain is a wide, flat area of land either side of a river in its lower course. The flood plain is formed by both erosion and deposition. Lateral (sideways) erosion is caused by meanders eroding on the outside of their bends. This makes the valley floor wide and flat. When the river floods, the flood water spreads out on the valley floor, slows down and deposits the sediments it was carrying. This can be seen in Figure 7.

Levees

Levees are natural embankments of sediment along the banks of a river. They are formed along rivers that carry a large load and occasionally flood. In times of flood, water and sediment come out of the channel as the river overflows its banks. The water immediately loses velocity and energy as it leaves the channel and so the largest sediment is deposited first, on the banks. Repeated flooding causes these banks to get higher, forming levees. You can see these alongside the river in Figure 7. Sometimes natural levees are built up further and reinforced by humans to protect settlements, industry and transport links on the flood plain from future floods.

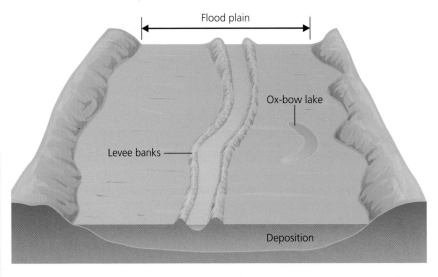

Figure 7: Flood plains and natural levees

The impact of weathering, mass movement and geology on river valley shape and sediment load

Weathering is the breakdown of rocks and minerals by physical and chemical processes. The rocks and minerals can then be moved by the processes of erosion. The main agents of erosion are wind, water, ice and gravity. Erosion involves the movement of material, whereas weathering takes place with little or no movement. In addition, on the sides of river valleys the processes of mass movement take place – transferring broken rock, minerals, soil and vegetation down the slope as a result of gravity.

The weathering and mass movement processes act on the valley sides, usually making them less steep over time, as material is moved from the top of the slope to the bottom. However, the river itself erodes its own channel, by wearing away the bed and/or banks, as well as possibly eroding the base of the valley side, making it steeper – and perhaps leading to mass movement.

Weathering can happen in many ways. Some common ones affecting river valleys are:

- Freeze-thaw – this happens when rainwater enters cracks or gaps in a rock and then freezes if temperatures drop below zero degrees. The water expands as it turns to ice and this exerts pressure on the rock, causing it to break down into smaller pieces.

- Rainfall – this happens because rain is slightly acidic. If the air is polluted by factories and vehicles, it can become very acidic. When rain falls on rocks, the acid in it can react with weak minerals causing them to dissolve, and the rock to decay.

- Biological weathering – the roots of plants, especially trees, can grow into cracks in a rock and split the rock apart as you can see in Figure 8.

There are several different processes of mass movement. In river valleys, two of the main forms of mass movement are:

- Soil creep – individual particles of soil slowly move down slope under gravity and collect at bottom of the valley sides. The river may then erode this material.

- Slumping – happens when the bottom of a valley side is eroded by the river. This makes the slope steeper and the valley side material can slide downwards in a rotational manner, often triggered by saturation due to rain, which both 'lubricates' the rock and makes it much heavier.

Geology influences river valley shape and sediment load in a number of ways: Where a river flows over bands of resistant rocks the valley sides tend to be steep, and the sediment load is small, because erosion is slow. Where a river flows over bands of less resistant rocks the valley sides tend to be gentle and the sediment load is large because erosion is rapid.

Figure 8: Roots of trees can grow into cracks in rocks and split them apart

ResultsPlus
Exam Tip

⚠ Good quality examination answers usually show a clear understanding of processes. Don't just name the process; explain how it works.

Why do rivers flood and how can flooding be managed?

Causes of flooding

A river flood is when a river overflows its banks and water spreads out on to the land alongside the river channel. Rivers usually flood as a result of a combination of causes. These are often a mixture of physical and human factors which, together, can result in large amounts of water getting into the river very quickly – so quickly that it fills up and overflows.

Flooding in Tewkesbury in July 2007

Figure 9 shows the rainfall in the Tewkesbury area in July 2007, together with the discharge of the river Severn from 17 to 27 July. This diagram is called a **hydrograph**. It shows the relationship between rainfall and river flow. The rising limb on the hydrograph shows the rapid increase in water in the river after the heavy rain of 19 and 20 July. The lag time is the difference between the time of the heaviest rainfall and the maximum level of the river. The falling limb is when there is still some rainwater reaching the river but in smaller and smaller amounts.

Point A on Figure 9 shows where some rainwater falls straight into the river. Point B is when a lot of water reaches the river by run-off on the surface. Point C is when some rainfall reaches the river more slowly because it has flowed through the soil and rocks. The hydrograph and Figure 10 show how the heavy rain combined with other factors to flood the area around Tewkesbury. This was a particularly severe flood which cut off the town from the surrounding area. Large numbers of people were forced to leave their homes and move to shelters above the flood waters. When the waters subsided, homes were found to be badly damaged. Carpets, kitchen units, cookers, fires, fridges and central heating boilers all had to be scrapped.

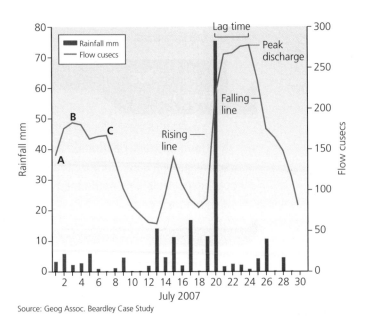

Source: Geog Assoc. Beardley Case Study

Figure 9: A flood hydrograph of the River Severn in July 2007

Figure 10: Flooding in Tewkesbury in July 2007

Physical factors that contribute to flooding

- Intense rainfall – When rain falls too fast to fully allow its infiltration into the ground, it flows quickly across the surface and into the channel. This happened in Figure 11.

- Snow melt – In some places lots of snow falls during the winter months, but when temperatures rise above zero in the spring all the snow that has built up over months suddenly melts.

- **Impermeable** rocks – Some rock types, like granite, do not let water enter the ground and so rainwater runs off across the surface into the channel.

Figure 11: Flooding in Hull, July 2007

Human factors that contribute to flooding

- **Deforestation** – Vegetation collects, stores and uses water from a drainage basin. The less vegetation there is, the more water can reach the channel.

- **Urbanisation** – In towns and cities, rainwater will not infiltrate the hard, man-made surfaces like concrete and tarmac and so it runs off into the channel.

- Global climate change – Increasing global temperatures, partly due to human activities such as burning fossil fuels, can cause more melting of ice in glaciers and, in places, more rainfall and more frequent storms.

Effects of flooding on people and the environment

There are many different and wide-ranging effects that floods can have on people and the environment (see Figure 12). Some of these are very immediate, while others can happen much later – and last much longer.

Activity 3

Read the news report below on flooding in Wales in October 2008.

FLOODS HIT HOMES, SCHOOL AND LAND

Firefighters have been dealing with calls about flooding in several areas of Wales after heavy rain.

They went to Llanelltyd, in Dollgellau, to help pump out a house with 1ft of water in the ground floor and were in the Lovesgrove area of Aberystwyth.

Firefighters were also called to Ty Newydd Farm at Felin Fach, Lampeter, to help rescue 20 sheep which were stuck in flood water.

There were reports of a flooded cellar at Cyfarthfa High School in Merthyr.

In Aberystwyth, crews spent the night pumping water which severely affected Fron Frith Lodge, and a high-volume pump had to be brought into operation.

http://news.bbc.co.uk/1/hi/wales/7691694.stm

(a) What caused this flooding?

(b) Identify three impacts of the flooding.

(c) How did the firefighters respond to this flood event?

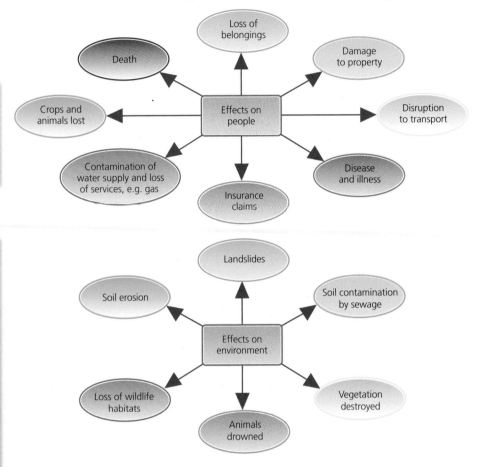

Figure 12: Effects of floods on people and the environment

The flooding that affected parts of northern England and the Midlands in the summer of 2007 caused the loss of 13 lives and over £2 billion worth of damage. In Lincolnshire, 40% of the pea crop was damaged, and in north Wales 30 tonnes of debris and earth blocked the only road out of Barland near Presteigne on 23 July.

Floods in Mozambique in Africa during February 2007 led to 121,000 people being made homeless. In Sudan, at least 12,000 livestock and 16,000 chickens were lost, while outbreaks of acute diarrhoea killed dozens of people as water supplies became contaminated.

Prediction and prevention of the effects of river flooding

Forecasting

Many rivers, especially those that have a history of flooding, are monitored by the Environment Agency. If river levels rise to potentially dangerous levels, they are able to warn people in areas of risk and, if necessary, evacuate people from their homes to a safer place. They have also produced maps, which are available online, that show areas at different levels of risk.

Computer simulation models are now being used to help predict flooding. In Texas, for example, in response to a prediction from Rice University, the Texas Medical Centre was evacuated before a large flood occurred in 2001.

The Environment Agency provides flood warnings online 24 hours a day. From this page you can view warnings in force in each of our eight regions covering England and Wales. You can also search for your local area and its current warning status using the panel on the right. The information is updated every 15 minutes. For further information about a Flood Warning in a particular area, you can call Floodline 24 hours a day on 0845 988 1188. Extracts from the Environment Agency website: www.environment-agency.gov.uk

Building design

There are things that people can do to their property that will make it easier and cheaper to clean up after a flood.

The Environment Agency advice includes:

- Lay ceramic tiles on your ground floor and use rugs instead of fitted carpets.

- Raise the height of electrical sockets to 1.5 metres above ground floor level.

- Fit stainless steel or plastic kitchens instead of chipboard ones.

- Position any main parts of a heating or ventilation system, like a boiler, upstairs.

- Fit non-return valves to all drains and water inlet pipes.

- Replace wooden window frames and doors with synthetic ones.

In many low-income countries, wooden buildings are built on stilts so that the living areas are above the **flood risk** level (see Figure 13).

Figure 13: House on stilts in Bangladesh

Planning

The local government is responsible for giving planning permission for new buildings. In recent years, many new buildings have been built on the flood plains of rivers, putting these properties at significant risk. Now, it is widely accepted that this should only be done if protection measures are put in place. Land use zoning is often used. This means flood risk areas are only used for activities that would not suffer too much from a flood. These include parks and playing fields.

Figure 14 shows how the town of Nome in Alaska has planned its land uses so that most of the area at risk of an extreme, '1 in 100 year' flood event is left as open space or is used for leisure and recreation.

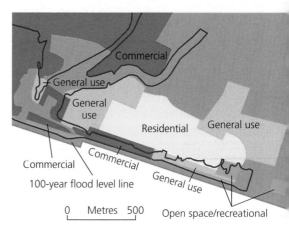

Figure 14: Land use in Nome, Alaska, showing how it has been planned to cope with possible future flooding

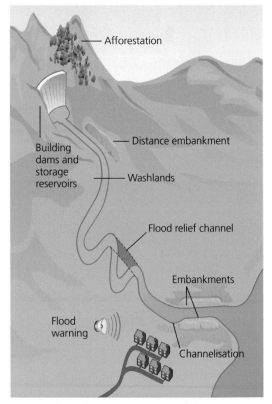

Figure 15: Flood prevention methods (Harcourt and Warren 2001)

Predicting and preventing river flooding

Education

Local governments or government agencies need to let people know about flood risk and what to do in a flood. Otherwise, emergency plans and strategies are not effective. This is where education is really important. Governments try to educate people by:

- Sending leaflets through the post

- Advertising in newspapers and on television

- Posting information on websites

- Offering helpline telephone numbers

- Having drills and training exercises.

The National Weather Service in the USA issues advice on what to do before, during and after flood events, such as:

- Store drinking water in clean bathtubs and in various containers – the water service may be interrupted.

- Keep a stock of food that requires little cooking and no refrigeration – electric power may be interrupted.

- Keep first aid supplies on hand.

- Assemble a disaster supplies kit containing: first aid kit, canned food and can opener, bottled water, rubber boots, rubber gloves, Weather Radio, battery-powered radio, flashlight, and extra batteries. [Weather Radios are issued free to all those who live in flood risk areas. These are permanently tuned into the weather service broadcasts.]

Flood management – from hard engineering to integrated and sustainable approaches

There is a wide range of different engineering methods that can be used to try to control rivers in the UK. In the past, the traditional approach was to use **hard engineering** methods to prevent the floods. This tended to ignore the impact of these 'solution' on all the other features of the river, such as the wildlife, or its use for leisure and tourism. So now the approach is more an **integrated river management** one, looking at all aspects of the river – at both its physical geography and its human geography. This new approach also emphasises the importance of trying to find sustainable solutions to the problems of river floods – solutions which do not damage the environment. Some of these solutions involve hard engineering, but most involve soft engineering, which is thought to be more environmentally friendly. Each method has advantages and disadvantages, as shown on page 97.

Hard engineering methods

	Advantages	Disadvantages
Embankments (levees) – these are high banks built on or near riverbanks.	They stop water from spreading into areas where it could cause problems, such as housing. They can be earth and grass banks, which blend in with the environment.	Floodwater may go over the top. They can burst under pressure, possibly causing even greater damage.
Channelisation – this involves deepening and/ or straightening the river.	This allows more water to run through the channel more quickly, taking it away from places at risk.	More water is taken further downstream, where another town or place at risk might lie. They do not look natural.
Dams – these can be built upstream to regulate the flow of water.	Water is held back during times of heavy rain or snowmelt. They can also be used for water supply, recreation and HEP.	They can be an eyesore and are very expensive to build. If they burst the damage would be devastating.
Flood relief channels – extra channels can be built next to rivers or leading from them.	The relief channels can accommodate the surplus water from the river so that it won't overflow its banks.	They can be unsightly and may not be needed very often. Costs can be high.

Soft engineering methods

	Advantages	Disadvantages
Washlands – these are areas on the flood plain that are allowed to flood.	This gives a safe place for floodwater to go. Inexpensive and leaves the natural environment unspoilt.	Flood plain cannot be used for other things.
Afforestation – trees are planted in the drainage basin.	Trees intercept rainfall and take water out of the soil. This reduces the amount reaching rivers. Wooded areas look attractive and provide wildlife habitats.	The land cannot be used for other activities, such as farming.
Land use zoning – governments allocate areas of land to different uses, according to their level of flood risk.	Major building projects are allocated to low risk areas. Open space for leisure and recreation is placed in high-risk areas because flooding would be less costly for them.	These may not be the best places for the different activities in terms of public accessibility.
Flood warning systems – rivers are carefully watched and if the levels are rising, places downstream can be warned.	People living in towns or villages downstream have a chance to prepare or to evacuate.	Sometimes it is not possible to give very much warning and so it is hard to save possessions.

Decision-making skills

What factors are used to decide the type of engineering method that is used?

Skills Builder 1

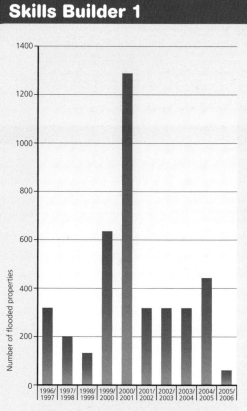

Figure 16: The number of flooded properties in the south-west of England between 1996/1997 and 2005/2006

Study Figure 16.

(a) In which years were the (i) highest and (ii) lowest number of properties flooded?

(b) Suggest two physical reasons why an area may have severe floods.

(c) Explain how any two hard engineering methods can be used to protect properties from being flooded.

Watch out!

■ Most management schemes involve a combination of different methods, rather than just one. In Figure 15 you can see how a number of different methods can be used in different places in a drainage basin and along a river.

Hard engineering and flooding in York

York is a city which is prone to flooding by the rivers Ouse and Foss which meet in the city. Over the past 20 years hard engineering works have been put in place to protect homes and families. Most of the works consist of raised embankments, some of which are made of earth and some of concrete supported by steel sheeting. In addition there is a natural flood plain upstream of York, called Clifton Ings, which can store 2.3 million cubic litres of water. This lowers the peak flood level in the city by 150 mm. The banks around the Ings were raised in 1982 and sluice controls built for letting flood water in and out of the Ings. All this hard engineering has been very expensive and it has had a significant impact on the environment – it has not benefited wildlife, for example. Further, despite all these works flood water does still overflow the embankments, causing damage to homes, offices and factories. This happened in 2000, leaving the city with a clean-up bill of £1.3 million, not including the cost of damage to people's homes and businesses.

A new system of flood defences was introduced for a trial in 2008. These are temporary hollow plastic barriers, called 'Aquabarriers'. They have holes in one side which allow flood water in and so increase their stability during the flood. Once the flood recedes the water simply drains away out of the holes. These are much cheaper than traditional hard engineering solutions to the problem and can be easily moved.

Figure 17: Aquabarriers – a new system of flood defence

Alternative approaches on the River Skerne near Darlington

A different approach to flood prevention has been developed on the River Skerne in north-east England (Figure 18). In the past, the river had meandered across a flood plain, but over the 200 years before 1998 the river had been straightened and deepened. The flood plain had then been used as a tip for industrial waste. The river was simply a straight urban drain used to carry flood waters through a built-up area. However, from 1998 the river was restored to its natural meandering state by using earthmoving equipment to create large bends in the river. This enables the river to flow from side to side and for water to run off on to the flood plain during flood times.

In addition, a wide range of wetland habitats was created alongside the river. The banks of the meanders were protected with coconut fibre rolls and willow stakes to prevent excessive erosion until the tree roots had become established. Over 20,000 m³ of industrial spoil were removed from the flood plain in order to allow the river to overflow its banks but still protect surrounding housing. The spoil was also used to raise and reshape the valley sides. Over 20,000 trees and shrubs were planted, as well as wildflowers and bulbs. As a result, water quality has improved in the river, flooding has been reduced, and the wildlife – such as swans, fish, dragonfly and even water voles – has increased in the area. This scheme is a much cheaper and more environmentally friendly way of protecting the houses and buildings in the area than traditional hard engineering methods.

Figure 18: Aerial view of the River Skerne

Rivers are vital sources of water, but they can bring destruction and suffering when they flood. Managing rivers so that humans and ecosystems can enjoy the benefits rivers bring, but not the costs, is a difficult challenge.

You should know...

- ☐ Key terms linked to river systems. such as drainage basin
- ☐ What the long profile of a river is
- ☐ How river characteristics change along the long profile
- ☐ How weathering, erosion, transport and deposition cause river landforms
- ☐ How to describe landforms of upper, mid, and lower course rivers
- ☐ How the landforms are formed
- ☐ How sediment changes as it is moved along a river
- ☐ How flooding is caused
- ☐ How human and physical factors increase flood risk
- ☐ To describe and explain hydrograph shape
- ☐ How climate change could increase flood risk
- ☐ The impacts of flooding, for example on Tewkesbury in 2007
- ☐ How hard engineering schemes might prevent flooding
- ☐ How integrated approaches might reduce flood risk
- ☐ How to evaluate different types of flood management scheme.

Key terms

Deforestation	Impermeable	Ox-bow lake
Deposition	Integrated river	River cliff
Drainage basin	management	Sediment
Erosion	Interlocking	Slip-off slope
Flood plain	spurs	Upper course
Flood risk	Levee	Urbanisation
Geology	Long profile	Waterfall
Hard	Lower course	Weathering
engineering	Meander	
Hydrograph	Mid course	

Which key terms match the following definitions?

A The relatively flat area forming the valley floor on either side of a river channel, which is sometimes flooded

B Not allowing water to pass through

C A natural embankment of sediment along the banks of a river

D The bend formed in a river as it winds across the landscape

E An arc-shaped lake which has been cut off from a meandering river

F The dropping of material that was being carried by a moving force

G The development and growth of towns or cities

H Areas of high land which stick out into a steep-sided valleys

To check your answers, look at the glossary on page 313

ResultsPlus
Maximise your marks

Foundation Question: Using a named example, describe the impacts of river flooding. (6 marks)

Student answer (achieving Level 2)	Examiner comments	Build a better answer (achieving Level 3)
Flooding happened in Tewkesbury in summer 2007.	• *Flooding happened in...* This is a good, specific example.	Flooding happened in Tewkesbury in summer 2007.
There was an unusual amount of rainfall for June and rivers burst their banks.	• *There was an...* Although this is correct, the question is about impacts not causes.	There were over 5 inches of rain and the rivers Severn and Avon burst their banks.
Lots of homes were flooded.	• *Lots of homes...* This mentions an impact, but does not state how many homes were affected.	1,800 homes were flooded and 10,000 had their water supply cut off.
Tewkesbury was cut off for a while.	• *Tewkesbury was cut...* This outlines an impact but could be more specific.	The 4 major roads into Tewkesbury were flooded for a day.
Insurance claims were high and insurance will rise in the future.	• *Insurance claims were...* This is a good, extended point.	Insurance claims were high and insurance will rise in the future. One year later, 800 people had not moved back into their homes.

Overall comment: The student achieved Level 2 because although the answer was correct, some details were a bit vague. Try to include facts and figures in your answer.

Higher Question: Explain how human and physical factors can increase flood risk. (6 marks)

Student answer (achieving Level 2)	Examiner comments	Build a better answer (achieving Level 3)
Flood risks are basically caused by too much rain, like the 5 inches that fell on Tewkesbury in 2007.	• *Flood risks are...* This is a good example with details on rainfall.	Flood risks are basically caused by too much rain, like the 5 inches that fell on Tewkesbury in 2007.
If the ground's already saturated, run-off happens rather than infiltration, making floods worse.	• *If the ground...* This includes correct geographical terminology.	If the ground is already saturated, run-off happens rather than infiltration, making floods worse.
Steep slopes like at Boscastle also increase run-off and flood risk.	• *Steep slopes like...* Another good example and physical factor are provided here.	Steep slopes like at Boscastle also increase run-off and flood risk.
Humans can make flooding worse by urban areas and impermeable surfaces.	• *Humans can make...* A human factor is detailed here, which is the second part of the question.	Humans can make flooding worse by urban areas and impermeable surfaces.
Deforestation is a natural factor.	• *Deforestation is a...* This is incorrect as deforestation is a human factor.	Deforestation is a human factor that reduces infiltration.
It reduces run-off and infiltration.	• *It reduces run-off...* This is unclear as there is confusion over which process is reduced.	It leads to more rapid run-off and a much steeper hydrograph.

Overall comment: The student uses good geographical terminology, has some balance of human and physical factors, and uses an example which is good. The answer achieved Level 2, but only a few minor changes would be needed to score Level 3.

Chapter 7 Oceans on the edge

102

Objectives

- Describe the impacts of human activities on marine ecosystems such as mangrove forest.

- Explain how marine food webs work, and why they become damaged.

- Understand that climate change brings new and often unpredictable stresses to oceans.

Results Plus
Build Better Answers

Study the distribution of mangrove swamp shown in Figure 2. Identify the main populated areas where mangrove swamp grows. (3 marks)

■ **Basic answers** (0–1 marks)
Name only one or two areas, or are very imprecise, writing things like 'there is a lot on the right hand side of the map'.

● **Good answers** (2 marks)
Correctly identify Asia and Central America as named locations.

▲ **Excellent answers** (3 marks)
Also provide more specific details of Asian locations (naming the west coast of India and Indonesia) or mentioning north-west and south-east Africa.

How and why are some ecosystems threatened with destruction?

The term ecosystem describes a grouping of plants and animals that is linked with its local physical environment – through use of soil nutrients, for example. The oceans, covering two-thirds of our planet, are home to distinctive **marine ecosystem** communities composed of fish, aquatic plants and sea birds – as well as tiny but very important organisms such as krill and plankton.

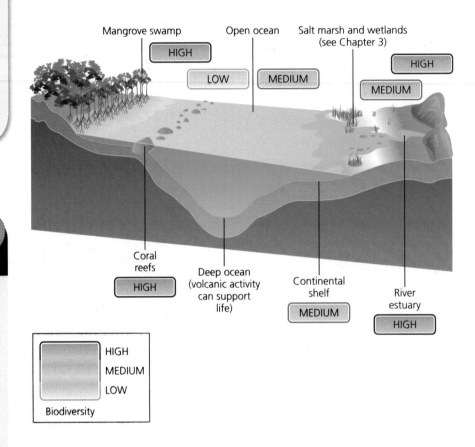

Figure 1: Biodiversity in the oceans' ecosystems

Coral reef ecosystems have an incredible level of **biodiversity** (their variety of plant and animal species). Although they cover less than 1% of the Earth's surface, they are home to 25% of all marine fish species. Biodiversity is also high in waters close to the edges of the world land masses – above the **continental shelf**. There, species enjoy shallow warm water, enriched with silt nutrients from river **estuaries**. Even in the deep ocean, where light cannot penetrate, unique ecosystems are found. Underwater volcanic activity can create densely populated sites of plant and animal life. Here, often at great depth, life has evolved that can survive truly extreme conditions of heat and pressure.

Investigating how mangrove swamp is used

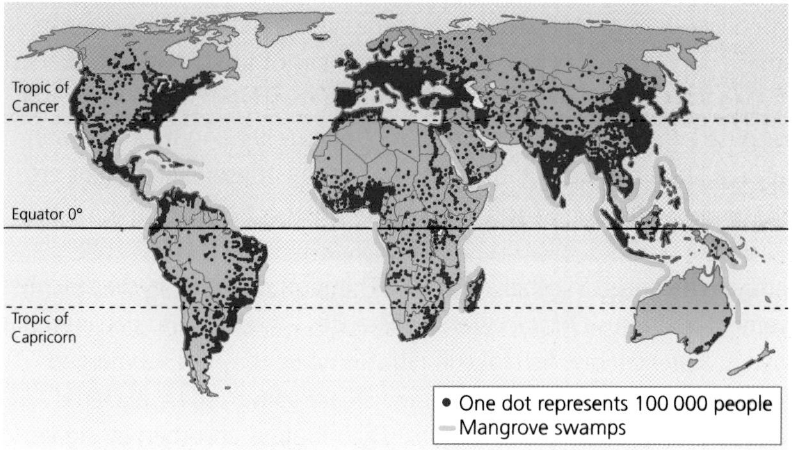

Figure 2: The distribution of mangrove swamp and populated areas

One dot represents 100 000 people
Mangrove swamps

Mangroves are areas of swampy forest, found in estuaries and along marine shorelines in around 120 tropical and subtropical countries. Many are densely populated, especially in coastal Asia (Figure 2). Mangrove plants have evolved to tolerate daily tidal flooding and high salinity. Long twisting roots anchor the trees against a constantly ebbing and flowing tide (Figure 3). The roots trap mud, making a **habitat** for lobster and prawn.

These species used to be fished in **sustainable** ways that preserved the **mangrove swamp**. However, prawns can be more intensively harvested from ponds dug in mud on cleared land. Global demand for prawns (worth £30 billion a year) has caused widespread removal of mangrove vegetation. Mangrove swamp naturally covered 200,000 km² of the Earth's surface but only half that area now remains, often because of this prawn **aquaculture**. Vietnam's Mekong Delta is a typical site where mangrove trees have been removed, leaving a flat muddy plain studded with blue plastic-lined ponds.

Mangrove removal creates problems of coastal erosion and loss of nursery grounds for fish. Crocodiles, snakes, tigers, deer, otters, dolphins and birds all lose an important habitat. Carbon dioxide stored over centuries in the rich mud beneath the swamp is released when the trees are removed.

Mangrove swamp is also nature's defence against tsunamis. The enormous ocean wave that struck the coast of south-east Asia in 2004 killed 230,000 people. Where they were still in existence, mangrove trees shielded lives and property. In Thailand, recent conversion of mangrove habitat into prawn farms and tourist resorts close to Phuket contributed significantly to the catastrophic losses experienced there.

Prawn aquaculture can also cause serious **pollution** problems. Antibiotics and pesticides used in the prawn pools frequently leak into the delicate ecosystem of neighbouring areas.

Figure 3: Mangrove forest

Quick notes (Mangrove swamp):
- Human use of this environment was originally sustainable.
- Because of commercial pressures, **unsustainable** exploitation now takes place.
- Vegetation removal and pollution damage local ecosystems and settlements.

Skills Builder 1

Study Figure 3

(a) Name one adaptation shown that helps the trees survive in a wet tidal environment.

(b) Describe two ways in which undisturbed mangrove forest can support human activities.

Unsustainable use and disturbance of marine ecosystems

Ecosystems are made up of different animal and plant populations, all of whom are dependent on one another in some way. A natural balance exists among these populations. For example, the number of top carnivores such as sharks is always relatively low compared with fish they feed on, such as tuna. This is because big hunting sharks use a lot of energy chasing their prey and each must eat large numbers of tuna to stay healthy. These relationships are shown in the illustration of a **food web** for an open ocean (Figure 4).

Ecosystems are balanced in other ways too. The term **nutrient cycle** describes the movement and re-use of important substances – such as nitrogen – within an ecosystem. For example, fish take in nitrates when they eat submerged plants and algae. Waste products from the fish are converted by bacteria into ammonia and eventually into nitrates. The nitrate is absorbed by algae, restoring the original balance. The entire cycle can then begin again.

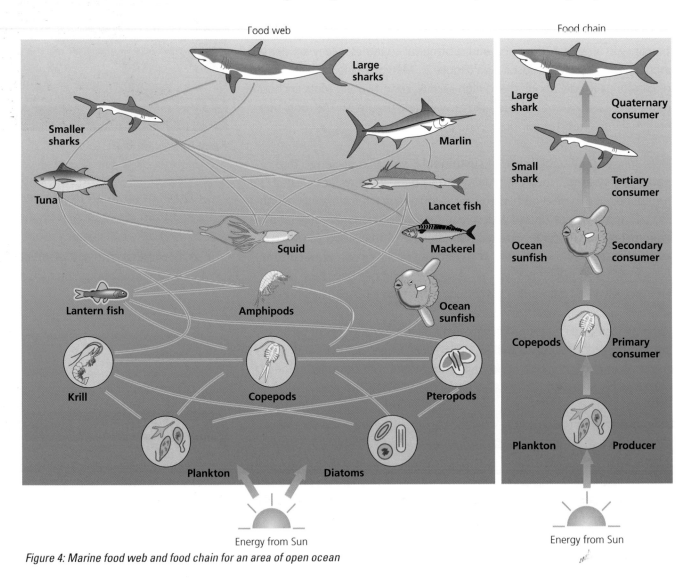

Figure 4: Marine food web and food chain for an area of open ocean

Food web disruption

Overfishing by humans interferes with the natural balance – the equilibrium – of ocean animal populations. Excessive hunting of a particular fish species such as tuna triggers a series of impacts. First, humans directly reduce the number of tuna. Next, this action impacts indirectly upon sharks. These top carnivores can no longer find enough tuna to meet their own needs and numbers of sharks fall too. In contrast, numbers of organisms lower down the food chain might actually increase at first – mackerel numbers, for example, could boom with fewer tuna preying on them.

If fishing ceases, the original balance may later be restored. In some cases, however, overfishing has been known to cause permanent species loss.

Worldwide, scientists say 5 million ocean species have yet to be discovered – and we are wiping them out before we have found them.

Cod fishing	North America's cod population has never recovered from overfishing in Newfoundland during the 1970s.
Baiji white dolphin	Some large marine species, including the Baiji white dolphin, were hunted to **extinction** during the twentieth century.
Whales and sharks	Humpback whale and great white shark populations are unlikely to ever regain their original sizes.

Another human impact that brings dire results to estuaries and oceans is **eutrophication**. This process occurs when excessive nutrients are added to a body of water. Nitrate fertilisers often get carried by rain **run-off** from farmland into rivers or over cliff edges into coastal water. In contrast to toxic pollutants that kill, the result here is one of 'over-feeding'. The nutrient-enriched waters initially experience a sudden growth in marine life. Tiny organisms flourish, creating an explosion of life called an algal bloom.

For a variety of reasons, the presence of so much algae uses up most of the water's oxygen. Fish and crustacean (crab and prawn) species suffocate in the de-oxygenated water. Eutrophic conditions can be found all over the world, with Japan and the Gulf of Mexico particularly badly affected (Figure 5 on page 106). The North Sea is another 'nitrate hotspot' where lobster populations have been lost due to a lack of oxygen.

ResultsPlus
Build Better Answers

Study Figure 4. Describe how an increased demand for squid in restaurants would impact on this marine ecosystem. (4 marks)

■ **Basic answers** (0–1 marks)
Simply state there would be fewer squid left in the ecosystem.

● **Good answers** (2 marks)
Also identify linked effects, such as fewer tuna or a growth in lantern fish.

▲ **Excellent answers** (3–4 marks)
Also make explicit mention of the distinction between direct and indirect impacts.

Activity 1

Why are there so few big fierce animals?

This deceptively simple question is a useful way to start discussing the workings of food webs. Think about how there are losses of energy at each level of the food chain (kinetic energy from movement, as well as heat losses). Also consider the physical losses of matter that take place due to urine and excretion.

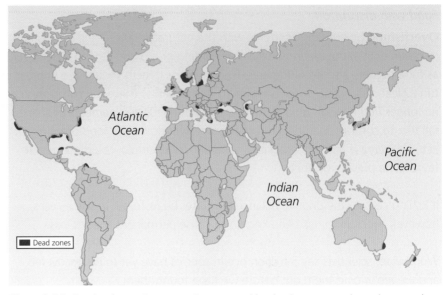

Figure 5: 'Marine dead zones' – areas where eutrophication has occurred on a large scale

Activity 2

Study Figure 6.

1. Figure 6 shows the UK enjoying a warm current. What is this current's name?

2. Study the cold ocean current flowing south from Greenland. It is fed by cold meltwater running off Greenland's ice sheets, especially in summer. What do you think will happen to the temperature of water surrounding the UK if the Earth gets significantly warmer?

3. How could this impact on life in the UK?

Closely linked with eutrophication is the problem of **siltation**. Run-off from the land also washes soil into estuaries and coastal waters. This is common in parts of south-east Asia that have been badly deforested. After heavy rain, underwater plant communities become buried in silt. Seagrasses are deprived of the sunlight they need for photosynthesis. These plants are the starting point for the entire marine food web and if they die then so too do the fish.

Climate change and the oceans

Climate change has started to create a warmer world. However, we should not expect to see a uniform increase in sea temperatures as a result. The world's oceans are dynamic systems where significant flows of warm and cool water can be identified (Figure 6). These flows could be disrupted or change direction because of climate change. For example, one impact of a major global temperature rise could be wide-scale melting of ice in Antarctica and Greenland. Billions of gallons of ice-cold water would be dumped into the world's oceans, leaving some seas cooler than before.

Complex physical factors such as ocean currents make it hard to predict exactly how marine life will be affected by global warming. However, a number of early impacts are already visible. Scientists have identified some stresses that climate change seems already to be bringing to oceans and their ecosystems (see table on page 107).

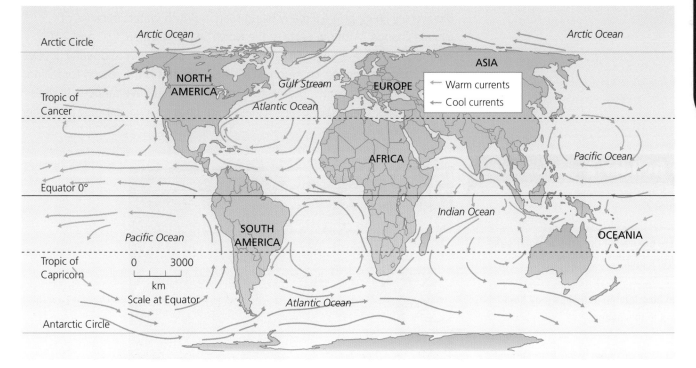

Figure 6: The world's major warm and cool ocean currents

Warmer water	Some computer models suggest that fish in tropical seas will face a famine if their water gets too warm. Microscopic plankton that fish feed on become stressed and less productive in warmer waters. This is because warmed-up surface water tends to mix less well with the colder nutrient-rich water below. Fewer nutrients coming to the surface means less plankton growing. Food shortages then develop throughout entire food webs, causing reductions in population size at every level of the food chain.
More acidic water and bleaching	The majority of the world's coral reefs are in danger of being killed off by climate change. This is because the oceans absorb some of the extra carbon dioxide that humans are putting into the atmosphere. This helps slow down the actual rate of global warming, but is bad news for marine life. Carbon dioxide dissolves in water to form carbonic acid. Small rises in acidity can seriously damage coral reefs, which then appear bleached after losing their vibrant colour. Australia's Great Barrier Reef may lose 95% of its living coral by 2050 if current trends continue.
Higher sea levels	Melting ice sheets on land masses will bring **sea-level rises**. Coastal marine ecosystems such as mangrove swamp or UK salt marshes would become permanently submerged. Unique maritime communities might disappear altogether in some locations if a significant global sea level rise of 10 cm or more takes place. Recent events in Antarctica already suggest that ice sheets there are becoming less stable as a result of warmer temperatures. The collapse of the Larsen B ice shelf in 2002 worried many scientists who believe land ice is now moving into the sea at an accelerated rate. (See Figure 7 on page 108).

Biodiversity changes are already being seen in British waters because of climate change. Water in parts of the North Sea is between 1 °C and 2 °C warmer than 20 years ago. During this time, the sandeel, a cold-water species of fish, has become less common. Scientists think this is the reason for falling populations of guillemots, puffins and other birds that have sandeels in their diet. However, there have been increased sightings of tuna, stingray and other warm water fish species.

Skills Builder 2

Study Figure 7.

(a) Estimate the size of the area of polar sea ice that broke apart.

(b) Explain how human activities may have contributed to this event.

(c) Describe the possible impacts of melting polar land ice on two named marine ecosystems.

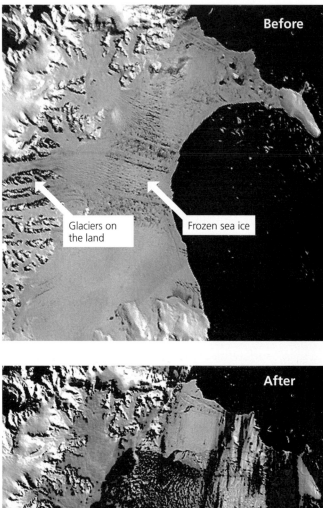

Before

Glaciers on the land

Frozen sea ice

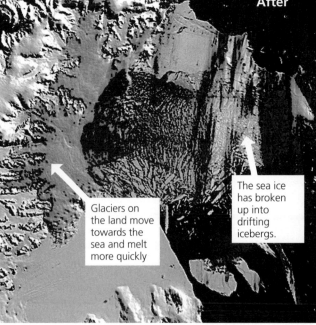

After

The sea ice has broken up into drifting icebergs.

Glaciers on the land move towards the sea and melt more quickly

0   50 km

Figure 7: A major break-up of the Larsen B ice shelf in the Antarctic, over just five weeks in 2002

How should ecosystems be managed sustainably?

Pressure in the Firth of Clyde

The Firth of Clyde is a 60 km stretch of sea water along Scotland's west coast (Figure 8). The waters of the Clyde are home to 40,000 animal and microbe species. Nature-lovers regularly catch glimpses of seals, harbour porpoises and large basking sharks. Leatherback turtles and killer whales are also sometimes seen. However, the water is used in a number of ways that put pressure on this marine wildlife:

● Fishing – This is still an important source of local income, along with kelp (seaweed) harvesting. Thanks to a warm ocean current called the Gulf Stream, commercial fish and crustacean population sizes are very high. However, over-fishing has caused numbers of species – like cod – to crash. Some fishing takes place using 'active' methods, where ships drive in circles, closing their nets around shoals of fish. There are also 'passive' methods, where pots and traps are laid along the seabed.

● Tourism and leisure – Falling incomes from fishing and farming have led local businesses to try and make more of tourism and the leisure opportunities offered by the water and coastline. The Firth of Clyde is now the UK's second-largest yachting centre. Snorkelling and kayaking are also popular activities. However, such activities disturb wildlife.

● Sewage disposal – In the past, on-land sewage treatment facilities were limited. Waste from toilets often flowed straight into the sea, damaging sea life. Thankfully, this is less of a problem now due to tougher new laws.

● Military testing – The last Ice Age left the Firth of Clyde with deeply eroded seafloor valleys. This makes a perfect testing ground for the Royal Navy's nuclear submarines. A serious accident would devastate the ecosystem.

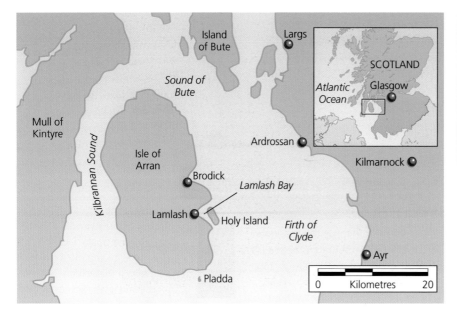

Figure 8: The Firth of Clyde

Scallop fishing in Lamlash Bay

Much of the Firth of Clyde has been completely over-fished. Lamlash Bay has been especially badly affected. Scallops used to thrive here. In restaurants, these shellfish are an expensive seafood delicacy. They live on the seafloor and fishermen harvest them, using heavy dredging machinery made up of metal chains and rollers. This equipment scours the seabed and in so doing destroys the maerl – a pink-coloured cousin of coral. Maerl is an important nursery habitat for many fish species such as cod, plaice and haddock. With the maerl mostly gone, these fish species have all but vanished too.

The first people to notice change were Lamlash residents, some of whom enjoy scuba diving. They saw places where the seafloor now looked like the surface of the moon! A dead wasteland replaced the previously colourful and species-rich environment. Along with other concerned locals, the divers formed an organisation called COAST (Community of Arran Seabed Trust). The 1,400-strong group started campaigning for a no-take zone to be established in Lamlash Bay as part of a sustainability plan. No take zones have helped regenerate fish stocks in other parts of the world by offering species a protected spawning ground. COAST's wider aims were to:

● Improve the marine environment, and reverse the maerl and fish decline

● Sustain the livelihood of people dependent on tourism, as well as fishing

● Increase the popularity of the area as a diving and tourist destination.

In 2008, the Scottish government made part of Lamlash Bay a no-take zone. All fishing within the specified area has been banned while scientific trials are carried out. Tests will establish whether the seabed can regenerate naturally. The remainder of the bay is now a management area where scallops can still be fished, but only in less destructive ways.

My family have been fishing for generations and it's the only work I know. If I can't fish, how am I going to pay for Christmas this year? Plus the sea is there for us to use – I don't see the point in keeping people like me away from doing honest work.' *Clyde scallop fisherman*

'I can't fish for fun much anymore – there's little left in the bay for me to catch. We should give the fish time to recover and leave them alone for a while.' *Amateur angler, Lamlash*

'It's a challenge managing places like this. We need to keep jobs, otherwise all the young people move out. That can leave us with a lot of independent elderly and no one paying the taxes we need to give them care. We know fishing jobs will be lost. But maybe some of the fishermen can refit their boats to take tourists on trips.' *Council Officer, Ayr*

'I retired here for the peace and the views. I feel sorry for the fishermen, but I approve of any attempt to try and limit industrial-type activities. Mind you, I hope this doesn't attract too many visitors – we don't want the roads getting too busy.' *Pensioner, Lamlash*

'It's a disgrace the way the fishermen have been allowed to ruin the sea life over in Lamlash Bay. The no-fish zone is the best news we've had in years. Finally our wildlife will be restored.' *School teacher, Ayr*

'The no-fish zone is good news for me. Our firm is going to invest in some new glass-bottomed boats. Lamlash is getting a lot of publicity and plenty of visitors will pay to see the wildlife. We've got porpoises and seals, and even basking sharks sometimes swim in these waters.' *Boat tour operator, Arran*

Figure 9: A view of Lamlash Bay, now a no-take zone

Local views on managing the Firth of Clyde

Many outsiders, both working and retired, have moved to places bordering the Firth of Clyde so that they can enjoy the sea views. They want to see the water treated well and some would like it to be kept completely free from commercial exploitation. In contrast, other people – such as local fishermen – rely on the Firth of Clyde for their livelihood. Some see laws like the no-fish zone in Lamlash Bay as a step taken in the wrong direction.

Future plans for the Firth of Clyde

The recent establishment of a no-fish zone in Lamlash Bay has shown that it is possible to make major adjustments to how local waters are managed. However, even bigger changes could soon be under way in the Firth of Clyde.

First, it is one of ten stretches of Scottish water that may soon be designated as a Coastal and Marine Park (CMP). The aim would be to ensure that coastal and marine-based activities are managed in sustainable ways to bring long-term economic benefits to people, while protecting the environment. The CMP, if established, will be run along the lines of Britain's existing National Parks.

This would mean much closer monitoring of commercial activities like fishing and sea-bed drilling for oil, gas or other minerals (which are all very important to the Scottish economy).

Secondly, the Scottish Marine Bill is a new set of laws being written to help manage future conflicts in Scottish waters. For example:

- Scotland's government wants 31% of electricity to come from renewable sources by 2011. Major tidal and off-shore wind resources can be exploited to assist with this. However, big new off-shore developments could interfere with navigation for ships. Critics say that off-shore wind farms with unsightly turbines – although designed to help the environment – will ruin the look of local landscapes (see Figure 10).

- Habitats of species like dolphins and other marine mammals need protecting from pollution. However, rising energy prices mean that companies want to exploit the remaining oil and gas resources found under the seabed – activities that can pollute.

Can compromises be reached that will keep everyone happy? Only time will tell.

Figure 10: Off-shore wind turbines

Build Better Answers

Study Figure 10. Explain one advantage this site brings for building wind power turbines. (2 marks)

■ **Basic answers** (0 marks)
State that they are in the sea.

● **Good answers** (1 mark)
State that the advantage is 'they are at sea and not on land'. This response only gains one mark, as it does not explain why this is a good thing.

▲ **Excellent answers** (2 marks)
Explain that a marine site will not be too close to anyone's home, unlike a land site. This response gains full marks.

Activity 6

1. Should people living in the Faroe Islands be allowed to keep killing Pilot whales if it is an important part of their cultural tradition?

2. How do we balance the global need to preserve species with the rights of local people to embrace their heritage?

IWC

The International Whaling Convention (IWC) was established in 1946 to oversee the management of the whaling industry worldwide. In 1986 it issued an indefinite ban on commercial whale hunting.

UNCLOS

The United Nations Convention on the Law of the Sea (UNCLOS) requires that the 156 nations who have signed it must follow the IWC guidelines.

CITES

The Convention on International Trade in Endangered Species of Wild Fauna and Flora (CITES) gives global protection to all of the great whales.

Sustainable management at the global scale

Global actions are needed to tackle pollution and to save threatened species from overfishing and extinction. International organisations already play an important role in helping countries to reach agreements. For example, the United Nations Food and Agriculture Organisation (FAO) regulates the management of deep sea fisheries. In addition, individuals can do their bit by making shopping choices that do not support unsustainable fishing. For example, many shoppers avoid buying tinned tuna that has been caught in nets that also may have trapped dolphins. Since the 1980s, many tuna products have carried the label 'dolphin-friendly', to show that no dolphins were harmed in the process of catching the fish.

Protecting endangered marine species

International laws exist to protect endangered species. Many species of whale were hunted almost to extinction during the twentieth century, bringing a public outcry in many countries. Now whale populations are being helped to recover by three global agreements (see table on left).

It is not just whales that have been helped by new international laws. CITES has helped protect other species too. In 2006, sturgeon fish were added to its endangered list and can no longer be hunted legally in the wild (sturgeon eggs are used to make caviar, an expensive delicacy).

However, international agreements are not always fully effective:

- Japan has a long history of defying international whaling laws and continues to slaughter whales.

- Norway has objected to proposals for the south Pacific to be made into a whale sanctuary.

- Each year in the Faroe Islands, around a thousand Pilot whales are massacred after they run aground in shallow water. Local people see the hunt as an important part of their culture.

Tackling pollution

Shipping is an enormous industry, with over 90% of all trade between countries involving sea travel. However, sub-standard ships and illegal activities cause marine pollution – especially along busy shipping lanes. Vessels are expected to follow international rules set out by UNCLOS. One success has been the retirement of single-hulled oil tankers after the single-hulled *Prestige* went down off the coast of Spain in 2002, causing great damage to sea species. A global phase-out of these old tankers is under way.

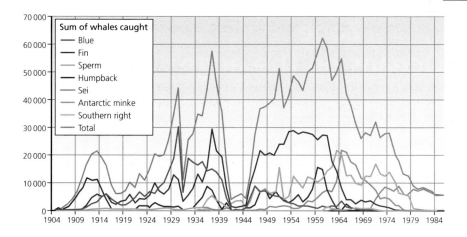

Figure 13: Twentieth-century whale harvests in the Antarctic Ocean

It is also now illegal for ships that have recently delivered oil to use seawater to wash out their tanks. Empty tanks are sure to still have plenty of oil stuck to the sides. Flushing tanks clean causes significant oil pollution (and brings in marine species to become 'stowaways', moving around the world inside the ships). Flushing is therefore banned, although monitoring what actually happens on ships is hard once they are out at sea.

A far more difficult problem to tackle on a global scale is the growth of the 'Pacific Garbage Patches'. These are enormous rubbish-strewn regions of the north Pacific. The scale of the phenomenon was first recognised by researchers in 1999. They counted one million pieces of floating plastic per square mile, most of it in the form of tiny fragments. These fragments are the remains of plastic bottles and bags that have been broken down by **attrition** and other erosion processes. They are captured and kept in place by a circulatory ocean current called the North Pacific Gyre. The current's flow creates giant pools of 'rubbish soup' (Figure 14). This rubbish has been flushed into the ocean by run-off and sewer discharge from thousands of different cities all over the world. With so many origins, it becomes a very difficult pollution problem to tackle at source.

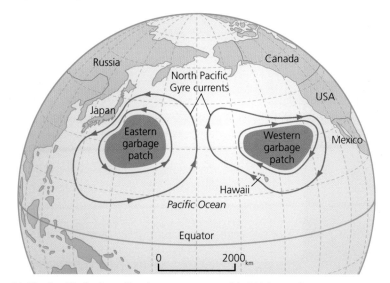

Figure 14: The Pacific Garbage Patches – two areas of 'rubbish soup'

Skills Builder 4

Study Figure 13.

(a) In what year were most Blue whales caught?

(b) What happened between 1939 and 1945?

(c) Describe changes in the numbers of Sei whales caught each year.

(d) Explain why fewer whales of any species were caught after the 1970s.

Results Plus
Build Better Answers

Study Figure 14. Describe the size and location of these areas of 'rubbish soup'. (3 marks)

■ **Basic answers** (0–1 marks)
Ignore part of the question and only describe either the size or location of the garbage patches.

● **Good answers** (2 marks)
State that two large patches are present, one on each side of the Pacific.

▲ **Excellent answers** (3 marks)
Also use the scale to estimate the size of the patches – or are far more precise about the location (e.g. proximity to the USA).

Know Zone
Oceans on the edge

The oceans are a very important resource. Millions of people depend on oceans for food and income. Human actions are rapidly degrading some parts of the oceans, increasing the threat of extinction and resource collapse. Climate change is a long-term threat on the health of the oceans.

You should know...

- ☐ What different types of marine ecosystem there are
- ☐ The distribution of marine ecosystems globally
- ☐ How humans use mangrove swamps for resources
- ☐ Which human activities, like overfishing, are degrading marine ecosystems
- ☐ How these human activities can lead to damage and destruction
- ☐ What nutrient cycles and food webs are
- ☐ How humans can damage marine food webs and nutrient cycles
- ☐ How climate change might damage marine ecosystems
- ☐ That marine resources are under increasing pressure
- ☐ How this pressure affects ecosystems locally, such as the Firth of Clyde
- ☐ Why, locally, people's views on their ecosystems can lead to conflict
- ☐ What sustainable management means
- ☐ How sustainable management can protect ecosystems like the Firth of Clyde and Coral Triangle
- ☐ How global actions such as CITIES and the IWC can help improve global sustainability.

Key terms

Aquaculture
Biodiversity
Bleaching
Climate change
Continental shelf
Coral reef
Estuaries
Eutrophication
Extinction
Food web
Habitat

Mangrove swamp
Marine ecosystem
Nutrient cycle
Overfishing
Pollution
Run-off
Sea level rise
Siltation
Sustainable
Unsustainable

Which key terms match the following definitions?

A An illustration of the grouping of animals and plants found in an ecosystem, showing the sources of food for each organism

B The deposition of silt (sediment) in rivers and harbours

C The number and variety of living species found in a specific area

D A tidal swamp dominated by mangrove trees and shrubs that can survive in the salty and muddy conditions found along tropical coastlines

E The loss of oxygen in water after too much nutrient enrichment has taken place

F Hard stony ridge, just above or below the surface of the sea, formed by the external skeletons of millions of tiny creatures called polyps

G Long-term changes in global atmospheric conditions

H Water that flows directly over the land towards rivers or the sea after heavy rainfall

To check your answers, look at the glossary on page 313

Foundation Question: Using examples, describe the threats facing marine ecosystems. (6 marks)

Student answer ● (achieving Level 2)	Examiner comments	Build a better answer △ (achieving Level 3)
Marine ecosystems, like coral reefs, can be overfished. They might suffer from bad fishing types, like dynamite fishing, which destroys coral. Divers break off bits of coral. This is happening a lot in the coral triangle area due to lots of tourists.	• *Marine ecosystems like...* The student names an ecosystem and a threat. • *They might suffer...* This is an extension of the first point and provides a detailed type of fishing. • *Divers break off...* is another correct threat and gains a mark, but is a bit vague on what divers do. • *This is happening...* The student uses a named example, linked to a threat, which is good.	Marine ecosystems, like coral reefs, can be overfished. They might suffer from bad fishing types, like dynamite fishing, which destroys coral. Divers break off bits if coral to collect as souvenirs. In the Mekong delta area of Thailand they are expanding aquaculture. Mangroves are being cut down and replaced with ponds used to farm shrimps.

Overall comment: The student would have scored additional marks if they had been more specific about the threat from divers, and remembered that a range of examples are needed for a longer answer question.

Higher Question: Using named examples, explain the short- and long-term threats facing marine ecosystems. (6 marks)

Student answer ● (achieving Level 2)	Examiner comments	Build a better answer △ (achieving Level 3)
Climate change is a short-term threat to marine ecosytems as climate changes quickly from day to day. Ecosystems like coral reef can be affected by rising water temperatures and this causes coral reef to become bleached. Another short-term threat is from overfishing, such as what happened in the Firth of Clyde. Fisherman caught too many fish so fish stocks fell and the catch fell too. This is not sustainable. A long-term threat would be like tourism.	• *Climate change is...* This part of the answer is incorrect because climate change is a long-term threat. • *Ecosystems like coral...* This is good because it explains why bleaching happens, but there is no example given. • *Another short-term threat...* This is good because a threat and an example are given. • *Fishermen caught too many...* This part of the answer uses good terminology and extends the point above by adding detail. • *This is not sustainable.* This is correct, but it could be linked more carefully to the example above. • *A long-term threat...* The threat of tourism is not explained here.	Climate change is a long-term threat, as water temperatures gradually rise due to global warming. Ecosystems like coral reef can be affected by rising water temperatures and this causes coral reef to become bleached. Another short-term threat is from overfishing, such as what happened in the Firth of Clyde. Fisherman caught too many fish so fish stocks fell and the catch fell too. Fishing like this is not sustainable as long-term fish stocks could disappear. Tourists often break coral off for souvenirs when diving, but coral might recover so this is a short-term threat.

Overall comment: The student used some good terminology and some examples of threats and places. However, they did not really cover both short-term and long-term threats, and made some errors.

Chapter 8 Extreme climates

118

Objectives

- Describe the characteristics of one named extreme climatic environment.

- Explain how ecosystems and people have adapted to life in that environment.

- Understand that people living there make a unique and valuable contribution to world culture.

What are the challenges of extreme climates?

The most important **extreme climates** are those where it is extremely cold (**polar**) or extremely hot and dry (**hot arid**). These two types can be subdivided further, as in the table below.

Polar	**Glacial** – ice-covered places, e.g. Greenland
	Tundra – places with frozen soils, e.g. Alaska
Hot arid	**Deserts** – truly arid places with less than 250 mm rain a year, e.g. Sahara
	Drylands – semi-arid places with 250–500 mm rain a year, e.g. Sahel

Places with extreme climates experience temperature or rainfall conditions that limit **flora** (plant) growth and **fauna** (animal) numbers. Without the aid of technology, human populations living in these places are limited too. Where extremely high or low temperatures are found, people's day-to-day existence is threatened. In contrast to life in a more **temperate climate** like that of the UK, special measures must constantly be taken just to stay alive. Places with an extreme climate have a low **carrying capacity**. This means that the number of people they can support without the aid of technology is lower than in other places where temperatures are more moderate and rainfall supplies are larger and more regular.

Of course, once money and technology becomes available, anything is possible. For example, 5 million people live in Las Vegas, in the Nevada desert – just 300 km from Death Valley, one of the driest (and deadliest) places on Earth. Anyone left without water in Death Valley would die from dehydration within a day, if not hours. It is the water piped from the River Colorado that keeps life going in Las Vegas (see Chapter 4).

Deserts and drylands

A truly arid climate is one that receives less than 250 mm rainfall each year. Arid regions are also known as deserts. Large areas of the Earth's surface are arid, including the Sahara and the Australian, Arabian and Kalahari deserts. Most are located in the Tropics, for reasons explained in Chapter 3. Key extreme desert facts include:

Hottest place on Earth
A temperature of 58 °C was once recorded in the North Sahara at a place called El Azizia in Libya. Death Valley in Nevada (57 °C) comes a close second.

Biggest desert
The Sahara measures 9 million km². Its biggest dunes are 150 m high. It manages to support 2 million people at its edges and in towns along caravan trails.

Driest place on Earth
There are parts of Chile's Atacama desert that have not seen rain for 400 years. In many places there, average rainfall is just 1 mm per year.

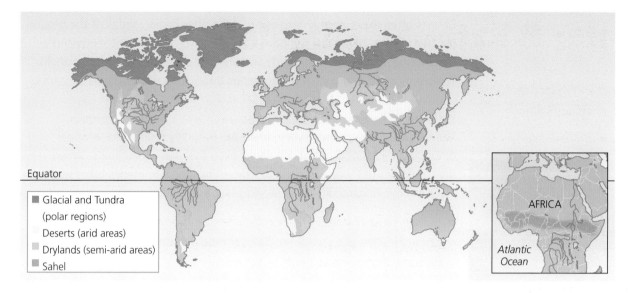

Figure 2: Deserts, drylands, and polar regions

On the borders of deserts there are semi-arid areas – also called drylands – where annual rainfall ranges from 250 mm to 500 mm. For example, the Sahel is a long strip of drylands that borders the south of the Sahara. It includes parts of Sudan, Chad, Burkina Faso, Niger and five other countries (Figure 2).

Compared with the Sahara, conditions in the Sahel are less testing for humans. Natural water supplies are sufficient for around 50 million people to live here – provided that their lives are well adapted to the conditions. Permanent, settled lifestyles exist side by side with nomadic (migratory) behaviour. Along the Sahel's northern edge, the nomadic Tuareg people live a migratory lifestyle. They live in small tribes consisting of 30–100 family members, who keep camels, goats, cattle and chickens which graze the land.

Desert and dryland flora and fauna

In the hottest desert regions, few animals can survive apart from tough scorpions and small reptiles. In dryland areas, with a greater supply of water, biodiversity rises as tough grasses, shrubs, cacti and hardy trees begin to form the basis of a larger food web. The plants and animals have evolved ingenious **adaptations** to help them deal with a shortage of water:

● Grasses with extensive root systems that can penetrate as far as 50 metres into the ground and can also reach sideways.

● Trees that store water. Acacia trees, for example, have short, fat trunks that act as reservoirs for excess water.

● Seeds that only germinate after heavy rain. Plants immediately produce brightly coloured flowers to attract insects.

● Insects that can collect moisture from the air. By standing still, some beetles let early morning mist condense as water droplets on their bodies which then runs into their mouths.

Skills Builder 1

Study Figure 2.

(a) Describe the distribution of deserts.

(b) Using examples, describe how plants and animal communities have adapted to survive the extreme conditions found in deserts and drylands.

(c) State and explain one change that these communities are vulnerable to.

Deserts often experience an explosion of life after heavy rainfall. If the ground has been baked hard then temporary lakes will form. Dormant seeds and hibernating animals awaken. Birds and animals are attracted from far afield. Temporary plant and animal communities thrive for a while (Figure 3).

Vulnerability

The species in desert and dryland areas are naturally resilient to seasonal or longer periods of drought. However, even they have their limits. When dry conditions get worse in drylands, desertification sometimes takes place, revealing the vulnerability of semi-arid plant and animal communities.

- Human causes of desertification include over-grazing by goats and cattle. This has been made worse over time by restrictions placed on the movement of nomadic herds by settled farmers.

- Physical changes include the occurrence of long-term and cyclical droughts (see Chapter 4, page 58).

Figure 3: A desert blooms

Extreme cold

Cold polar climates are found in areas of high **latitude** – especially inland areas, far from the warming influence of the sea (Figure 4). Places where the land is partly or completely covered with ice are called **glacial** regions. 98% of Antarctica is covered with snow and ice. The coldest ice-covered glacial regions such as central Antarctica do not support any life at all. In the northern hemisphere, lying south of the glacial regions are the **tundra** regions. They are not directly beneath ice but do experience very cold weather for most of the year, leaving the ground beneath permanently frozen. This gives them a distinctive appearance and environment, as you will see. Around 20% of the Earth's land surface is tundra. Key extreme cold facts include:

Coldest temperature ever recorded	Antarctica has a recorded low of -89 °C. The coldest temperatures for populated continents are -81 °C in Yukon, Canada and -55 °C in Siberia, Russia.
Darkest settlement	Alert Bay, on the north-east coast of Ellesmere Island, Canada is 82.5° north. It receives no direct sunlight for 50 days of each year.
Deepest ice	In parts of Antarctica, the massive sheet of ice is nearly 5 kilometres deep.

Figure 4: A view of the Alaskan tundra

In the high latitudes of the northern hemisphere, conditions are cold and icy but less extreme than in Antarctica. Daytime temperatures at the surface of Alaskan glaciers can reach as high as 12 °C in summer. Daytime temperatures keep skiers warm and the reflected sunlight even brings sunburn.

The tundra landscape (Figure 4) is found south of the glacial regions in the northern hemisphere. Much of Alaska – a separate US state to the west of Canada – is tundra. In populated regions, such as the northern town of Barrow, the Sun never rises above the horizon for up to one month of the year and temperatures may fall to –50 °C. Sunlight here is so weak that the growing season is generally under three months. Native people hunt extensively during the most brightly lit months, building up meat stores for the year, or working alongside settlers to mine and extract resources, including oil.

Glacial and tundra flora and fauna

Away from the deepest ice sheets, small tough alpine plants grow on the exposed rock surfaces that can be found poking through the snow in glacial areas. Along with lichen and moss, they form the basis for simple food webs. Plants survive strong winds by growing close to the ground in crevices and cracks. Spiders and tiny mice can also scratch out a living on glacial mountain sides.

Tundra food webs are far more complex, with around 1,700 species of vegetation and animals. A range of adaptations help plants and animals to survive here:

Water-loving plants can thrive	In the flat low-lying tundra regions, surface layers of soil melt in summer, leaving many areas waterlogged and creating perfect conditions for sedges and moss (Figure 4).
White fur as camouflage	Plants form the basis for a food chain with many plant-eaters and meat-eaters present. Some, like snowshoe rabbits and snowy owls, have evolved white fur and feathers so that they cannot easily be seen against winter snow.
Caribou have two layers of fur	The caribou is a type of deer that has adapted to survive the bitter tundra cold. They have two layers of fur to keep the heat in and large hooves to help them travel over soggy ground.

Vulnerability

In both glacial and tundra regions, we find that flora and fauna can be vulnerable to sudden environmental changes:

- Bursting lakes – Catastrophic changes occur when a glacial meltwater lake bursts, which is likely to become more common due to climate change and warmer temperatures in polar regions. Meltwater running off a glacier sometimes builds up behind a natural barrier, such as a ridge or bank of soil. Failure of the barrier can release a surge of meltwater, destroying local ecosystems.

- **Solifluction** – Tundra regions are vulnerable to seasonal soil melting. This can trigger a process caused **solifluction**. Rather like a landslide, whole sections of a slope start to move under gravity. The surface vegetation gets rolled beneath the moving mass of soil, like the tracks on a tank. Again, a warming atmosphere increases the likelihood of this happening. Heat from buildings can also cause melting.

Skills Builder 2

Study Figures 2 and 4.

(a) Describe the distribution of glacial and tundra regions.

(b) Using examples, describe how glacial or tundra plants and animal communities have adapted to survive the extreme conditions found in these places.

(c) State and explain one change that a named polar ecosystem is vulnerable to.

ResultsPlus
Watch out!

■ Polar bears are not part of a typical polar ecosystem. Their habitat is actually the ice that covers the Arctic sea, where they hunt seals. They rarely hunt on the land and are not present on the tundra, unlike grizzly bears that do hunt in places like Alaska.

Living with extreme cold

Many of the Earth's coldest regions show evidence of long-term settlement, despite the challenging environmental conditions found there. In Chapter 2, we learned about the ancient historical activities of some groups of people living in the world's far north. During the last Ice Age, people were migrating out of Asia and into North America across ice and land bridges, spreading their culture through regions that remain cold today. More recently, Scandinavian Vikings were exploring Greenland during the Medieval Warm Period, around the tenth and eleventh centuries.

Today, people can be found living in all the cold regions of the Arctic Circle. In Russia, we find nomadic reindeer herders called the Khanty people. In Alaska, native Inupiat and Yup'ik tribes have been joined by more recent settlers. The adaptations to cold conditions seen in Alaskan settlements are thus a mixture of ancient and modern technologies:

Buildings	High-pitched steep roofs allow snow to slide off, while triple-glazed windows keep the heat indoors. In areas of **permafrost**, houses are raised on stilts to prevent their heat from melting the frozen ground beneath (Figure 5). Ice loses volume when it melts, causing the land to sink and subside, which would damage the building. Even worse, melting the ice might trigger solifluction flows.
Transport	Roads can also suffer from permafrost melting, causing cracks to develop. In order to prevent this, roads are built on 1–2 metre thick gravel pads that stop heat transfering from vehicles to the frozen soil beneath.
Farming methods	Native Alaskans have always relied on hunting and fishing for food rather than farming the land, because of the permafrost and very short growing season for crops.
Clothing	Traditionally, Inupiat and Yup'ik people favoured coats made of caribou skin (with the double lining of fur) and sealskin boots. Goose down was used as a lining. Now they also wear modern man-made textiles, like Gore-tex.
Energy use	Energy use is high for people living in cold environments, especially at high latitudes where sunlight is extremely limited for part of the year. In parts of Alaska, geothermal heat provides energy and hot water, a beneficial side-effect of volcanic activity along this part of America's Pacific coast.

Exploring polar cultures

Many of Alaska's native people are part of a larger ethnic group called the Inuit who are spread throughout the Arctic Circle – in Canada, Alaska, Russia and Scandinavia. In all these places, unique cultures have survived into the modern age.

Figure 5: An Alaskan house built on stilts to avoid melting the permafrost

Yup'ik masks	The Yup'ik community in south-western Alaska are well known for their carved masks, which are a unique feature of their storytelling culture.
Inupiat art	You may be familiar with the art of totem poles. Tlingit village in Alaska attracts tourists to see an impressive collection of poles that have been collected there.
Inupiat whaling ceremony	Some aspects of polar culture are controversial. In Alaska, the Inupiat are still allowed by US law to hunt and kill bowhead whales. Communal sharing of freshly cut whale meat is an important ritual that the Inupiat people believe helps bind their culture together.

Other cold places such as Iceland and Greenland have their own highly distinctive cultures. Life in a challenging environment has evolved over time so that practices born in hardship have become treasured traditions. For example, in Iceland, people still love to eat 'rotten shark'.

Iceland's 'rotten shark'

Until recently, Iceland was regarded as one of the least developed countries of the northern hemisphere. It has a harsh glacial, stormy and volcanic environment. However, thanks to investment in banking and hi-tech research, the Iceland economy has made enormous progress over the past century.

Despite this economic advancement, Icelandic people still like to share 'rotten shark' at parties. The meat of the Greenland Shark is naturally poisonous because the shark contains fluids that allow it to live in cold water without freezing. To make the meat safe, Icelanders used to bury chunks of the meat for months at a time before eating. During this time, the fluids drain from the shark, making it safe.

The decomposition that takes places also gives the meat a very strong ammonia smell that many people find disgusting. Although modern Icelanders no longer need to rely on it for survival, many have learned to enjoy the taste of 'rotten shark'. They view it as part of their heritage. You can watch a short film about this at: http://video.nationalgeographic.com/video/player/places/culture-places/food/iceland_rottensharkmeat.html

Figure 6: Many people in Iceland enjoy eating 'rotten shark'

Quick notes (Iceland's 'rotten shark'):
- Extreme climates gave rise in the past to unique and interesting cultures.
- Originally, many cultural practices developed through necessity.
- These cultures live on today, through tradition and familiarity – and sometimes to attract tourists.

Living with aridity

The word 'arid' might at first bring to mind inhospitable places where culture is unlikely to blossom. But remember that the first world's first great ancient civilisations – the Egyptian pharaohs and the Arabian and Persian empires – all flourished in places we classify as arid and semi-arid regions.

Desert and dryland cultures flourish today, especially when newer technologies have been introduced alongside older ones. In the Middle East, desert cities like Dubai and Abu Dhabi have air-conditioned buildings and enjoy desalinised water supplies paid for with oil money (Figure 7). Whether in the Middle East, or the Sahel, common adaptations to the arid climate include:

Buildings	Flat roofs collect water, while walls are painted white to reflect sunlight and keep buildings cool. In poorer areas, windows are small to reduce light and keep temperatures low. Where there is money, there are bigger windows – and air-conditioning.
Transport	In the early twentieth century, camels were still the dominant transport for Middle Eastern and African desert tribes. Today, modern roads allow motorised transport for those with money. Rich people in the Gulf states drive powerful off-road vehicles in areas where sand frequently covers over the roads.
Farming methods	Water is incredibly precious in deserts and drylands. The ancient Egyptians relied on irrigation from the Nile to help their farming. The Nile is now dammed at Aswan, giving Egypt reliable water supplies for its crops. Where irrigation is not available, traditional **nomadic pastoralism** is still practised, with goats and sheep being moved to new places when they begin to put too much stress on local vegetation.
Clothing	Middle Eastern tribes of Bedouin people favour loose-fitting clothing, often white to reflect sunlight. To avoid sunburn and protect the wearer from wind and sand, heads and faces are often covered with head-scarves called *kufiyya*. Clothes were originally woven from sheep or goat wool.
Energy use	For traditional societies, energy use has often been low, thanks to the high temperatures. Wood fires keep nomads warm inside their tents at night. Where oil has been found in the Middle East – and where wealth is now high – energy use has, of course, soared, in response to rising demand for air-conditioning and refrigeration in permanent settlements.

ResultsPlus
Build Better Answers

Describe how people's clothing helps them survive in the area of extreme climate you have studied. (3 marks)

■ **Basic answers** (0–1 marks)
State that animal skins are worn.

● **Good answers** (2 marks)
Also mention something about the nature of the animal skin (for example that caribou skin has two layers of fur and is especially adapted to a cold climate).

▲ **Excellent answers** (3 marks)
Not only describe how a traditional type of clothing helps keep people warm, but say that modern outdoor textiles are now used as well.

Figure 7: The Dubai skyline

Exploring arid cultures

Some of the world's desert and dryland cultures have given us truly amazing cultural and archaeological artefacts:

- **Egypt's Pyramids at Giza** – Perhaps the greatest wonder of the ancient world, Egypt's pyramids lie in the desert, an important reminder that harsh climates have never stopped human progress.

- **Australia's Aboriginal Culture** – Australia's desert is home to the native tribes of Aboriginal people, with a rich history of ceremonial songs and paintings that stretch back 40,000 years.

- **South America's Nazca Lines** – In the desert of Peru, giant shapes have been 'drawn' in the desert. They were constructed around 1,500 years ago, but they can only be seen properly from high above – by aeroplane – making them one of the world's greatest archaeological mysteries.

The Sahel region that you have been learning about in this chapter is also home to many rich cultures and traditions:

- The use of instruments such as the Hausa flute and gourd rattle has produced a distinctive musical culture not found anywhere else (Figure 8). For instance, Mali's Dogon people use music and dance in a festival, called a 'Dama', to honour the dead.

- The city of Timbuktu in Mali was an ancient place of learning and is said to have been home to 25,000 scholars in the sixteenth century.

- In nearby Niger, the Tuareg tribe are excellent craftsmen, renowned for their indigo cloth, gold and silver jewellery and carved wooden masks.

However, culture clashes are also found in the Sahel. In Sudan's Darfur region, for example, the recent wars between tribes have resulted in 200,000 people being killed and around 2 million displaced since 2003. Other conflicts are currently underway in Chad, Niger, Mali and Senegal. Too often, the same problem lies at the root of the conflicts – different groups of nomadic cattle-herders and settled farmers are fighting over limited grass, water and soil.

ResultsPlus
Build Better Answers

Describe how building styles are adapted to the extreme climate you have studied. (3 marks)

■ **Basic answers** (0–1 marks)
State that buildings are white in arid places.

● **Good answers** (2 marks)
Also say something about why this is the case (because white reflects sunlight, keeping temperatures lower inside).

▲ **Excellent answers** (3 marks)
Not only describe a traditional type of adaptation, but also say that modern houses have air-conditioning (especially where oil money is found).

Figure 8: Musicians of Mali's Bobo tribe playing traditional instruments

Objectives

- Describe the human and physical threats faced by one extreme environment.

- ◉ Explain how climate change could further threaten this environment and its people.

- ◉ Understand that a range of local and global management strategies can help protect this environment.

How can extreme environments be managed and protected from the threats they face?

Human threats to extreme climates

Extreme climates give rise to physical environments that are extremely sensitive to change. Human actions often threaten to change local conditions and disturb the delicate balance found in extreme environments in a number of ways. The tables below and opposite show examples of the impacts in Alaska and in the Sahel.

Pollution occurs because of **resource exploitation**, often in previously pristine wilderness environments. One reason for this is that valuable oil underlies many of the world's arid and polar regions. **Land degradation** sometimes results from poor management, accelerating desertification in drylands and permafrost melting in some polar areas.

It is not just the physical geography that is affected. Tourism and settlement by in-migrants can result in **cultural dilution**, because outside influences cause a culture to lose its unique characteristics. Ancient languages become lost – and replaced by English or Spanish for instance.

Figure 9: Striking a pose for the tourist gaze

	Impacts in the Sahel
Pollution	One theory suggests that air pollution in North America and Europe may have made drought and desertification worse in the Sahel region since the 1960s. Clouds of sulphur particles emitted from power stations and factories may have affected patterns of cloud formation in ways that have left the Sahel without its normal level of rainfall for several decades, resulting in a sustained drought.
Land degradation	Desertification is an enormous problem. Traditional nomadic herders find they cannot migrate with their cattle as easily as in the past, for various reasons. International food-growing companies have seized the best land. National boundaries established in the colonial era interfere with age-old migration routes. Civil war and political instability today drive people into marginal lands. In too many cases, the result is overgrazing, removing the marginal vegetation that was present. Soil erosion can then take place when the wind blows.
Cultural dilution	The Sahel countries were colonised by European powers in previous centuries. Many have experienced disruption and civil war since independence. As such, it is not surprising to find that some native languages and art have been lost and forgotten. Today, foreign tourists are a vital source of income, and tourism continues the process of cultural dilution and loss. In the case of the Dogon people of Mali, they sometimes perform their sacred Dama funeral ritual for the amusement of tourists. Some people think that cultures like the Dogon should not be exploited like this (Figure 9).

How can extreme environments be managed and protected from the threats they face?

127

	Impacts in Alaska
Pollution	Alaska has suffered badly from the extraction of oil in the North Slope area. In 1989, an oil tanker, *Exxon Valdez*, ran aground on the Alaskan coast while transporting oil to market. Only 15% of the 11 million spilled gallons was ever recovered, and 5,000 sea otters and many seals and eagles were killed. More recently, a broken pipeline in 2006 spilled 200,000 gallons of oil in the fragile North Slope region.
Land degradation	Millions of square kilometres of permafrost have been damaged because people didn't take account of the sensitive soil conditions. Wide-scale melting can be found around warm urban areas such as Fairbanks and Barrow. However, the greatest human threat comes from global warming. Temperatures are rising in Alaska, perhaps by as much as 5 °C over the past century. Scientists are also reporting an increased frequency of landslides in the permafrost.
Cultural dilution	In the past, 20 native languages were widely spoken in Alaska. As elsewhere in North America, European languages such as English have been adopted by the youngest generations of tribes (Figure 10). In-migration by new settlers since the 1700s has brought new languages. Native names were replaced by names like Peter and John after missionaries visited. In the 1970s, American schooling insisted on classroom use of English. Now, some languages, such as Eyak, have lost their last speaker, while others are on the verge of dying out.

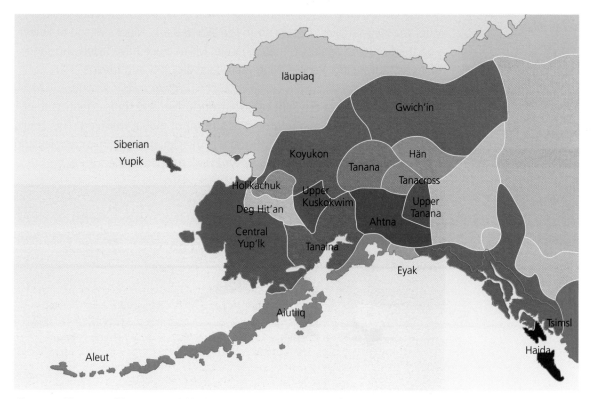

Figure 10: Threatened languages of Alaska

Climate change and extreme environments

Climates and environments everywhere face change and challenge because of global warming, as we learned in Chapter 2. Events under way in arid and polar regions are especially concerning. The additional stress in arid and dryland regions endangers some of the poorest and most vulnerable people on the planet. Impacts in polar regions also threaten people and ecosystems living there – while melting polar ice, in turn, brings a sea-level rise that threatens societies on a worldwide scale.

Climate change and desertification

Some predicted impacts of climate change for the Sahel drylands include:

Increased aridity	Drylands where rainfall is already low could turn into fully arid desert as the twenty-first century progresses. Most predicted changes of weather show the Sahel rain belt being pulled either further north or south, away from populated areas. Dramatic drying is likely.
Nomads losing their livelihood	Nomadic herders in the northern Sahel will be unable to find enough vegetation for their cattle to graze.
Sand dune threats	Old inactive sand dune systems in the Sahel could re-activate and begin to move again, threatening crops and grazing land.
Climate change refugees	Millions of people, unable to find enough water and land to grow food or feed their cattle, may be forced to migrate elsewhere. Will already over-populated coastal cities in Africa be able to cope with an even greater influx of migrants?

Reports suggest that people are already feeling the effects of climate change. In some parts of the Sahel, women now walk as far as 25 km a day to fetch water for their families. According to the World Bank, Niger's village of Feteye is on the front line against climate change. Millet crops have failed and sand dunes are advancing. People living there face drought and famine.

However, evidence for the Sahel is still inconclusive, and some scientists think there is a possibility of rain returning as a result of global warming. This would lead instead to the 'greening' of the Sahel. The climatic system is complex and there remains much uncertainty over what will happen in the future.

Figure 11: The creeping desert in Niger

How can extreme environments be managed and protected from the threats they face?

129

Climate change and polar melting

Polar regions face their own climate change threats, including:

Changes in sea ice cover	Arctic and Antarctic sea ice are melting. Scientists are recording record low levels in summer. This is changing sea temperatures and threatens the decline of species like polar bears that hunt seals on sea ice.
Land glacier retreat	Most of the world's glaciers are now shrinking in size and scientists blame a warmer climate. All over the world, the pattern seems to be the same. Tourism in icy regions such as Canada is threatened.
Permafrost melting	In Alaska and Siberia, up to 40% of permafrost is expected to thaw if climate change predictions are right. As we have already learned, this poses many threats to people and the tundra ecosystem. Temperatures in parts of Alaska have risen by 5 °C over the past century. Permanent, rather than seasonal, melting of permafrost is already triggering long-term flooding of some parts of the tundra.
Species migration	One possible beneficial effect is that the treeline could move further north as temperatures get warmer. Areas of tundra might be colonised by coniferous forest, bringing increased animal and bird life.

In south-west Alaska, the Yup'ik people believe they are already suffering the effects of climate change. Described by one newspaper as 'America's first community of global warming refugees', the Yup'ik no longer take their dog teams safely out on to Bering Sea pack ice in search of fish. The ice has lost half of its thickness, making it unsafe. In the village of Newtok, all the buildings are sinking and tilting as the permafrost beneath melts. Whereas once the Yup'ik drove their building piles 3 metres deep into the ground, now it must be 4 metres. It is clearly becoming warmer. According to scientists at NASA, temperatures in Newtok have risen by 4 °C since the 1960s – and by as much as 10 °C in winter months.

For complex reasons, global warming is felt most by people living in polar regions. The loss of snow cover also sets off a vicious circle for the Yup'ik. With less bright white snow to reflect the sunlight, the ground warms and melts even more.

Activity 1

1. Will the world see millions of climate change refugees in your lifetime?

2. Where will they go?

3. Should countries like the UK offer them homes?

4. This may turn out to be one of the biggest issues of the twenty-first century. What are your views?

Decision-making skills

Which do you consider to be the major threats facing either polar or arid areas? To help you with your answer, use the tables on page 126 or 127 and 128 or 129.

Figure 12: A retreating glacier – the Rhone Glacier in 1996 and 2000

Sustainable management of extreme environments

Extreme environments are under threat due to a combination of local and global pressures:

- Global warming is changing world temperatures, rainfall patterns and global sea level. Temperature changes are especially high near the poles, where warming of several degrees has been recorded since the 1950s.

- Local population growth, land-use changes (often linked to the actions of global companies) and political problems (such as the civil war in Sudan's Darfur region) threaten environments and ecosystems.

How can we ensure that polar and arid places have a sustainable future? How can we ensure that future generations of communities living there today can still enjoy and use these special environments? How can we make sure that biodiversity levels remain the same and species are not lost?

In Chapter 4 you read about the use of **intermediate technology** to help aid water supplies in drought-stressed parts of Tanzania. If your chosen extreme environment is arid regions, then you should now re-read that section (pages 124–125). The WaterAid website contains details of many more projects, plenty of which are located in the Sahel region, especially Burkina Faso and Mali (www.wateraid.org).

One of the most exciting developments for arid areas is the move towards greater use of solar power. In America, Australia and the Middle East, enormous areas of desert are already being used to collect sunlight in photocells (Figure 13). In the near future, countries with the greatest amounts of intense sunlight may come to be regarded as the world's most resource-rich places – rather than those with the most oil, as is the case today.

Figure 13: Rows of solar panels quietly produce energy in America's Mojave Desert

Sustainable management of polar regions

Iceland is an excellent example of a cold region where every effort has been made to use sustainable energy supplies. The island is located along the mid-Atlantic ridge, a zone of volcanic activity. Thanks to the up-welling of magma beneath Iceland, groundwater becomes heated and sometimes pressurised to form steam. This steam can be harnessed to drive turbines that generate electricity (Figure 14, page 131) while hot water can be used directly to heat buildings. It is a renewable energy resource that can be used to heat and light greenhouses, allowing Icelanders to grow fresh fruit and vegetables and heat water for fish farms all through the year.

How can extreme environments be managed and protected from the threats they face?

131

In Alaska, similar technology has been installed in many places located close to the coastal tectonic plate boundary that forms part of the Pacific 'Ring of Fire'. Unlike Iceland, this is a plate boundary where dangerous, unpredictable and explosive activity sometimes takes place. However, it still produces beneficial resources such as hot springs and renewable energy. The first Alaskan volcano to be tapped for geothermal power is Mount Spurr, a tall, snow-capped and steep-sided volcano in the Aleutian Volcanic Arc. A tourist resort situated at Chena Hot Springs near Fairbanks is entirely powered by geothermal power.

Another example of sustainable living, which we have already seen, is the construction of houses on stilts (Figure 5, page 122) – an adaptation which helps preserve the permafrost in its pristine state for future generations.

Global actions to protect extreme environments

Polar and arid regions are widely regarded as special places. International efforts have been made to try and ensure they have a sustainable future. The best known of these is the 1961 Antarctic Treaty. It has become one of the most successful international agreements of all time, restricting commercial exploitation of the Antarctic continent. Following on from this, the 1998 Protocol on Environmental Protection to the Antarctic Treaty is one of the toughest sets of rules for any environment in the world. Under the agreement, no new activities are allowed in Antarctica until their potential impacts on the environment have been properly assessed and minimised. Tourist boat operators taking visitors there have to follow incredibly strict guidelines.

As climate change begins to damage the Arctic, we are starting to see global efforts focused on this polar environment as well. In 2008, the US placed Arctic polar bears on its endangered species list. If it really wants to help the bears, however, the US still needs to do far more to curb its own CO_2 emissions. Only global action on climate change will stop more Arctic ice from being lost, depriving the bears of their hunting ground.

The threat posed to the world by ice melting makes global action to protect polar environments absolutely essential. Loss of arid environments is perhaps seen as a less urgent concern by many. However, efforts are still made to unite the international community and help to halt desertification and 2006 was the 'International Year of the Deserts and Desertification'. According to the UN, the aim was 'to get the message across that desertification is a major threat to humanity, compounded by both climate change and loss of biological diversity. **Land degradation** affects one-third of the planet's land surface and around one billion people in over a hundred countries.'

Figure 14: Geothermal energy generation in Iceland

Skills Builder 3

Study either Figure 13 or Figure 14.

(a) Describe how energy is being generated in the extreme environment shown.

(b) Explain why this is a type of renewable energy.

Decision-making skills

Suggest what is meant by a sustainable future for extreme environments and explain how renewable energy can assist with this.

For either polar or arid environments, identify the features of what a sustainable future may look like. Explain how renewable energy can assist with this.

Extreme climates

Some places are just very difficult to live in. Extreme heat and intense cold make survival a major challenge for plants, animals and humans. Extreme climate regions are often untouched, but they are facing growing threats to their very survival.

You should know...

You can choose to study either polar regions or hot arid areas. For either choice you need to know:

☐ What is meant by an extreme climate

☐ Facts and figures for your chosen climate

☐ Where your extreme climate can be found

☐ How plants and animals have adapted to live there

☐ How people adapt and cope with life in an extreme climate

☐ The value and uniqueness of the people who live there

☐ How climate change could alter your extreme climate zone

☐ How your extreme climate zone could be affected by other threats

☐ How local people are adapting to the threats they face

☐ How global actions might protect extreme environments.

Key terms

Adaptation	Land degradation
Carrying capacity	Latitude
Cultural dilution	Nomadic pastoralism
Desertification	Permafrost
Exploitation	Polar region
Extreme climate	Pollution
Fauna	Solifluction
Flora	Temperate climate
Glacial region	Tundra
Hot arid regions	
Intermediate technology	

Which key terms match the following definitions?

A The position of a place north or south of the Equator, expressed in degrees

B Plants

C The maximum number of people that can be supported by the resources and technology of a given area

D A climate that is not extreme (in terms of heat, cold, dryness or wetness)

E Permanently frozen ground, found in polar (glacial and tundra) regions

F The spread of desert conditions into what were semi-arid areas, caused by natural climate change or by human activities, such as overgrazing and deforestation

G Animals

H The movement downhill of soggy soil when the ground layer beneath is frozen

To check your answers, look at the glossary on page 313

Foundation question: Describe how plants and animals are adapted to either polar or hot arid extreme environments. (3 marks)

Student answer ⬤ (achieving 2 marks)	Examiner comments	Build a better answer △ (achieving 3 marks)
In polar places like the Arctic plants have very small leaves. *They grow close to the ground.* *Maybe only a few inches high.*	• *In polar places...* scores 1 mark. It is a good example of an adaptation. • *They grow close...* This also scores a mark as it is another clear adaptation. • *Maybe only a...* As the question asks for plants and animals, another point on plants is not needed and does not score a mark.	In polar places like the Arctic plants have very small leaves. They grow close to the ground. Animals usually have thick, white fur, such as the Arctic Fox.

Overall comment: The candidates should have spotted that the question asked for plants and animals. Read questions carefully and underline or highlight key words.

- -

Higher question: Explain how plants are adapted to either polar or hot arid extreme environments. (3 marks)

Student answer ⬤ (achieving 2 marks)	Examiner comments	Build a better answer △ (achieving 3 marks)
Plants need to have shallow roots. *Some plants grow close to the ground.* *Polar regions are cold and dry. Plants have small leaves to limit water loss.*	• *Plants need to...* This is a correct point but is a description rather than an explanation. • *Some plants grow...* This is another correct point but, like the first, it describes rather than explains. Together with the first point, it scores 1 mark. • *Polar regions are...* This is a correct answer and it explains why, so scores 1 mark.	Plants need shallow roots to avoid the permafrost. Plants grow close to the ground to avoid wind damage. Polar regions are cold and dry. Plants have small leaves to limit water loss.

Overall comment: The student tended to describe rather than explain. The last point was linked to a reason, but the first two were not.

Unit 2 People and the planet

Your course

This unit focuses on human geography and the topics link together to build an overall understanding of how humans interact with the planet. You will learn about how populations grow and change, where people live and work, and how they exploit and use resources. There are three sections:

Section A topics are compulsory and introduce you to the main aspects of how people live on our planet. You will study **all** topics:

- Topic 1 (Chapter 9): Population dynamics

- Topic 2 (Chapter 10): Consuming resources

- Topic 3 (Chapter 11): Living spaces

- Topic 4 (Chapter 12): Making a living

Section B will cover how people interact with different aspects of our planet on a small scale and you will study **one** topic:

- Topic 5 (Chapter 13): Changing cities

- Topic 6 (Chapter 14): Changing countryside

Section C will cover how people interact with different aspects of our planet on a large scale and you will study **one** topic:

- Topic 7 (Chapter 15): Development dilemmas

- Topic 8 (Chapter 16): World of work

Your assessment

- You will sit a 1-hour written exam worth a total of 50 marks

- There will be a variety of question types: short answer, graphical and extended answer, which you will practice throughout the chapters that you study. You will answer **all** the questions in Section A, **one** question from Section B and one question from Section C.

- **Section A** contains questions on the compulsory topics.

- **Section B** contains questions on the two small-scale topics.

- **Section C** contains questions on the two large-scale topics.

Remember to answer the questions for the topics that you have studied in class!

Study the photograph of Las Vegas.

(a) Explain the term 'sustainable city'.

(b) Describe two ways of making cities more sustainable.

Chapter 9 Population dynamics

Objectives

- Learn how the rate of global population growth is changing.

- Recognise that the structure of population changes with time.

- Understand the link between population change and development.

ResultsPlus
Watch out!

■ Remember that although the rate of global population growth is slowing down, the total population is still increasing, but not as quickly as it was a few years ago.

How and why is population changing in different parts of the world?

Global population growth

During 2008 at least another 65 million people were added to the world's population. This annual increase was less than during the 1980s and 1990s. Population growth rates have fallen from 2.1% per year to around 1.95 per cent. Besides showing the upward curve of global population since 1800, Figure 1 shows how long it has taken for the world's population to increase by 1 billion. It took 118 years from 1804 to double from 1 to 2 billion. Since then, the length of time has shortened considerably to a mere 12 years between 1987 and 1999.

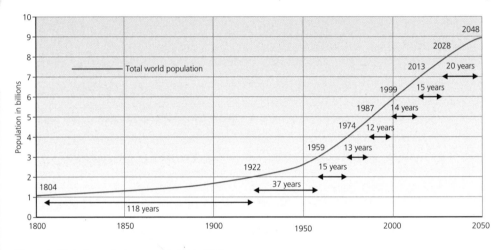

Figure 1: Global population growth, 1800–2050

What the graph also shows is that the length of time it will take to add further billions will start to increase, from 12 to 14 to 20 years. In short, the rate of population growth is predicted to slow down. The interesting question is – given this slowing down – when exactly will the point of **zero population growth** be reached? There is much disagreement about the likely date. Some have suggested it may be as early as the 2020s. Others say it will not be before 2060.

No matter where you are in the world, population change (growth and decline) is produced by two processes – **natural change** and **migration** change. Figure 2 shows that natural change depends on the **birth rate** and the **death rate**. If there are more births than deaths, population will increase. If there are more deaths than births, population will decrease.

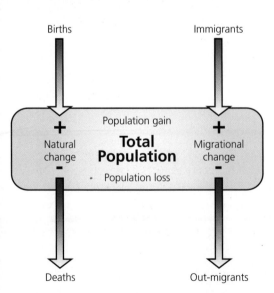

Figure 2: Elements of population change

How and why is population changing in different parts of the world?

137

Migration is the movement of people into and out of an area or country. If **immigrants** (incomers) exceed **emigrants** (outgoers), there will be a gain in population. If the situation is reversed, there will be a loss of population. But, of course, any change in the total size of the global population results only from natural change. Migration simply helps change the distribution of that global population.

So the overall rate of global population growth depends on the difference between the birth and death rates. The more the birth rate exceeds the death rate, the higher the rate of **natural increase**. Remember too that a rise in the rate of natural increase has a multiplier effect and adds even more population growth. This is shown in Figure 3.

The most important factor responsible for the rising rate of population growth has been the fall in the death rate, as shown in the table on the right. (You will also see that the trend in the birth rate has been downward too, working in the opposite direction.) A whole range of factors has contributed to the fall in mortality. They include:

⦿ The development of modern medicines

⦿ The introduction of vaccination and immunisation programmes

⦿ Better healthcare – more doctors, nurses and hospitals

⦿ More hygienic housing

⦿ Cleaner drinking water and better sewage disposal

⦿ Better diet.

A reduction in the **infant mortality** rate has also contributed to this lowering of the overall death rate.

Forecasting future population figures is a difficult business, because there are often surprise changes. Who really knows what is likely to happen to the global birth and death rates? Will the spread of birth control and worries about population pressure on the world's resources continue to lower the birth rate? What about the HIV/AIDS pandemic? Will this cause the death rate to rise significantly? Is there some new highly infectious and lethal disease just around the corner? What will happen if there is an outbreak of nuclear war? Only time will tell.

Changes in the global birth, death and infant mortality rates, 1800–2050

Year	Birth rate per 1,000 people	Death rate per 1,000 people	Infant mortality rate per 1,000 live births
1800	40	35	no data
1850	40	34	no data
1900	37	28	no data
1950	37	20	126
2000	23	9	57
2050	14	10	10

Activity 1

Study the table above.

1. What changes do you think have helped to lower the global infant mortality rate?

2. How might the infant mortality rate differ between developing and developed countries?

3. Calculate the natural increase rates for 1950 and 2050. What does the difference between the rates tell us about future population change?

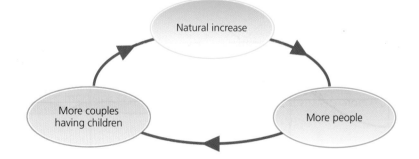

Figure 3: The population multiplier effect

Objectives

- Learn how populations are managed.

- Recognise that there is a need to manage populations.

- Understand the reasons for the management of population change and migration.

Activity 3

1. Name two countries suffering from overpopulation. What is your evidence for this?

2. Can you identify two countries that might be suffering from underpopulation? What is your evidence for this?

Decision-making skills

Some countries are underpopulated. Assess the likely success of the following ideas to increase population:

- Tax credits
- A 'birth bounty' – a payment for each child born
- Opportunities for parental leave for both parents for up to a year after the birth of a child
- Free nursery care for all up to the age of three
- Guaranteed employment at the same level on return to work following maternity/paternity leave

Also think about how migration policies could help to solve underpopulation.

How far can population change and migration be managed sustainably?

In order to answer the question – How far can population change and migration be managed sustainably? – we first need to ask another question: Why is it necessary for governments to manage populations, particularly their numbers and movements?

Managing populations

One of the duties of government is to monitor what is happening to a country's population. Is it growing in number or declining? Is it changing? And, most importantly, is it changing in a way that is likely to lead to problems? The key to answering this last question lies in the balance between the resources of a country and its population. As shown in Figure 8, if population outweighs resources, then all sorts of problems, known collectively as overpopulation, are likely to arise. Feeding all the people is an obvious challenge. Can enough food be produced to ensure that people do not suffer from malnutrition or starvation? Unemployment is another challenge – can sufficient work be found so that people are able to support their families? Can sufficient decent housing be provided so that people do not have to live in slums or shanty towns?

Conversely, but much less commonly, the balance may tip in favour of resources. This is known as underpopulation. Generally speaking, its problems and challenges are much less demanding than those of overpopulation. Providing services and exploiting resources in underpopulated areas are two of the challenges.

An optimum population exists when resources and population are equally balanced. Achieving this sustainable situation is probably the main aim of most governments.

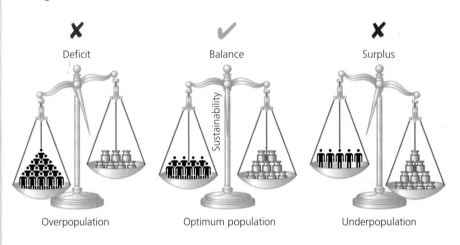

Figure 8: Three different relationships between population and resources

There are many examples of governments deciding that they need to manage their populations. In the majority of cases, the reason has been the need to control numbers – to stop them passing the **tipping point** that leads to overpopulation. So what does a government do to control population numbers? The target of any action is most often the birth rate. If this is lowered, then in time the whole population will become less. The birth rate is usually lowered by encouraging birth control (Figure 9) and making it expensive to have too many children. However, it is important to understand that there are some religions that are strongly against such policies.

When the aim of government action is to increase population, again the target is likely to be the birth rate. Couples may be offered money or other benefits to have more children. But there is another possible action. That is to encourage migrants to come to the country, particularly young adults, who are likely to want to have children eventually.

We will now take a look at two very different case studies – one about dealing with population growth, the other about population decline. They are also different in terms of size. China is the most populous country in the world – and one of the largest, while Singapore is one of the smallest. But they do have one thing in common – most of their people are of the same ethnic origin – Chinese. China covers an area of 9.6 million km² and has a population of 1,320 million. Singapore occupies a small island of just 639 km² and has a population of 4.8 million.

Figure 9: Birth control devices

Decision-making skills

Argue the case *for* and *against* population control. You should consider it from an environmental, economic, political, social and cultural viewpoint.

Case study: China's one-child policy

Faced with a high rate of growth in its already huge population, the Chinese government introduced 'voluntary' schemes to control birth rate as early as the 1970s. With the birth rate already falling fast, it introduced its 'one-child policy' (Figure 10). For nearly twenty years after that, no couple was supposed to have more than one child, and those who did were penalised in various ways. All couples were closely monitored by female health workers who were trusted members of the Communist Party. Couples with only one child were given a 'one-child certificate' entitling them to such benefits as cash bonuses, longer maternity leave, better childcare, and preferential access to housing. In return for the certificate, couples would have to pledge that they would not have more children. Unmarried young people were persuaded to postpone marriage. Couples without children were advised to 'wait their turn'. Women with 'unauthorised' pregnancies were pressured to have abortions. Those who already had a child were urged to use contraception or undergo sterilisation. Couples with more than one child were virtually forced to be sterilised.

Since 1996 the policy has been relaxed a little, particularly in rural areas. Birth rate fell from 34 per 1,000 in 1970 to 13 per 1,000 in 2008, and the annual population growth rate fell from 2.4% to 0.6%. Even so, the total population has grown from 996 million in 1980 to 1,320 million today. The brake has certainly been put on population growth, and as the cutback in children works its way up the population pyramid, so its effect will become stronger.

One thing seems very clear. The policy has been much more effective in urban areas than the countryside. In cities, finding enough living space for a family of three is difficult. Raising a child there is much more expensive. In rural areas, however, there is always the need for an extra pair hands to help on the family farm. In short, there are two very different attitudes towards children.

China's one-child policy remains very controversial. Population growth fell very rapidly before it was introduced in 1981 as a result of charges in Chinese society, land reform and, no doubt, a 'voluntary' policy that may not have been entirely voluntary in practice.

The one-child policy had a number of unwanted consequences. The Chinese tradition is to prefer sons. So as couples are limited to having only one child, there has been widespread sex-selective abortion. If you look closely at the age bars in Figure 6 on page 140, you will see there are more males than females below the age of 45 years. There are now 120 males to every 100 females. This is having consequences:

- Parents 'spoil' their 'one-boy' child and as a result he tends to be obese, demanding and delinquent. They are referred to as 'little emperors'!
- Because of the increasing shortage of women of marrying age, bartering for brides and 'bride kidnapping' have become common in rural areas and prostitution has increased in the cities.
- There may well be a future shortage of labour, particularly if life expectancy remains low and illness rates remain high.

Figure 10: Promoting the happy image of a 'one child' policy

Case study quick notes:
- Population management needs tough government.
- Population management can also be tough on people.

How far can population change and migration be managed sustainably?

145

Case study: Singapore's 'Have three or more' policy

Since the mid-1960s, the Singapore government has controlled the size of its population. First, it wanted to reduce the rate of population growth, because it was worried that the small island would soon become overpopulated. This policy was so successful that in the mid-1980s the government was forced to completely reverse the policy. The old family planning slogan of 'Stop at two' was replaced by 'Have three or more – if you can afford it'. Instead of penalising couples for having more than two children, they now introduced a whole new set of incentives to encourage them to do just that. These include:

- Tax rebates for the third child and subsequent children
- Cheap nurseries
- Preferential access to the best schools
- Spacious apartments.

Pregnant women are offered special counselling to discourage 'abortions of convenience' or sterilisation after the birth of one or two children.

Case study quick notes:
- Governments are able to control population numbers in a variety of ways.
- Control is usually achieved by a 'stick and carrot' approach.

Figure 11: Singarore's 'Have three or more – if you can afford it' policy offers incentives to have lots of children

Managing migration

Governments are particularly interested in migration. Is the number of people entering the country (immigrants) exceeding the number of people leaving (emigrants)? If it is, then the situation (net in-migration) will be adding to any growth in population coming from natural increase. This situation might then begin to ring alarm bells – about possible overpopulation, unemployment and housing shortages. So, in this situation, some sort of action will need to be taken to check the inflow of migrants. If the situation is the complete opposite, then a government might take action to stop it, particularly if the loss of people involves a 'brain drain' of its best workers.

So whether or not a government decides to manage or control migration depends on what its impacts are likely to be. Those impacts can be both positive and negative. Let us start by looking at the immigration into the UK that took place between 1950 and 1975.

Activity 4

Read the case study on page 144 and look back at Figure 6 on page 140. Can you pick out any evidence of the one-child policy in China?

Activity 5

Try and find out if the population of the UK has ever been managed. Is it being managed today?

Figure 12: Immigrants from Jamaica arriving in the UK in 1948

Decision-making skills

Draw a quadrant (as shown) which looks at the positives and negatives of migration into the UK for both the host country and the source country.

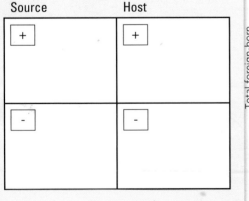

Case Study – The UK opens its doors

The UK's post-war immigrants came mainly from colonies in the Caribbean, and from what had been the Indian Empire (India, Pakistan and Bangladesh). Immigration was encouraged by an Act of Parliament which gave all Commonwealth (ex-colonial) citizens free entry into the UK. The first ship to bring in immigrants from Jamaica docked at Tilbury (Essex) in June 1948.

It is estimated that during the 1950s and 1960s over a quarter of a million immigrants came from the Caribbean. Roughly the same number came from what had been the Indian Empire. By 1971, there were over 1 million immigrants from Commonwealth countries. The new settlers took up a variety of jobs. Many found work in textile factories and steelworks. Many drove buses or worked on the railways. Later arrivals, particularly from India, opened corner shops and restaurants or ran Post Offices. By the 1970s, the UK had more than enough labour, and controls were introduced to reduce the migrant arrivals.

Figure 13 shows the number of UK residents who were born abroad. Despite the controls on immigration, clearly the number has been steadily rising. So too has the percentage of the UK population that they represent.

Enough time has passed for us to see and understand the impacts of this post-war immigration. Its positive economic impacts were:

- It met the shortage of unskilled and semi-skilled labour.

- It played an important part in the post-war reconstruction of the country.

On the negative side:

- Public money had to be spent on meeting the everyday needs of the immigrants and their families – housing, schools, healthcare, etc.

- When the economy went into recession in the 1970s, these immigrants added to the burden of unemployment.

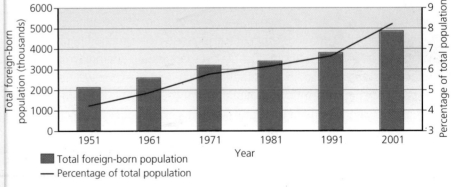

Figure 13: UK residents born abroad, 1951–2001

How far can population change and migration be managed sustainably?

147

As for the social impacts, it is clear that the immigration created great tensions (Figure 14). The UK was not used to having sizeable ethnic groups in its population. There was hostility towards the immigrants, and they were discriminated against – and abused. As a consequence, they tended to settle in particular areas, for personal security reasons. Better to live with people from the same ethnic group than run the risk of being victimised by white neighbours. Most often, immigrants became segregated, to form 'ghettos' in areas of rundown housing in the inner areas of towns and cities. Clearly, there were some serious negative social impacts.

Slowly, the social situation has changed. Discrimination has been made illegal. UK law now states that all citizens, regardless of ethnicity, should enjoy equal opportunities. Slowly, most white people have come to realise that they are not threatened by immigrants. They have come to realise that there are positives. Ethnic groups add to the country's skill base and culture. The offspring of the original immigrants have made their way in the UK, and many now occupy well-paid and responsible jobs. They have moved into areas of better housing. They represent the country in a range of sports. They have seats in Parliament. They are now truly UK citizens. The situation is still not one of complete harmony, and many still live in poverty. But the situation is much better than it was 40 years ago.

Since the 1970s the immigration of New Commonwealth citizens has been subject to some form of government control. In the late 1990s the economic situation in the UK began to change. A period of economic boom saw the country short of labour once again. This time, the search for willing workers turned to Eastern Europe.

Case study – East European workers come to the UK

In 2004 the East European states of the Czech Republic, Estonia, Hungary, Latvia, Lithuania, Poland, Slovakia and Slovenia joined the EU. Since then, many of their citizens have come to work in the UK. Figure 15 shows the push and pull factors. In most cases, these **economic migrants** intend to stay only until they feel they have made enough money to take home.

Figure 14: Inner city race riots

Skills Builder 3

Take a look at a town or city you know. Do members of different ethnic groups live in particular areas? Draw a simple sketch map to show these areas.

Quick notes (the UK opens its doors): The UK's need for labour led to it becoming a multi-ethnic society.

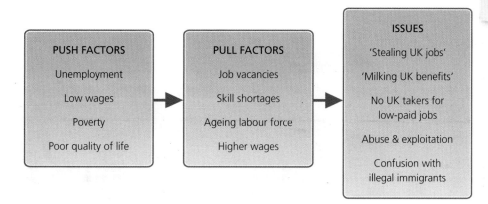

PUSH FACTORS	PULL FACTORS	ISSUES
Unemployment	Job vacancies	'Stealing UK jobs'
Low wages	Skill shortages	'Milking UK benefits'
Poverty	Ageing labour force	No UK takers for low-paid jobs
Poor quality of life	Higher wages	Abuse & exploitation
		Confusion with illegal immigrants

Figure 15: The push factors, pull factors and the issues of East European workers in the UK

Activity 6

Explain in no more than 150 words why economic migrants from Eastern Europe have been attracted to the UK.

Employment of East European migrants

Employment type	%
Hotels and catering	28
Factory worker	20
Farming	14
Food processing	5
Cleaning	5
Care assistant	3
Sales assistant	3

Quick notes
(Eastern European workers):
The 'boom and bust' cycles that most countries go through mean alternating shortages and surpluses of labour. The economic migrant can play a useful part in smoothing out the differences.

Figure 16 shows that over half of the East European migrants come from Poland, the largest of the new member states. The vast majority of migrants are young and single, with over 80% of them aged between 18 and 34.

Some UK newspapers (and citizens) take a very negative attitude towards these economic migrants. They are accused of depriving UK workers of jobs and taking advantage of our state benefits system. Figure 15 (page 147) shows these as two of a number of issues relating to these economic migrants. But the critics choose to ignore four important facts:

- The migrants contribute to the UK's economy by the taxes they pay.

- The jobs that many of them take up are mainly low-paid. Such jobs are often avoided by UK workers.

- The migrants have a strong work ethic, which can directly benefit employers. They are efficient and polite. Sadly, there are employers who unfairly exploit these qualities.

- Less than 5% of them receive any sort of state benefit.

Many of these workers are now returning home – persuaded by the economic recession and better employment prospects back home. Also important is the changing exchange rate. With the pound sterling falling in value, workers are not able to send or take home so much money.

One of the advantages of belonging to the European Union is that workers are free to move between member countries. All they need is a passport or national identity card.

While it is becoming easier to migrate within the EU, it is becoming more difficult for migrants to enter it from other parts of the world (Figure 17). In order to enter the UK from outside the EU, for example, you need a visa. There are various types – visitor, business and working holiday – and they are usually valid for less than a year.

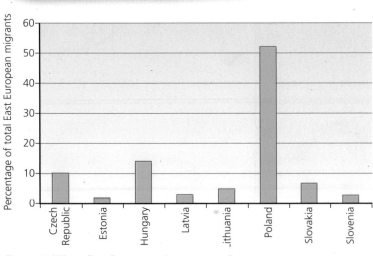

Figure 16: Where East European migrants come from

Figure 17: UK border control

If you wish to come to the UK to work and settle down, then you have to go through the points-based system, which was introduced in 2008. It allows British businesses to recruit the skills they need from abroad. Only migrants with those skills will be able to come and live in the UK. The system recognises five tiers of migrants:

● Tier 1 Highly skilled workers – scientists and entrepreneurs, for example

● Tier 2 Skilled workers with a job offer – teachers and nurses, for example

● Tier 3 Low skilled workers filling specific temporary labour shortages – construction workers for a particular project, for example

● Tier 4 Students

● Tier 5 Youth mobility and temporary workers – musicians coming to play in a concert or people recruited to help with specific projects, for example.

As national boundaries are tightened, the volume of illegal immigration increases. There are two 'porous' frontiers through which most illegal immigrants enter the EU – in the Mediterranean countries (see Figure 18) and along the eastern border with Belarus, Russia and the Ukraine.

149

Decision-making skills

Argue the case *for* and *against* the idea of the skills list. What do you think about the make-up of the tiers? Do you think that the government has got its priorities right? Are there any people that we need who are not included, i.e. ageing population?

Skills Builder 3

Study Figure 18.

Describe the routes being taken by illegal immigrants into the European Union.

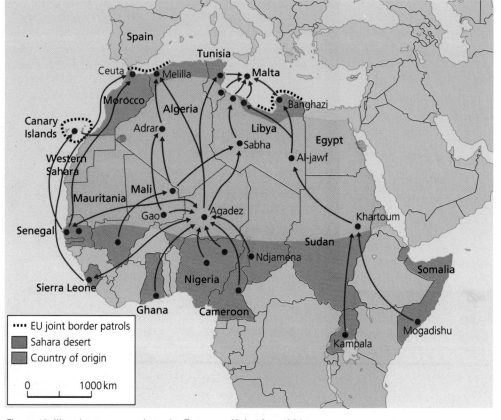

Figure 18: Illegal entry routes in to the European Union from Africa

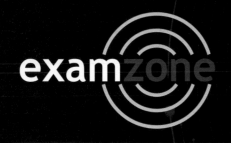

Population is always changing. You can study the effects of both natural change from births and deaths, and migration. This change has important implications so countries try to manage both population and migration.

You should know...

☐ How the world's population is still growing at a decreasing rate

☐ What causes population to grow or decline

☐ Why some countries have more rapid growth than others

☐ How birth rates and death rates can change population

☐ Why these rates vary around the world

☐ How immigration and emigration can affect population growth

☐ Why infant mortality rate varies across countries

☐ How you can fit countries pattern of population change into the Demographic Transition Model

☐ How you can draw population pyramids to show population structure

☐ What is meant by sustainable or optimum population growth

☐ How countries try to manage growth to avoid over and under population

☐ How and why countries manage migration

☐ How migration can have both benefits and costs for source and host

☐ Why economic migration is becoming so widespread

Key terms

Ageing population	Immigrants	pyramid
Birth rate	Infant mortality rate	Population structure
Death rate	Life expectancy	Tipping point
Development	Migration	Youthful
Economic migrant	Natural change	population
Emigrants	Natural increase	Zero population growth
	Population	

Which key terms match the following definitions?

A Economic and social progress that leads to an improvement in the quality of life for an increasing proportion of the population

B A person leaving a country or region to live somewhere else (for at least a year)

C The change (an increase or a decrease) in population numbers resulting from the difference between the birth and death rates over one year

D The number of births per 1,000 people in a year

E A person arriving in a country or region to live (for at least a year)

F The point at which the momentum of a change becomes unstoppable

G Diagrammatic way of showing the age and sex structure of a population

H The average number of years a person might be expected to live

To check your answers, look at the glossary on page 313

ResultsPlus
Maximise your marks

Foundation Question: Describe the key features of Stage 1 of the Demographic Transition Model. (3 marks)

Student answer ● (achieving 2 marks)	Examiner comments	Build a better answer △ (achieving 3 marks)
This stage has a very high birth rate. It also has a very high death rate. This is because of the high amount of disease.	• *This stage has...* This is correct and scores 1 mark. • *It also has...* This is another correct statement and scores 1 mark. • *This is because...* This is an explanation rather than a description. As the question asks for a description, this part of the answer does not score any marks.	This stage has a very high birth rate. It also has a very high death rate. Both the birth rates and the death rates fluctuate from year to year.

Overall comment: The student should have circled the command word – *describe* – to make sure that they responded to the question being asked.

- -

Higher Question: Explain why Stage 1 of the Demographic Transition Model is called the 'High Fluctuating' phase. (3 marks)

Student answer ● (achieving 2 marks)	Examiner comments	Build a better answer △ (achieving 3 marks)
The very high birth rates go up and down because of infant mortality varying. The death rate is high and it fluctuates because of famines.	• *The very high...* Although this is a true statement, it is not really an explanation as it only implies fluctuation. • *The death rate...* This is a valid point as it explains why the death rate fluctuates.	The infant mortality fluctuates because of AIDS and poor hygiene, which leads to disease. The birth rate is high because there is no contraception. The death rate is high and it fluctuates because of famines.

Overall comment: The student attempts to explain in their answer, but the explanations both need to be developed.

Chapter 10 Consuming resources

Objectives

- Be able to define and give some examples of natural resources.

- Be able to explain why some countries have higher demand for resources than others.

- Understand that there are often considerable variations within countries.

Resource consumption in different parts of the world

Classifying resources

Resources can be divided into three types:

1. **Natural resources**
2. **Human resources** (the skills of a population in a society)
3. **Material or capital resources** (the goods and equipment already in place in a society).

In many contexts the term 'resources' is taken to mean natural resources, but it is useful to be aware that this is just a handy abbreviation. The term 'natural resources' is often defined as:

> *Those things found in the natural world that are of use to us, and that we have the technology and the willingness to use.*

Not all natural materials are of value to us at all times. Flints might have been useful to our ancestors but they have no real value today. Uranium was of little practical use before we developed the technology to use it for nuclear energy and nuclear weapons.

'Resources' also include those features of the natural environment that, in the past, were largely taken for granted. This includes water, air and the biosphere. For much of our history, we have simply used these as though it was our right to do so, without reflecting on whether or not they would always be available, let alone whether they were 'ours' to use. From dodos to giant moas, American bison and blue whales, we simply destroyed them for our own convenience – to extinction, in the case of the first two.

Availability

It is very useful to define resources in terms of their availability:

- **Non-renewable resources** – like coal, oil or diamonds – cannot be 'remade', because it would take millions of years for them to form again. They exist in a fixed amount that is gradually (and sometimes rapidly) being used up.

- Some **sustainable resources** – like wood – can be deliberately renewed. This means that they can be managed so that they are usable now but will also last into the future.

- Some sustainable resources – like solar power or wind power – renew themselves, and do not need to be managed i.e. renewables.

Figure 1: The tar-sand reserves of Alberta, Canada

Benefits and costs

One way of assessing resources is to look at the benefits that come from their use and compare these with the costs. This can be difficult because some of the costs and benefits can be indirect and even unintentional (see table). These factors are important today because we are facing a future in which 'cheap' oil is running out and **'alternative' energy** will need to fill an increasing gap between demand and supply. Important political decisions will need to be made.

Decision-making skills

Study the table below. Draw a scale from minus three to plus three. Plot each of the three resources listed by their environmental cost and economic benefit for the future.

The benefits and costs of three resources

Resource	Benefits	Costs
Tar-sand oil (non-renewable) There are enormous reserves of tar-sands, especially in Canada (see Figure 1), and they have attracted more attention since oil prices have risen rapidly.	• There are about 300 billion barrels of oil in the tar-sands (equivalent to Saudi-Arabian oil reserves). • Tar-oil extraction would make profits for the major oil companies. • There would be tax benefits for countries with tar-sands. • Dependence on the dangerous Middle Eastern region would be reduced. • Continued production of oil would avoid the costs of switching to other fuels, such as hydrogen.	• Heavy oils, such as those from tar-sands, produce up to three times more CO_2 than 'light' oils. • Tar-oil extraction uses vast quantities of water – up to six barrels for each barrel of oil. • Ancient spruce forests will need to be removed to scrape away the oil-rich sands, and pollution of groundwater and rivers will be inevitable. • 300 billion barrels is about five years' supply, thus it only delays the need to search for alternative technologies.
Biofuels (sustainable and **renewable**) There are several types of biofuel, but the best known are ethanol (often extracted from corn) and bio-diesel, extracted from crops such as soya or palm oil (see Figure 2).	• CO_2 is absorbed when they are grown – but CO_2 is released when they are used. • They can be grown in many different environments allowing every country the possibility of producing some of its own fuel. • Internal combustion engines need very little modification to cope with biofuels, so there is no need for costly changes to vehicle designs.	• They are not carbon neutral because the farming methods themselves release CO_2. • The amount of land needed to replace conventional fuels would be enormous. All the current food crop regions of Europe and the USA would need to be devoted to biofuels to meet current demand for fuel. • They would be grown as 'monocultures', reducing habitat variety. In some areas of the tropics palm oil is already replacing tropical rainforest.
Solar energy (sustainable – unlimited)	• Solar energy is unlimited. • It is environmentally friendly, with minimal carbon emissions after initial panel production. • New thin panels are being developed that will be much more efficient than present technology.	• It is intermittent. so back-up systems and new ways of storing electricity are needed. • Current production is tiny (see Figure 3). • It is relatively expensive. Present panels last about twenty years and, for household usage, do not even pay for themselves. • Huge areas of panels are required to produce significant amounts of energy.

Figure 2: A palm oil plantation in Malaysia

Source	Value
Coal	1,814.013
Natural gas	752.095
Nuclear	716.729
Hydroelectric	266.407
Petroleum liquids	40.349
Wood	35.981
Wind	23.512
Waste	22.124
Petroleum coke	18.389
Other gases	14.765
Geothermal	13.520
Other	5.151
Solar	0.503

Figure 3: The energy sources used to generate electricity in the USA, in billion kilowatt hours

Activity 1

1. Identify two other forms of 'alternative' energy.

2. Draw up a table to show the costs and benefits of developing them.

Note: You should use the Internet to conduct your research, but remember that this a very controversial topic and many websites are arguing strongly *for* or *against* the development of alternative energy sources, e.g. www.alternativeenergy.com.sg/Default. aspx .

Skills Builder 1

Study Figure 3 on page 153.

(a) Identify three renewable sources of energy for electricity production.

(b) Calculate the percentage of electricity produced in the USA using fossil fuels.

Skills Builder 2

Study Figure 4.

(a) Calculate the percentage of global diamond production that comes from developed countries.

(b) Using an atlas, locate the five largest diamond-producing countries.

A changing world of 'haves' and 'have nots'

It might seem obvious that countries with large quantities of natural resources would be more developed – and more successful – than those without those resources. After all, if the person sitting next to you has more gold, diamonds and oil than you have, you might suppose that they are wealthier than you are. And, of course, there are examples of countries – such as Saudi Arabia with its oil – where the wealth is a result of an abundant natural resource. But the picture is much more complicated than common sense might suggest. Take a look at Figure 4, which shows the world's major producers of diamonds.

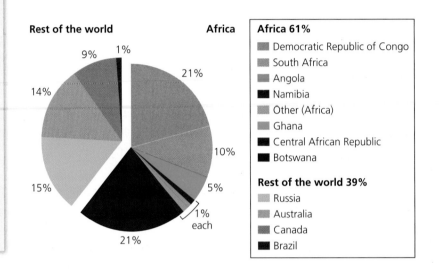

Figure 4: The world's leading diamond-producing nations

It is obvious that although there are a few developed countries that produce diamonds there are a large number of developing countries too, some of which are very poor indeed. The Democratic Republic of Congo (DRC) and Botswana are the world's two largest diamond-producing nations. But the DRC is 177th out of the 179 countries on the United Nations Human Development table. In the DRC, the average life expectancy is 45. Botswana is placed 126th on the list, with an average life expectancy of 48. So the world's two largest diamond-producing countries are poor countries, in one case extraordinarily so. It would be fair to conclude that, in both of these countries, the money being made out of diamonds isn't benefiting most of the people. And that is why common sense can mislead you. Countries are made up of different groups of people – different interest groups – and there are often considerable gaps in wealth between the rich and the poor. In the DRC, a very small number of people have done very well indeed as a result of the diamond mines – but most of the country's people have not.

Global wealth – just like resources – is also unevenly distributed. Figure 5 shows the size of each country in terms of the average wealth of its people rather than the true geographic size of the country.

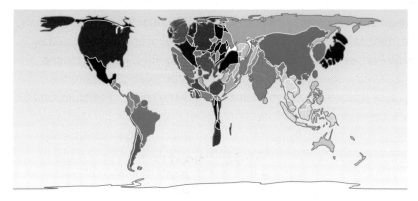

Figure 5: World wealth distribution in 2002, by country

It is obvious that Europe, the USA and Japan are the wealthiest global regions, and equally clear that South America and, above all, Africa are very poor indeed. In fact, the richest 20% of the human race (most of whom live in the three wealthy regions) consume over 86% of all global resources, whilst the poorest 20% consume only 1.3% of global resources. These inequalities have increased dramatically in recent years.

But this world of 'haves' and 'have nots' is not quite a simple as it appears. Over 20 million US citizens live on 'trailer parks', in what we would call 'caravans'. At the same time, there are a significant number of millionaires living in Mumbai, a city better known for its slums.

Figure 6 shows that as societies develop, it is highly likely that their impact on the environment will increase. One of the many ironies of the battle against global warming is that the societies that are least equipped to cope with its effects – Malawi and Mozambique, for example – are not generally responsible for the problem in the first place. Figure 6 shows that the present impact of both China and India is low. But these countries, with their huge populations, are expected to develop into the emerging superpowers, raising concerns that the scale of their future **consumption** and resource use may have a significant impact on the future of the planet.

Skills Builder 3

Study Figure 6.

(a) Identify the global 'average' ecological footprint, in hectares per person.

(b) Describe the relationship between human development rank and ecological footprint.

(c) Identify two countries that do not 'fit' the general relationship.

ResultsPlus
Watch out!

■ Whenever you write 'in developing countries…' or 'in the UK…', remember that there is lot of variation within countries. Try to avoid using a statement such as 'The UK is rich, so…' and replace it with 'Many people in the UK are relatively rich so…'.

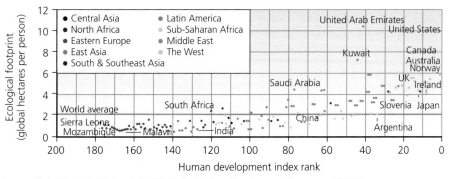

Source: Global Footprint Network (2006); United Nations Development Programme (2006).

Figure 6: The relationship between human welfare and ecological footprints

Activity 2

1. Using the Internet, define the term 'peak oil'.

www.peakoil.org

www.dieoff.org/l1.html

www.eia.doe.gov

www.cges.co.uk

www.iea.org

2. Suggest reasons why some countries face more challenges than others as oil production declines.

The uneven patterns of oil supply and oil demand

The distribution of oil is determined by nature, but the production of oil is affected by our technical ability to extract oil, some of which is found in very difficult environments. As with the tar-sand oil which we studied earlier, there are potential sources of oil in a number of locations that are not shown on the current reserves map (Figure 7). Current production is dominated by the Middle East, especially Saudi Arabia. Many countries, including some in the Middle East, have reached what is known as 'peak oil', where the production of relatively cheaply obtained oil has reached its maximum, and are now experiencing falling production. Figure 8 shows the per capita (per head) consumption of oil, by country. Consumption is largely controlled by the wealth of a country and its dependence on motor vehicles. 70% of oil production is used in transporting people and goods around, both within countries and between them. Oil is consumed at the rate of about 1,000 barrels (160,000 litres) a second.

The USA has less than 5% of the world's population but it consumes 25% of the oil. This is not just because the Americans are wealthy – there are other wealthy nations whose people use much less oil. It is the country's high standard of living – relying on air-conditioning, for example – and, more than anything, the dependence on cars that causes the massive consumption. This dependence – which is built into the American way of life – is caused by:

- The poorly developed public transport system – within cities and between cities

- The pattern of low-density urban settlements, requiring long journeys to work, school and the shops

- The long history of very low petrol prices, which is only recently coming to an end.

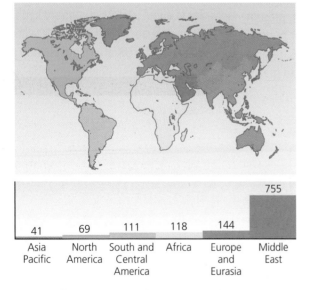

Figure 7: Oil reserves by global region, in billion barrels, 2007

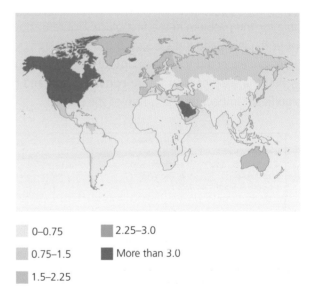

Figure 8: Consumption (in tonnes) of oil per head, by country, 2007

As a result, there are about 260 million cars in the USA – not far off one for every person in the country.

The USA situation is especially worrying because there are many people in the world who would like to share the 'American Dream' of low-density living with very high energy usage. Unfortunately, economic growth causes increases in the use of resources, which in turn increase the emission of greenhouse gases. The table below shows the scale of the environmental problem if countries like China and India move towards being more like America.

People and cars: the situation now (2008)

	Population	Number of cars per 100 people	Car ownership
USA	300 million	87	260 million
China	1,300 million	4	59 million
India	1,100 million	1	12 million

As China and India continue to grow rapidly (see Figure 9) it is unlikely that their citizens will give up dreams of owning a car and enjoying the type of lifestyle that is accepted as 'normal' in the USA and other developed countries.

It is also unlikely that these countries – and others in South and East Asia that are developing fast – will accept too many lectures about the need to slow down their rate of growth from countries that have already reached high levels of consumption. In fact, exactly the opposite is likely to happen as businesses seek to expand their markets in these global regions. One of the most striking examples is that of Tata Motors, a branch of the huge Indian corporation, that has begun production of a car priced at around £1,500. Every car bought and every bicycle or rickshaw abandoned will increase the demand for oil (and add to CO_2 emissions). With demand rising rapidly and global peak oil predicted by many to be very close (or even behind us), the future is best described as 'challenging'.

Although the sums are different and the global players may take on different roles, similar scenarios can also be described for several other non-renewable resources. For the first time in many years, the debate about the relationship between human beings and the planet's resources is back at the top of the agenda.

Outline how increasing car ownership might cause an increase in the demand for oil (2 marks)

■ **Basic answers** (0 marks)
Fail to identify the link between car usage and demand for oil.

● **Good answers** (1 mark)
Recognise that increasing car ownership is likely to increase demand for oil.

▲ **Excellent answers** (2 marks)
Also recognise this link and add some data to support the point.

Skills Builder 4

Study Figure 9.

Describe the growth of the Chinese and Indian economies shown in the figure.

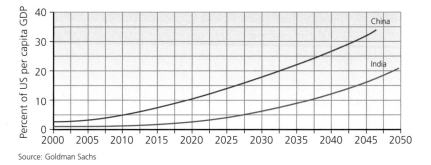

Source: Goldman Sachs

Figure 9: The growth of the economies in China and India compared with the US

Objectives

- Know that there are several different ways of explaining the link between population and resources.

- Be able to describe the main aspects of Malthus's and Boserup's theories.

- Explain the ways in which sustainable development is interpreted by different groups.

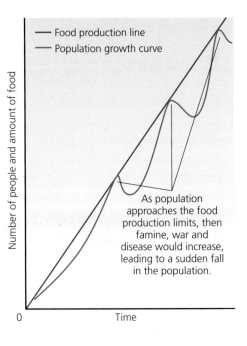

— Food production line
— Population growth curve

Number of people and amount of food

As population approaches the food production limits, then famine, war and disease would increase, leading to a sudden fall in the population.

0   Time

Figure 10: The relationship between food supply and population, according to Malthus

Skills Builder 5

Describe the growth of the population shown in Figure 10.

How sustainable is the pattern of supply and consumption?

The different theories on how far the world can cope

The world's population is predicted to reach 10–12 billion in the second half of this century. Are global resources going to be sufficient to support that number of people?

Although many experts are optimistic about the world's ability to produce enough food in total, there will still be the problem in the future – as now – of unequal distribution. Today, when the world's population is between 6 billion and 7 billion, there are many people – about 1.5 billion – who are never very far away from famine. About 4,000 babies (under the age of 1) die every day from a lack of fresh water and basic sanitation.

Are there too many people in the world? Is there a limit to the number of people that the world's resources can support? At first sight, these questions seem simple enough to answer but, in fact, they are two of the most controversial of all issues, and have been argued over for centuries. You don't need to take sides in the debate (although you can), but you do need to understand the different points of view.

The Malthusian theory

Thomas Malthus (1766–1834), an English economist, believed that because population increases in a different (and faster) way than food supply, there would inevitably come a time when the world could not cope. He argued that population would increase geometrically (1, 2, 4, 8, 16, 32, 64, 128, etc.) because two people would have four children, and those four would have eight, and so on. Food supply, on the other hand, would only increase arithmetically (1, 2, 3, 4, 5, 6, 7, 8, etc.) because improvements in farming practices could only come gradually. From these calculations, Malthus predicted in 1798 that population growth would outstrip food supply, leading first to tensions in society and ultimately to deaths from famine, war and disease – causing a sudden fall in the population. Figure 10 shows a series of these collapses. After each one, the population rises rapidly again until it hits the 'barrier' of the slowly increasing food supply – and collapses again.

Malthus wrote at a time when the rich owned most of the resources and enjoyed a life of comfort, whereas the poor had very little access to resources, especially land. He thought that it was wrong to give assistance to the poor because he believed that they were less able to understand these risks and thus limit population growth.

Malthus's predictions turned out to be wrong because food production increased rapidly. At the end of the nineteenth century there were more people who were, by and large, better off. If there was a relationship between people and resources, it was obviously not quite as Malthus had argued.

The pattern seemed to be repeated in the twentieth century, when population growth was more rapid than at any other time in human history (from about 1 billion to about 6 billion) and living standards and quality of life rose globally more than in any other century.

Those with Malthusian views today – the 'neo-Malthusians' – tend to be building on the 'common sense' idea that poverty is caused by there being too many people in the world. They look at images of famine victims, often in Africa, and assume that famines are the result of too many people rather than, as is often the case, poor distribution of food or too much land devoted to profitable crops for export – much as it was in Malthus's own day.

The Boserupian theory

Esther Boserup (1910–1999) a Danish economist, had a very different view from Malthus. She suggested that population growth has a positive impact on people that will enable them to cope, because as resources start to run out we are forced to 'invent' or innovate our way out of the problem (see Figure 11). This belief relates to the old saying 'necessity is the mother of invention' – the process of invention is really driven by basic need, rather than by curiosity or individual genius. Some believe that we 'invented' farming because we were starting to run out of hunted food. The same explanation might be suggested for the 'Green Revolution' or even genetic modification.

According to this idea, population growth becomes something positive and, what is more, very important to our development as a species. This rather reassuring and optimistic view has not gone unchallenged. The fact that we have invented our way out of food production problems in the past is no sure indicator that we can do so in the future. So, although for some, the 'more people equals more wealth' idea is confirmed by the past 200 years, for many others this ignores the environmental impact of population growth which, they say, will bring it all to a very abrupt and unpleasant halt.

The debate

A few years ago the debate about the growing global population tended to concentrate on the use of mineral resources – not just oil, but also metals. Recently, the debate about the future of the planet has shifted away from arguments about mineral reserves and tends to concentrate on the destruction of the planet's biosphere. Many commentators agree that we may very well find alternatives to minerals – even oil – but that our consumption patterns will damage the Earth's living resources beyond repair. For most of these commentators, population growth is only one element of this problem – and probably not the most important. This view is summarised by the 'IPAT' formula:

Impact = Population x Affluence x Technology

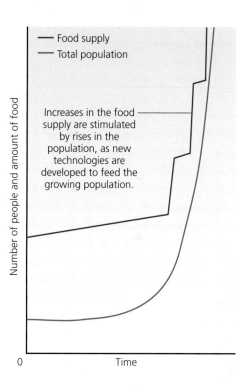

Figure 11: The relationship between food supply and population according to Boserup

Graph labels: Food supply / Total population. Y-axis: Number of people and amount of food. X-axis: Time (from 0). Annotation: Increases in the food supply are stimulated by rises in the population, as new technologies are developed to feed the growing population.

Skills Builder 6

Study Figure 11.

(a) Compare the growth of total population and food supply.

(b) Compare the growth rate of food production with that shown on Figure 10.

This formula suggests that the impact on the environment (I) is result not just of the number of people (P) but also their wealth or affluence (A) which affects how many resources they use and the type of technology available (T). This last factor – technology – is thought by some to be the potential answer to our problems, because we can find solutions to the challenges through invention and innovation, much as Esther Boserup suggested.

There is a vast amount of evidence that the world's biosphere is in trouble and that the prime cause is our increasing consumption rather than the growing population:

- In the past 25 years, 20% of the world's mangrove swamps have been destroyed, leaving the coast more and more vulnerable.
- 77% of the world's fish stocks are either over-fished or fished out completely.
- More than a billion people do not have access to fresh, drinkable water. The demand for water is likely to triple in the next 50 years.
- Energy consumption rises every year. In 2006, for example, we consumed in six weeks what we consumed over the whole of 1950.
- The biosphere's resource base has fallen by 40% since 1970.
- Between 15% and 37% of living species are predicted to disappear by 2050.
- 96% of deforestation takes place to make way for agriculture.
- In Europe, more than 40% of bird species and 30% of amphibians are threatened with extinction.
- 3 billion people depend on wood for heating and cooking and, as pressure mounts in rural areas, much of this fuel is not replaced by replanting.
- In 2004, UK farmers used 2.4 kg of pesticides per hectare against 0.5 kg in 1961.

Achieving sustainability – the challenge for resource consumption

Worries about the environment began in the 1960s when writers such as Rachel Carson pointed out that the uncontrolled use of pesticides was destroying bird life – and possibly affecting human health. She also accused the chemical industry of lying about the effects of its products. A few years later in 1972 a group of scientists known as the 'Club of Rome' published a book, *Limits to Growth*, that suggested that there was a problem in maintaining the growth rates of 'industrial output' and 'population' while also maintaining 'natural resources' and 'food production'. They revised their pessimistic forecast in 1999 (see Figure 12).

Skills Builder 7

Compare the trends in oil production and population shown in Figure 12.

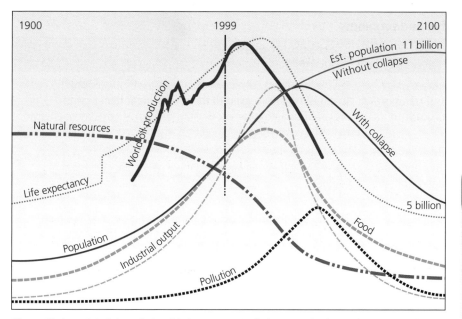

Figure 12: A model of the relationship between population growth, the environment and natural resources

Since the 1980s the idea of '**sustainable development**' has become very well known. In fact, the word 'sustainable' is now used by almost every global organisation – from governments to corporations, from schools to local authorities, from charities to churches. The word is used – and misused – so much that it is easy to lose track of what it really means. The 'official' definition of sustainable development is the one in Our Common Future, the report of the 1987 UN World Commission on Environment and Development. (This is sometimes called the 'Brundtland report', after the chairperson, former Norwegian Prime Minister, Gro Harlem Brundtland.) According to this definition:

> *Sustainable development is development that meets the needs of the present without compromising [limiting] the ability of future generations to meet their own needs.*

The report added that this could be broken into two parts:

● The concept of 'needs', in particular the essential needs of the world's poor, to which overriding priority should be given.

● The idea of limitations imposed by the state of technology and social organisation on the environment's ability to meet present and future needs.

Individual action to reduce resource demand

'Live simply' campaigns stress the need to change our own habits. To live sustainably means that we should live in a manner that preserves our environment and that we should take personal responsibility for the future of the planet. We should first reduce our impact on the Earth's resources, but this is not as simple as it first looks (see table on page 162).

Reducing our use of resources – benefits and problems

Ways of reducing our use of resources	Benefits	Costs and problems
Using local farmers' markets – do we really need imported food?	Supports local farmers and reduces the costs of transport and the '**carbon footprint**'.	Buying locally produced greenhouse-grown tomatoes in the spring is more damaging to the environment than buying imported tomatoes grown outside in Spain. It costs more energy to heat the greenhouses than to transport Spanish tomatoes.
Reduce the unnecessary luxuries in your life – how many pairs of shoes do you have?	Reduces the waste of resources in their manufacture and their transport, quite apart from the packaging.	What about the foreign workers who make these 'unnecessary' goods? Will they still have jobs if we all cut back on these luxuries?
Get on your bike	Cars are great polluters. The average journey by car is just 8.7 miles, and with 33 million cars on the UK roads there is enormous scope here to reduce CO_2 emissions.	The old and the infirm cannot jump on to bikes. It is also hard to imagine how the economy would continue to operate as effectively if we cut back on road freight.
Recycling and conservation	There are huge savings here as materials are reused and resources saved.	It isn't obvious how the rural peasants of Bangladesh benefit from our careful recycling. The impact on global warming is quite small and certainly doesn't improve the life of the poorest, especially if our trips to the recycling bins are in our 4x4s.

Decision-making skills

Study the above table. Carry out a scaled cost-benefit analysis of the four schemes. (Use a minus for costs and a plus for benefits).

Activity 4

(a) Using the following links:

- www.interfaceinc.com/
- http://uk.youtube.com/watch?v=RcRDUIbT4gw
- www.bigpicture.tv/videos/watch/e00da03b,

choose one of the 'seven goals' listed and write about 100 words explaining the details of the relevant policies.

(b) Having watched the YouTube link, explain how the policies that you described in (a) might help reduce the 'externalities' described by Ray Anderson, former Chief Executive of Interface Carpets.

Corporate action to reduce resource demand

Most companies have 'mission statements' that frequently talk in terms of striving for sustainability. Many of the promises don't translate into very much obvious action, but there are a number of honourable exceptions. One of the best known is Interface Carpets. This corporation Interface Carpets is the largest manufacturer of commercial carpets in the USA and its mission statement is very different from those of many other corporations:

Why is striving for sustainability so important?

Here's the problem in a nutshell. Industrialism developed in a different world from the one we live in today – fewer people, less material well-being, plentiful natural resources. What emerged was a highly productive, take–make–waste system that assumed infinite resources and infinite sinks for industrial wastes. Industry moves, mines, extracts, shovels, burns, wastes, pumps and disposes of four million pounds [1,800 tonnes] of material in order to provide one average, middle-class American family with their needs for a year. Today, the rate of material throughput is endangering our prosperity. At Interface, we recognise that we are part of the problem. What's the solution? We're not sure, but we have some ideas. We believe that there's a cure for resource waste that is profitable, creative and practical. Our vision is to lead the way to the next industrial revolution by becoming the first sustainable corporation.

Adapted from: http://www.interfaceinc.com/goals/sustainability_overview.html

The company has identified seven goals to improve its environmental performance:

1. Eliminating waste – in every area of business;

2. Benign emissions – eliminating toxic substances from products, vehicles and facilities;

3. Using renewable energy sources – solar, wind, biomass, geothermal, tidal, etc.

4. Closing the loop – redesigning processes and products to close the technical loop using recovered and bio-based materials;

5. Using resource-efficient transport – to reduce waste and emissions;

6. Creating a culture amongst stakeholders that integrates sustainability principles and improves their lives;

7. Creating a new business model that demonstrates the value of sustainability-based commerce.

Adapted from: www.interfacesustainability.com/seven.html

Will technology save the day?

There are still a number of people who believe that technology will come to the rescue. Building on the ideas of Esther Boserup and others, they see pressure as positive. In 1980, a famous bet took place between the neo-Malthusian Paul Ehrlich and another academic, Julian Simon. Simon said that we wouldn't run out of resources because we were clever enough to find substitutes – developing new technologies and new resources before they disappeared. Ehrlich forecast that resources would begin to run out. The two agreed that if resources were running out then their prices would rise and, having chosen a list of resources to monitor, the two decided to wait ten years and see what happened. Ten years later, the prices had fallen – and Simon won the bet. And he would have won it again should they have repeated it for the next decade (see table below). But critics are quick to point out that prices have risen very sharply since 2000. They also claim that concentration on metals misses the point about the destruction of the environment.

Key commodity prices chosen for the Ehrlich–Simon bet

	1990	2000	Change (%)
Copper	167	88	−47
Nickel	10,770	7,190	−33
Tin	368	242	−34
Tungsten	51	79	+55
Chromium	1,175	761	−35

The history of technological 'fixes' is a very mixed one. The problem of finding alternative fuels to replace 'cheap' oil was investigated at the beginning of this chapter. In reality it will need a supreme effort by governments, corporations and individual researchers to cope with 'peak oil'. It remains to be seen whether technology will always be able to come to the rescue.

ResultsPlus
Build Better Answers

Outline how technology might help solve shortages of resources. (2 marks)

■ **Basic answers** (0 marks)
Misunderstand the word 'technology' or 'resources'. Include some information about how consumption can be reduced.

● **Good answers** (1 mark)
Suggest that technology might lead to higher production, but do not offer any evidence or illustration.

▲ **Excellent answers** (2 marks)
Show that a particular technology (e.g. hydrogen cells) might replace a non-renewable resource such as oil.

Resource consumption is a major concern globally. There are different views as to how sustainable our consumption is as supplies of many resources, such as oil, are finite. There is also concern over the 'two-speed' world, with MEDCs consuming 80% of the resources but only containing 20% of the population.

You should know...

- [ ] How resources can be classified by type and availability
- [ ] How exploiting resources has both benefits and costs
- [ ] Why the supplies of some resources are running out
- [ ] How demand for resources varies between countries and is linked to their state of development
- [ ] How supplies of resources lead to a world of 'haves' and 'have nots' as countries have different levels of resource base
- [ ] Why there is concern over dwindling supplies and growing demands for oil
- [ ] That different theories exist about how far the world can cope with the current consumption of resources
- [ ] How we can become more sustainable in our resource consumption
- [ ] How technology may help solve shortages of resources

Key terms

Alternative energy	Malthusian theory	Sustainable development
Boserupian theory	Material resource	Sustainable resource
Carbon footprint	Natural resource	
Consumption	Non-renewable resource	
Human resource	Renewable resource	

Which key terms match the following definitions?

A Development that meets the needs of the present without compromising (limiting) the ability of future generations to meet their own needs

B The skills and abilities of the population

C A measurement of all the greenhouse gases we individually produce, through burning fossil fuels for electricity, transport, etc., expressed as tonnes (or kg) of carbon dioxide equivalent

D Those resources – like coal or oil – that cannot be 'remade', because it would take millions of years for them to form again

E Resources – such as wood – that can be renewed if we act to replace them as we use them

F Energy sources that provide an alternative to fossil fuels

G The using up of something

H A natural substance that humans choose to use

To check your answers, look at the glossary on page 313

Foundation Question: State **three** ways we can use resources more sustainably. (3 marks)

Student answer ● (achieving 2 marks)	Examiner comments	Build a better answer △ (achieving 3 marks)
We can recycle our waste.

We can recycle our mobile phones.

We can use bikes instead of cars. | • **We can recycle our waste** is correct and scores 1 mark.

• **We can recycle our mobile phones** does not gain any marks as it is an example of the first point, rather than a new point.

• **We can use...** scores 1 mark. However, it would be better to extend this answer and link it to resources. | We can recycle our waste.

We can buy local food and cut down on food miles, saving on aeroplane costs and fuel.

We can use bikes instead of cars so that we use less petrol. |

Overall comment: When you are asked to state three ways, make sure that you list three different ways. In this question, the student also needed to link their answer to the idea of sustainability to be sure of the mark.

- -

Higher Question: Explain how developing sustainable transport can lead to a reduction in resource use. (3 marks)

Student answer ● (achieving 2 marks)	Examiner comments	Build a better answer △ (achieving 3 marks)
The best way of not using scarce resources is to walk or cycle.

It is also useful to cut down on car use by developing good public transport, such as buses or cars.

Another way is to develop bus lanes. | • **The best way...** is a straightforward and correct explanation that scores 1 mark.

• **It is also...** This is also correct and is another explanation linked to resource use. It scores 1 mark.

• **Another way is...** This is potentially a good idea but it is not linked to an explanation so does not score any marks. | The best way of not using scarce resources is to walk or cycle.

It is also useful to cut down on car use by developing good public transport, such as buses or cars.

Another way is to develop bus lanes. These lanes make the buses faster and more popular and will save on fuel. |

Overall comment: The student demonstrated that they knew what sustainable transport was and linked their ideas to resource use. The only mistake was not to explain bus lanes.

Chapter 11 Living spaces

Objectives

- Know some of the things that might make a place seem attractive to people.

- Be able to explain why rural and urban areas are attractive to some groups but not to others.

- Understand why the attraction of rural and urban areas is related to the stage of development of a country.

What are the ingredients of good living spaces?

People have different views about what makes a good living space

We can all describe what we would consider to be an ideal place to live. Or at least we think that we can – in reality it is not as easy at it appears. The ideal location for a summer holiday may not be an ideal living space for the rest of the year. An ideal place to live when you own a car may not be quite so ideal if you don't.

Young people don't normally get the chance to choose where they live. Until they leave home, they are dependent on their parents and guardians, and have to live in the place chosen by them. But, because they become familiar with the local area and develop a network of friends, young people are usually content to stay where they are, even if others regard their area as unattractive (or worse). People become quite attached to their 'living spaces' and, even if they would like to live somewhere else, they may find it hard to break away.

Figure 1 shows a breakdown of living spaces in Scotland into six types. A survey of Scottish adults (see the table on the left) showed that, whatever type of area they lived in, they rated their own living spaces very highly. Even in large urban areas (cities), nearly 90% of people thought that their neighbourhood was very or fairly good. The survey also found that only 5% of people said that there was 'nothing they particularly liked' about their neighbourhood, whereas almost half said there was 'nothing they particularly disliked'.

Throughout Scotland, a neighbourhood's 'quiet and peaceful' aspect was identified most often as what made it attractive. Almost half the people in large urban areas identified this aspect, rising to some 75% in rural areas. The 'friendly people' in the neighbourhood was another factor which people across Scotland (some 30%) recognised as something they particularly liked. Although all agreed that it was the people and the peacefulness of an area that made it attractive, there were significant differences between rural and urban dwellers over the other attractive and unattractive features of their living spaces (see tables below and right).

How Scottish adults rated their own neighbourhoods in 2001

Type of area	Very or fairly good (%)	Very or fairly poor (%)
Large urban areas	89	10
Smaller urban areas	92	8
Accessible small towns	93	6
Remote small towns	95	4
Accessible rural areas	96	4
Remote rural areas	96	3

The views of Scottish residents

Urban likes	Urban dislikes
Convenient shops	Vandalism
Good public transport	Young people hanging about
Nice landscape/open spaces	Drug abuse

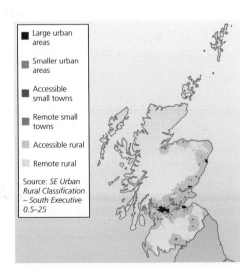

Large urban areas

Smaller urban areas

Accessible small towns

Remote small towns

Accessible rural

Remote rural

Source: SE Urban Rural Classification – South Executive 0.5–25

Figure 1: The different types of living space in Scotland

Rural likes	Rural dislikes
Nice landscape	Poor public transport
Safe area/low crime	Lack of leisure facilities
Good views	Poor local shops

It is clear from the Scottish survey that people are mostly happy where they are. Other research has shown that when A level students are asked where they would like to live in ten years' time, the most common response is generally their local region. In the UK for example, students in Bristol choose Bristol as their preferred living space for the future (giving London as their most frequent second choice). This pattern is repeated in many parts of the UK.

Factors that affect how we view different places

There are several factors which affect how we view different places and how we come to regard some as more attractive than others, including:

Age	Younger people, especially teenagers and young adults, often seek more variety in their social lives and more opportunities for socialising together. This tends to make cities more attractive than rural areas and large successful cities more attractive than failing cities.
Mobility	Personal mobility affects how isolated we are likely to be, both from one another and from important facilities such as schools, hospitals and shops. Rural areas obviously pose more problems unless households have private cars that are available and relatively cheap to run.
Cultural background	Our cultural background plays a large part in controlling how we 'see' places. As with images of attractive people, this has changed over time, and changes from place to place (Figure 2).
Knowledge and perception	What we take to be 'ideal' is often based on incomplete information. We tend to ignore what we don't know about local crime rates, local facilities and local job opportunities. Places can be 'sold' to us by clever branding and then turn out to be quite disappointing.
Economic status	We are likely to find places attractive that offer us security through offering us work.

Processes that affect the quality of living spaces

The processes that affect the quality of living spaces can be broadly divided into two groups:

- Natural processes that affect the environmental challenges and opportunities of an area.

- Human processes that control the evolution and development of that living space.

Figure 2: Cockington in Devon

Natural processes

Some natural environments pose more challenges than others. It would seem obvious that multiple challenges make an area both more 'difficult' and therefore less attractive. That is especially the case when people are able to make decisions for themselves about where they would like to live. But the same areas may also prove to be attractive to other people. As the Scottish survey suggests, all decisions about where to live are likely to be compromises – and not all people have the luxury of making such decisions.

Human processes

Human action, including the actions of governments, can have a profound effect on the quality of living spaces. Above all, the natural environment can be protected while also attempting to provide economic opportunities. Without economic opportunities almost all living spaces are likely to decline. A lack of jobs will mean out-migration, whether it be from declining cities such as Detroit in the USA or the remote fjords of Iceland.

People in different countries want different living spaces

Just as there are variations from time to time about what constitutes a good living space, so there are variations between countries. To some extent this is explained by how developed the countries are, and how much they are changing. In many developed countries, such as the UK, rural regions have often been seen as offering a better lifestyle than cities, giving rise to counterurbanisation – a movement back to rural regions. In the UK this process saw the decline in the population of large and medium-sized cities, while smaller market towns (and even small cities) grew. Rural villages also grew – although only if allowed to do so by the planners who control the supply of building land. In 2007 more than 105,000 people moved from urban to rural areas. Many were no doubt attracted by lower rates of crime, cheaper housing than the cities and the idea of a **rural idyll**.

Decision-making skills

1. Working in pairs, write down five features of your local area that you think are positive and five things that you think are negative.

2. Suggest two ways in which your list might be different if you were aged 40.

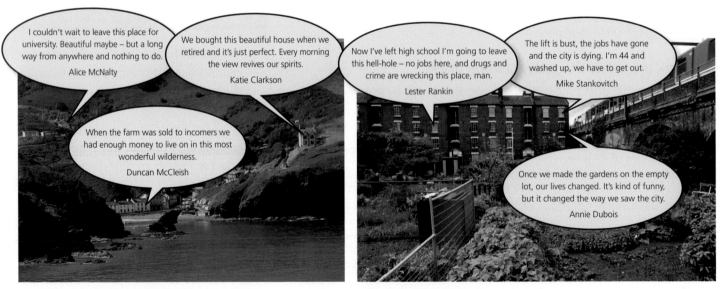

Figure 3: People living in rural areas can have different views about their surroundings

Figure 4: People living in urban areas can have different views about their surroundings

This concept has been much reinforced by TV programmes (*Heartbeat, The Darling Buds of May, Lark Rise to Candleford,* etc.) with their images of attractive English country villages populated with a rich stream of fascinating characters. In their efforts to rebrand and make themselves more attractive to both businesses and individuals, some rural local authorities have consciously traded on this very favourable image. However, Figures 3 and 4 show that people living in rural and urban areas also have different views about these areas.

Unfortunately the rural idyll does not prove to be quite so idyllic for all rural in-migrants. Some rural areas are so dominated by in-migrants that very little is left of the original rural character that made the areas so appealing.

The *State of the Countryside* report, published in 2008, estimates that the population of rural areas will increase by 18% within twenty years, compared with just 9% in urban areas.

Skills Builder 1

Study Figure 5.

Describe the changes in the rural and urban populations of China.

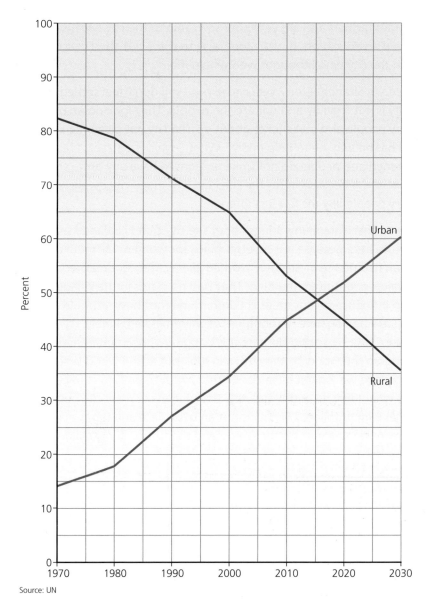

Source: UN

Figure 5: The past and predicted growth of China's urban and rural population

China: moving to the cities

Although urban to rural migration is a feature of many developed societies, on a global scale the most important movement is in the other direction. As Figure 5 shows, China will shortly become an 'urban society' – in which more than half of the population live in cities. The movement of people from rural, largely western and northern China, to the eastern and southern cities is driven by economic change in the country as a whole. The countryside offers fewer jobs and less variety of employment (and limited educational opportunities). In some countries, including both China and India, changes in farming have sometimes pushed people off the land, forcing them to find alternatives.

Migrants arriving in Chinese cities do not always find it easy to adjust, as this part of an interview with a young woman in Shanghai in 2004 shows:

Activity 4

Read the interview on the right.

Identify two advantages of urban life and two disadvantages that she mentions.

> In 2000, I came to Shanghai from Yunnan province and I miss it very much. I am 23 this year and when I came here I was only 16. My family are farmers and are very poor so I came to the city because I wanted an education. It was a long way to come but I had cousins here in the city who I thought would help me.
>
> I've been in Shanghai for 7 years now, but I still don't feel settled. The city is not friendly and nor are the people. They look down on country people and make jokes about us. The city has given me a job which has helped me to help myself. When I walk the streets I feel threatened and people do not smile or seem to have any time, and they mock me because of my background. For many city dwellers country people like me are just inferior.
>
> However, there are many things about the city that I like and I see my future here because it is a better place for women. My mum and my gran had no life at all. They worked all day, they got married, had children, got tired and died. They had no status in the village and no influence except on us children; that is not how it should be. I feel like a person here, although it is sometimes very lonely. Modern China is exciting because it is bringing opportunities and opening up the world. In this city I can look out at the world whereas in my village we looked inwards. All that I wish to do is to study and that is what I will do, night and day. That will make me a city person and make me fulfilled. This city is not an easy place and it has made me very miserable but it is the future and the future of my children, if I have them.

Rural–urban migrants are frequently young. As a result, the rural communities that they leave frequently have ageing populations, adding to the problems faced by such abandoned areas. There are other examples of how this affects living spaces.

Living spaces for the young in London

Age, ambition and aspirations can affect how individuals view various living spaces. Urban **regeneration** projects, for example, have frequently been built around the idea that young, single and mobile individuals can be attracted back into city centres. About 80% of England's population live in cities and towns of over 10,000 people. Despite the counterurbanisation process, we remain very much an urban society.

Redevelopments often provide small apartments and flats that are ideally suited to young, single people. The jobs that have replaced the old industrial jobs in the urban areas have been in finance and business services that are frequently dominated by younger people. As such, it is no surprise that the age structure of an inner city borough such as the London Borough of Camden is so distinctively different from the capital as a whole (and the country), as shown in Figure 6.

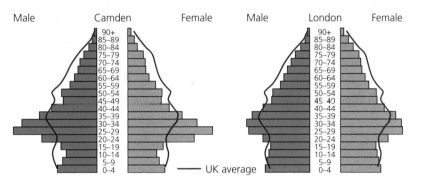

Figure 6: Population pyramids for London and Camden

The culture of inner city areas is frequently geared to the young adults that drive their economies. And the **inner city living spaces** that have proved attractive to the twenty-something generation are usually less appealing to other sectors of the population. Figure 7, for example, shows a futuristic design, with its small modern apartments, that is aimed at a 'young' market.

Living spaces for the retired

Most people do not move when they retire, but some do – and they have a freer choice of location. In recent years there have been increasing numbers retiring abroad, especially to Spain. Around 8% (900,000) of UK pensions are paid overseas. There has been a tradition of Britons residing in Spain since the 1950s, increasing rapidly in the 1980s. The British who 'retired' to Spain began to appear on British television in the early 1990s in documentaries, holiday programmes, dramas and comedies. The image of the 'Brits' in Spain is of people searching for paradise, living an extended holiday with other 'Brits', rarely taking part in local life or culture, refusing to learn Spanish, and trying to recreate England 'in the sun', complete with fish and chips and pubs.

Watch out!

Figure 7: 'Visage by Barratt', an apartment block in Swiss Cottage in the London Borough of Camden © Barratt

Activity 5

Which of the following age-groups would be most attracted to the property shown in Figure 8: people under 25, people aged 26–55, or people over 55? Justify your answer.

Decision-making skills

Think about people you know who have retired to a place in the sun. What factors do you consider to be the most important in influencing them to do this? If you don't know anybody then watch the TV show 'A Place in the Sun' to get an idea.

Many of these people are thinking of returning home. What factors may have changed to make them? Think about economic climate, healthcare, rising costs of aviation fuel, pound versus euro, family dynamics, land grabs, etc.

ResultsPlus
Build Better Answers

Explain two reasons why people who move overseas when they retire sometimes return home disappointed. (4 marks)

■ **Basic answers** (0–1 marks)
Make general statements about 'things not working out' without offering any detail as to why this is the case.

● **Good answers** (2 marks)
Identify at least one reason, such as ill-health, that might lead to people wishing to return.

▲ **Excellent answers** (3–4 marks)
Explain two reasons, such as ill-health and worries about language problems in 'foreign hospitals' and falling value of pensions as the value of the pound declines.

More recently the movement of UK citizens to rural rather than coastal Spain has appeared in popular culture through, for example, television programmes such as *A Place in the Sun*, which profile people 'opting out' in what are perceived to be idyllic rural locations (Figure 8).

Figure 8: A dream house in Spain

There have been a number of factors which have led to this type of retirement migration:

● Increasing numbers of people reaching retirement in good health

● Reduction in the average age of retirement

● Increasing income and savings

● The attractions of the climate of Spain, especially in the south

● Cheaper living costs

● Increasing property values in the UK

● Growth of budget airlines

● The growth of the internet, satellite TV, mobile phones and other ICT developments (e.g. Skype)

● The growth of a British **expatriate community**.

But the migrants are not all motivated by the same factors. They are a very varied group, with some returning home frequently to keep in touch, others returning permanently – disappointed and disillusioned – while others set up successful businesses and integrate into local culture. Recent changes in property values and in the relative value of sterling and the euro make the future of this type of international migration uncertain.

The growing demand for good living spaces

The pressures on the quality of living spaces

Although populations are rising quite slowly in most developed countries there are still pressures to build on **greenfield sites**. In most countries this pressure is a result of movements that take place within a country, as the people move from one area to another.

Figure 9 shows that in recent years there has been a significant shift of the population from the north of the UK to the south. Although the trend is not exclusively one-way, it is the southern counties that have had the most serious problem of finding space to accommodate a growing population.

Housing demand and supply in North Wiltshire

In 1996, according to the local planners, the county of Wiltshire required 60,000 more houses by 2016. This was necessary if the local economy was to continue to grow and attract new businesses. Economic growth attracts people – and people need somewhere to live. If houses were not built then the prices of existing houses would rise and this would discourage businesses from investing.

Local authorities do not build houses, but they control what goes where because they give (or withhold) the planning permission that is needed for almost every type of development. The three main policies applied in Wiltshire (and elsewhere) are to:

◉ Balance likely employment growth and the projected growth in the working population

◉ Reduce commuting flows (whether in or out)

◉ Concentrate new housing in the areas that are most self-contained – those that already have services such as schools and shops.

In North Wiltshire, population growth has steadily increased, encouraged by an increase in new housing. The land-use of the district, however, remains 81% rural, although only 19% of the population live in these rural areas, mostly in hamlets and small villages. About 60% of North Wiltshire's total population live in the six main towns – Chippenham, Calne, Corsham, Cricklade, Malmesbury and Wootton Bassett. Chippenham is the biggest, with 26% of North Wiltshire's total population, and is the key location for future growth and investment. It is also within commuting distance of London (75 minutes) and has attracted in-migration as a result.

Government policies, both nationally and regionally, emphasise the re-use of brownfield land and, since 1996, North Wiltshire has been able to direct 44% of its housing development to brownfield sites. The other 56%, however, has been built on greenfield land, and it is the local authority that decides which land can be used.

173

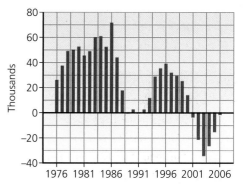

Figure 9: Migration from north to south in the UK, 1976–2006

Quick notes
(North Wiltshire):

- Rural areas face pressures from population growth.
- Economic growth inevitably means the loss of some greenfield land.
- Local authorities in the UK select some areas to be more suitable for development than others.
- Most development is restricted to the larger settlements.

ResultsPlus
Build Better Answers

Explain the pressures that can result from the demand for new housing in rural areas. (4 marks)

■ **Basic answers** (0–1 marks)
Use terms such as 'difficult' and 'hard' with reference to finding available land but do not explain why. Some students turn the question into 'why is there pressure' and talk about the causes and not the consequences.

● **Good answers** (2 marks)
Identify at least one pressure in a community, perhaps the pressure in 'greenfield sites' and the impact this has on existing residents.

▲ **Excellent answers** (3–4 marks)
Explain two pressures, such as the need to find greenfield sites and local resistance to this because of impact of their living space and, perhaps, the expansion of small towns and pressure on local services as population rises.

This process takes place in two stages:

Stage 1. Identification of 'key settlements', where growth is allowed because it can be sustainable (and meets the three main policies outlined in the bullet points on page 172). In North Wiltshire, the bulk of new development on greenfield land will take place in and around Chippenham.

Stage 2. Choice of individual sites, based on the following criteria:

- Policy restrictions – are the sites in or close to protected areas of special note, 'Areas of Natural Beauty' for example?

- Physical problems or limitations – this includes looking at drainage, flood risk and access, among other factors.

- Potential impacts – is there an impact on surrounding land-users?

- Availability of the land and desirability of the site for commercial development – can the land be purchased and would the houses be easy to sell?

The pressures in the USA

In a few developed countries population has grown, largely through immigration. The best-known example of this is the USA – a country that has historically grown by allowing in-migration. With a population density only one eighth of that of the UK (USA: 30 per km^2 , UK: 241 per km^2) the country has much more space and, with its extensive resources, more to gain from encouraging population growth. However, there are also important internal trends, with two that are particularly important:

- A shift from the older industrial cities of the north (e.g. Chicago) to the newer cities of the south and west (e.g. Phoenix) from 'Rust Belt' to 'Sun Belt'.

- The loss of population from inner city areas as industries, followed by people, have moved out into the surrounding rural areas.

In many areas of the south-west of the USA, cities have grown up in remote rural areas where only tiny settlements existed before. Their continued growth through **urban sprawl** consumes surrounding countryside very rapidly. From Orange County in California to Phoenix in Arizona the march of new housing estates is continuing, despite serious worries about how this type of growth can be sustained in an era of very much higher oil prices. One of the best known examples of how a once-rural area has been transformed is Las Vegas in the Mojave Desert.

Las Vegas and the Mojave Desert

Las Vegas is located in Clark County, at the southern tip of the state of Nevada. The most recent census shows that approximately 1.6 million people call Las Vegas home, but only a quarter of them were born there. It also shows that half a million of them live beyond the city limits, in the low-density urban sprawl that has spread out into the surrounding Mojave Desert.

The climate is extremely hot and dry. The occasional rainstorm in the mountains sometimes cause violent flash floods. It is these floods, over the years, that carved the deep canyons and deposited sediment in the dry valleys. Like most deserts, the plant life and animal life is scarce and easily damaged – this is a fragile environment.

The city of Las Vegas takes its name from the Spanish word for fertile valley. The springs that created this little oasis in an otherwise barren desert first attracted European settlers in the 1850s. But today Las Vegas is like a very big oasis. Palm trees line the streets, and the houses are surrounded by green grass and flowering shrubs. This illusion is created by the millions of litres of water imported from nearby Lake Mead, together with water taken from the region's groundwater – water that is not being renewed in what is now an extremely dry area.

Sometimes known as 'Sin City', sometimes as the 'Entertainment Capital of the World', Las Vegas has a reasonable claim to both titles. The city has an exceptional dependency on tourism – there were 35 million visitors in 2007 – and, with the ten largest hotels in the world, it competes for top global tourist destination with Orlando in Florida. It is also the fastest-growing urban area in the USA, and this growth is evident in the aerial images taken in 1972 and 2000 shown in Figure 10. This growth is vital to the local economy – and to Nevada, which has no state income tax and relies on gambling and other tourist spending for its tax revenues.

The growth of Las Vegas has had a significant impact on the environment – both in the city itself and in the surrounding rural area. These effects reduce the quality of the living space and make the area less attractive. Such growth, therefore is not sustainable:

- Las Vegas consumes more water per person than any other city in the world – 1,360 litres daily per capita, compared with 800 in Los Angeles, 600 in Tucson, or 420 in Oakland.

- Groundwater levels have declined more than 90 m in some portions of the valley.

- Off-road vehicles cause immense damage to the surrounding area of the Mojave desert and destroy local wildlife.

- The primary sources of water for desert plants and animals are the Colorado River and groundwater, both of which are under pressure from urban growth.

- Urban 'heat island' temperatures are as much as 5 °C above the local averages. The main contributory factor is the thousands of air-conditioning units.

- Las Vegas ties with New York City for fifth place in carbon monoxide pollution for US cities.

Figure 10: Aerial images of Las Vegas in 1972 (top) and 2000 (above).

Skills Builder 4

Study Figure 10. Describe the growth of Las Vegas between 1972 and 2000.

Quick notes (Las Vegas):
- Urban sprawl in the USA has turned many rural areas into urban landscapes.
- This growth has often been uncontrolled.
- Cities in dry and hot regions are especially likely to have a negative impact on the surrounding rural areas.

Figure 11: The urban sprawl of Las Vegas

Activity 6

Use the internet to research the causes and the consequences of the growth of Lagos (Nigeria) which is predicted to grow from 10.1 million (2003) to 17 million in 2015.

The attraction of cities

Urban areas have been attracting people – and growing bigger – for centuries. By definition, an urban area is a place where large numbers of people live in close proximity. And people generally do this willingly – they choose to live there – because they prefer this kind of living space. Cities can act like magnets, drawing people in. But being 'attractive' in that sense, does not necessarily make them attractive in the other sense – being pleasant to look at. Good living spaces may be both but, as with Las Vegas, attracting large numbers of people to live in one area not only places stresses on the surrounding rural areas but also on the city itself. This can have a very negative effect on the living space and, potentially, destroy the possibility of sustained future growth (Figure 11). It is this issue which preoccupies many urban planners.

There are two ends to every population movement, with **push** and **pull factors** affecting people's decisions. In some parts of the developing world, rural poverty is so bad that cities do not have to offer very much to be a better alternative. This is the case with Mumbai, Lagos, Jakarta and many other megacities in the developing world. In some cities, however, there are opportunities for the new arrivals. The table below summarises some of the causes and consequences of the growth of three megacities.

The growth of three megacities: causes and consequences

Predicted growth (millions)	Causes for past and future growth	Consequences
London 2003: 7.1 2015: 8.0	• Continued importance of London as a **global hub** for finance and business services. • Growth of other tertiary jobs to serve these global jobs. • Many poorer paid jobs are filled by immigrants from other EU countries. • Higher incomes than competing UK cities.	• Housing shortages, especially cheaper housing. • Strain on the public transport system. • Outward pressures on the surrounding areas, especially to the west and to the east.
Tokyo 2003: 35 2015: 36.2	• Significance as a **global city**. • Many Japanese companies moved their HQs there to be close to government – personal contact is very important in Japanese culture. • Planners continue to encourage growth by allowing the development of 'hub' cities in the Greater Tokyo area.	• Some of the most expensive property in the world. • Enormous problems of managing waste, pollution and urban transport systems.
Mumbai 2003: 17.4 2015: 22.6	• Creation of a significant finance and business sector has created jobs in the CBD. • The changes taking place in the Indian countryside – leaving many small farmers without land – are a major push factor.	• The creation of very large slums, as in Dharavi, home to over 600,000 people. • One third of the population do not have access to fresh drinking water. 2 million do not have access to a toilet.

Strategies to make future living spaces more sustainable

The idea of **sustainable living spaces** is quite new. Cities have always required 'support' from the land around them, because they can rarely grow much of their own food, or provide much of their own fuel or materials. But some cities have a much larger ecological footprint than others. Las Vegas is an example of a city that makes very little effort to be 'sustainable' – it only exists because of modern technology (bringing in water and using air conditioning to make life bearable). In fact, with its extravagant lights and spectacular fountains, the city seems to symbolise the conspicuous consumption of energy and resources – right in the middle of one of the most severe environments on the Earth.

Other cities, however, have tried a number of strategies to minimise their environmental impact, so that they can be more sustainable. One of the best examples of this is Singapore, an island 'city-state' in south Asia.

Making Singapore more sustainable

The original British colony of Singapore was situated on the southern coast of the island – the area that is now the CBD of the Republic of Singapore, founded in 1965. Before independence, the rest of the island had been mostly farms and tropical forest, but the new government has constructed many new towns, so that today the island is almost entirely built-up. This has brought very challenging problems in handling traffic and waste, which are likely to become even more pressing as the country continues to enjoy high rates of economic growth. The Urban Redevelopment Authority, which is responsible for implementing efficient land use, tries to maintain convenient transport and minimise pollution. Its attempts to make the city more sustainable are often quite imaginative:

Figure 12: Singapore

Quick notes (Singapore):
- Population: 4,260,000
- 640 km²
- Density: 6,650/km²
- Ex-British colony
- Strong, authoritarian government
- Rapid economic growth – one of the Asian 'tiger' economies
- GDP per capita: $31,000 – 20th in the world.

- To reduce strain on the Central Area, several regional centres have been developed, each containing a concentrated commercial district, so that travelling across the island to shop is unnecessary.

- Light industry is distributed around the island in industrial estates that are easily accessible to housing areas. Only industries that produce close to nil pollution are allowed.

- Heavy industry is concentrated on Jurong Island – a small separate island to the south-west.

- A Mass Rapid Transit network handles all public transport – an integrated system of railways (underground and overground), buses and taxis.

- Two-thirds of all journeys are made by public transport.

- Singapore implemented the world's first congestion charging system in 1975. Today, with 'Electronic Road Pricing', almost all vehicles on the island have electronic readers that debit bank accounts directly when they pass through one of the 27 entrances into the CBD. (There are also park and ride schemes.)

- Singapore is one of the most expensive places in the world to own a car. Prospective car owners must first have a Certificate of Entitlement – costing as much as £6,000 – and then pay high import costs and duties.

- The island has one of the most efficient waste management systems in the world. In 2007, a total of 5.6 million tonnes of rubbish was generated, of which 3.0 million tonnes was recycled and the rest disposed of at incineration plants and landfill. There are plans for new minimum-emission incineration plants that produce electricity in the process.

- Recycling by residents is encouraged by easily accessible centralised sites, each serving about five apartment blocks. Door-to-door collection of recyclable materials is planned.

Like other successful cities, Singapore is unlikely to reduce its environmental footprint below the global average of 2.7 ha but, with a figure of 4.2 ha, it compares very favourably with the UK (5.6 ha) and the USA (9.6 ha). (There is no figure available for the environmental footprint of Las Vegas – but it would be shockingly high.)

Sustainable cities

There are many different ways of making cities more sustainable – some of them local and some city-wide. One of the many organisations that advise and gather information about these projects is the Danish organisation Sustainable Cities (http://sustainablecities.dk/city-projects/cases). The following examples are based on some of their material.

Barcelona (Spain)

Like other European cities, Barcelona has developed a scheme by which bikes are made available as an alternative to cars. The service is called 'Bicing', and anybody with an address in the region can buy a year's membership for £30. This allows them to pick up a bike from one of the 400 specially made 'Bicing stations', and drop it off at another. The stations are spread all over the inner city – about 400 m apart – and most of them are close to another type of public transport, such as buses or the metro. The bike can be used for two hours at a time per day. The first half hour is free and then it costs 50 cents for every half hour. Today there are 6,000 bikes and more than 175,000 Bicing members, making 50,000 trips a day on average.

Havana (Cuba)

When the Soviet Union collapsed in 1989, Cuba lost its most important trading partner. The country was forced to take extreme measures simply to feed itself, and from that necessity grew the world's best-known example of urban agriculture – Havana's popular gardens (huertos populares) – Havana's gardens which use every scrap of available ground in the city (Figure 13). Cuban statistics are difficult to get but in 1995 it was estimated that there were 26,600 popular gardens in the city, ranging in size from a few square metres (used by one person) right up to three hectares (subdivided into smaller plots and shared by as many as 70 people).

Figure 13: A Cuban works in a popular garden in Alamar, east of Havana.

A wide selection of produce is cultivated, depending on the soil, family needs and market availability. Principles of **organic agriculture** are followed, with minimal external inputs (such as chemical fertilisers). Instead, gardeners use organic fertilisers – chicken or cow manure or compost from household food waste – which are low-cost, readily available, and environmentally sustainable.

Reykjavik (Iceland)

Reykjavik has made the most of its position near a constructive tectonic boundary. Since 1930 its buildings have been heated by the natural hot water distributed by the world's largest and most sophisticated geothermal district heating system. Today, geothermal energy powers the entire city – with an electricity distribution network harnessing 750 MW of thermal power from steam, and a water distribution system generating 60 million cubic metres of hot water per year. In 2006, 26.5% of electricity generation in Iceland derived from geothermal energy, 73.4% from hydro power, and just 0.1% from fossil fuels. The use of geothermal energy has massively reduced Reykjavik's dependence on fossil fuels – making it one of the cleanest cities in the world and saving up to 4 million tonnes of CO_2 emissions every year. Geothermal energy (Figure 14) has also contributed to Iceland's transformation from a relatively poor country to one that enjoys a very high standard of living.

Masdar (United Arab Emirates)

The cities that have grown up rapidly along the Arabian Gulf are not generally known for their 'sustainable' vision. On the contrary, they have a reputation for very expensive and extravagant building projects. And in Abu Dhabi, the capital of the United Arab Emirates, for example, the air-conditioning systems consume vast amounts of energy coping with the summer temperatures, which average over 40 °C. Locals drive instead of walking, because the streets are considered too hot to walk on – in fact, there is often no pavement provided.

To tackle these issues, a new settlement for 50,000 people – Masdar – is to be built near Abu Dhabi airport, about 17 km outside the capital. It is claimed that it will be the world's first 'zero carbon, zero waste' city. Part of Masdar is going to be covered with a solar roof, which will generate energy – and provide shade. Masdar will draw on traditional Arabic architecture, using wind towers to funnel air through the city as 'natural air conditioning'. (Wind towers are attached to the tops of buildings and suck cool air in and down from the roof, pushing the warmer air out.) The new city will be surrounded by a wall to protect it from the hot desert air, and splashing fountains in courtyards will dampen the dry heat. The buildings will be no more than five storeys high, and huddled close together on narrow streets, with rooftops covered with solar generators and street-level 'solar canopies' providing shade. Masdar will be car-free, but public transport links will never be more than 200 m away. It is hoped that the shaded walkways and narrow streets will create a pedestrian-friendly environment, despite the area's extreme climate.

Figure 14: Geothermal energy generation in Iceland

ResultsPlus
Build Better Answers

Explain two ways in which cities might be made more sustainable in the future. (4 marks)

■ **Basic answers** (0–1 marks)
Include a general comment about how being more sustainable is desirable but do not explain what might be done to achieve this.

● **Good answers** (2 marks)
Identify one method, such as reducing dependence on cars by development of public transport or schemes such as bike hire.

▲ **Excellent answers** (3–4 marks)
Explain two ways that cities might be made more sustainable, such as increasing provisions of public transport so reducing car usage and energy consumption and, perhaps, improving waste disposal by collection schemes and recycling to reduce landfill and incineration.

Know Zone
Living Spaces

People have very different ideas, depending on factors such as age, and cultural background as to what are good and bad living spaces. Popular choices have led to rapid growth and development of unsustainable towns and cities. Future living spaces are now being designed to be more sustainable.

You should know...

☐ What qualities make good or bad living spaces

☐ Why people have different views about what is good or bad when choosing a living space

☐ What factors are likely to influence their opinions

☐ Why many people in developed countries dream of living in a rural area

☐ Why large numbers of people especially in developing countries are migrating to cities to live

☐ How those migrants find life in the cit and what sorts of living spaces they have

☐ Why some people, especially those who are retired want to live in 'a place in the sun'

☐ What are the pressures and impacts on the environment of popular and growing living spaces

☐ How some living spaces have become unsustainable

☐ How decision makers are developing plans to make existing settlements more sustainable

☐ Examples of sustainable schemes in a range of cities, which reduce their eco-footprint

☐ How architects and planners are designing new eco cities and towns to be very sustainable with zero emissions

Key terms

Accessible rural areas

Cultural background

Economic status

Expatriate community

Global city

Global hub

Greenfield sites

Inner city living space

Organic agriculture

Pull factor

Push factor

Redevelopment

Regeneration

Remote rural areas

Rural idyll

Sustainable living space

Urban sprawl

Which key terms match the following definitions?

A An overseas community made up of non-nationals, e.g. the British living in Spain

B A piece of land that has not been built on before, but is now being considered for development

C The common perception that rural areas are quiet and attractive – and therefore good places to live in

D Countryside within easy reach of urban areas

E A major centre of global communications, such as an international airport

F Something that attracts people to a location

G Development that aims to stimulate growth in areas that have experienced decline

H Rural areas that are distant from and thus little affected by urban areas and their populations

To check your answers, look at the glossary on page 313

ResultsPlus
Maximise your marks

Foundation Question: State **three** reasons why people in developed countries want to live in rural areas. (3 marks)

Student answer ● (achieving 2 marks)	Examiner comments	Build a better answer △ (achieving 3 marks)
People see the country as a much cleaner and safer environment to bring up their children or to retire to. Many people find the countryside much less crowded than the urban areas they live in. These urban areas are very stressful. And the price of housing is very high in urban areas and they are very congested.	• **People see the...** This is an excellent beginning and scores 1 mark. • **Many people find...** This is just about focused on rural areas and also scores 1 mark. • **And the price...** In this part of the answer the student has moved the focus to urban areas and is not looking at why people want to live in rural areas.	People see the country as a much cleaner and safer environment to bring up their children or to retire to. Many people find the countryside much less crowded than the urban areas they live in. These urban areas are very stressful. The price of housing in many rural areas is cheaper than in inner urban areas and people can buy houses or cottages with land.

Overall comment: This question requires details of the *pull factors* of rural areas. The material relating to the *push factors* of urban areas was not required for this question.

- -

Higher Question: Explain how both push **and** pull factors are encouraging counterurbanisation. (3 marks)

Student answer ■ (achieving 0 marks)	Examiner comments	Build a better answer △ (achieving 3 marks)
The pull factors are the opportunity to live in a vibrant city with good services and entertainment. The countryside is increasingly becoming a place without services and employment. In urban areas there are huge numbers of jobs in service industries. The countryside is better for retirees who like the quiet environment.	• **The pull factors...** The student has misunderstood the term counterurbanisation and has confused it with urbanisation. This part of the answer is incorrect so does not score any marks. • **The countryside is...** This part of the answer is also irrelevant to the question so does not score any marks. • **In urban areas...** This also strays from the main point of the question so does not score any marks. In addition, the point about retirees may be a relevant factor, but it is not linked to the question.	The main push factors are often a mirror image of each other. The main push factors are the stress, high cost of housing, lack of space and noise, especially in inner city areas. The main pull factors are the opportunity for a place in the country with land, space and in a tranquil environment free from crime and noise.

Overall comment: The student misunderstood the idea of counterurbanisation, so the pull and push factors were both incorrect.

Chapter 12 Making a living

182

How and why is work changing in different places?

The balance between employment sectors

Lots of people have to work (parents looking after children, school pupils) but here we are only talking about paid work. So employment is paid work.

There are lots of different types of employment, and the various jobs are usually grouped into four sectors:

- **Primary employment**. These are jobs that involve getting food and raw materials from the land and the sea. So people who work in farming, mining and fishing work in the primary sector.

- **Secondary employment**. These are jobs that involve making or manufacturing things. So people who work in factories making steel, cars or textiles all work in the secondary sector.

- **Tertiary employment**. These are jobs that provide services, such as health, education and transport. So people who work in schools and hospitals or who drive buses or lorries all work in the tertiary sector.

- **Quaternary employment**. These are jobs that are in the high-tech service industries – usually based on information, or in the consultation, research and development sectors of companies. So people who work in research departments or laboratories work in the quaternary sector (Figure 1).

The **employment structure** of a place means how the workforce is divided up between the four main employment sectors. These employment sectors change over time – and they change from place to place.

Figure 1: People working in a high-tech industry

Changes over time in the employment structure

In 1851 about a quarter of people in Britain worked in the primary sector – as farmers, foresters, miners or fishermen. The largest employment group was the secondary sector with about half of all employed people being involved in manufacturing. The remaining quarter of the working population were employed in the service sector, as lawyers, administrators, shopkeepers, bankers and similar jobs.

Today, however, things have changed. Now only about 2% of the working population are in the primary sector, and only 27% are in the secondary sector. The tertiary sector is by far the biggest employer today – with 68% – and the remaining 3% are in the quaternary sector.

The Clark Fisher model of changing employment

The Clark Fisher model is a simple model which shows how the stages in which the different employment sectors change as the economy of a country grows:

- **Pre-Industrial Stage** – You can see from Figure 2 on page 184, in this first stage, the primary sector of farming, mining and fishing is the most important. Britain before 1700 would be an example of this employment pattern.

- **Industrial Stage** – Then, as time passes, manufacturing industries start to grow and develop. New factories spring up in many different places and towns start to grow and expand. So the secondary sector grows and becomes more important. At the same time, the tertiary sector also starts to grow – but not as fast as the secondary sector. The tertiary sector grows to provide services such as transport, water, electricity and finance for the growing population. Britain after 1850 had this employment structure.

- **Post-Industrial Stage** – In this stage the tertiary sector becomes more important than the secondary sector, which tends to die away. As more people start to live in towns, there is a growing demand for various services – from health and education to transport and finance. Britain in the 1950s had this employment pattern.

- As more and more countries have developed their tertiary sector, countries like the UK and the USA have had to develop their quaternary sector. This is the most recent sector to have emerged in many developed countries and its growth is based on research and development, as well as information technology. The quaternary sector provides research expertise for the development of a wide range of new products and services.

Results Plus
Watch out!

■ This model is based on what happened in developed countries like Britain. It may not work in the same way for developing countries which may bypass some part of the model. For example developing nations might encourage tourism in their country and so bypass the Industrial Stage.

Results Plus
Exam Tip

▲ Most descriptions of the growth of towns and the parallel growth of industry are based on what happened in Britain in the nineteenth and twentieth centuries. These processes of industrial growth and urban growth were spread over a long time. The growth of industry and towns in developing countries is taking place much more quickly so the results may be different. This is a good point to mention in an exam answer.

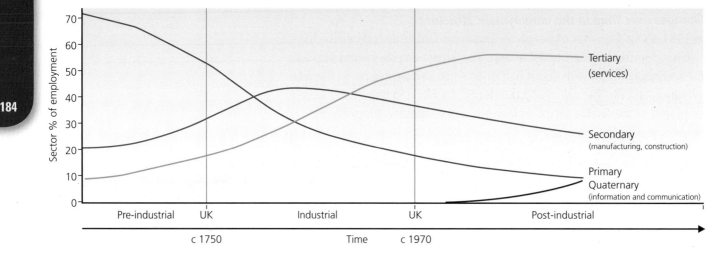

Figure 2: The Clark Fisher model of changing employment

Skills Builder 1

Study Figure 3.

(a) What percentage of people in the USA work in the primary sector?

(b) What percentage of people in China work in the secondary sector?

(c) Using the data in Figure 3, draw pie charts for each country. Compare the graphs for Bangladesh and the USA under the headings: 'Primary employment', 'Secondary employment' and 'Tertiary employment'.

Activity 1

1. Why are the numbers of people in quaternary activities included in the numbers for tertiary activities?

2. Why is farming difficult in Bangladesh?

3. Where does China sell many of its manufactured goods?

4. Why do the numbers of people in tertiary activities increase over time?

Changes in the employment structure from place to place

Bangladesh, China and the USA are three countries with very different employment structures, as Figure 3 shows. (Because the quaternary sector numbers are so small they are included in the tertiary sector in these figures.) These three countries show how the employment structure of a country changes as it develops its industries and its services.

Country	Primary %	Secondary %	Tertiary %
Bangladesh	63	11	26
China	43	25	32
USA	3	23	75

Figure 3: The employment structure of Bangladesh, China and the USA, 2007

BANGLADESH LIVES OR DIES BY FARMING!

Bangladesh is a developing country with most of its people working on the land as farmers. So it has a lot of people in the primary sector of employment. It is a country which tries to produce enough food for all its people. However farming is difficult here, partly because of the environment . For example there are sometimes serious floods which cover the land and destroy the crops. If this happens there can be a failure of the harvest and people may go hungry or even die. So everyone depends on farming. But Bangladesh is working hard to develop its industry. There are more and more people in its secondary sector making the country's own manufactured goods, such as textiles and clothing. But progress is slow because many people are poor. The tertiary sector is large in Bangladesh because a lit of people work in government departments which administer the country.

CHINA BECOMES THE WORKSHOP OF THE WORLD!

China used to be a poor country at an early stage of development, and farming is still important because it wants to feed as many of its own people as possible. But in the past ten years the country has become one of the most important centres of manufacturing industry in the world. So it is now an industrialising country. Since 2000 there has been a massive growth of new firms, making everything from steel, ships and cars to toys, pharmaceuticals, drugs, paint, clocks, clothing and computers. China exports most of these goods to richer countries such as Japan, the USA and the UK. Therefore the secondary sector in China is very important. This has helped to create more wealth, so the tertiary sector services – such as health, transport and education – have all grown too, to provide for the growing numbers of people in factories or on farms.

DOES THE USA DEPEND TOO MUCH ON SERVICE INDUSTRIES?

Recently it emerged that the USA has over-three quarters of its workers in the tertiary sector – which can make the economy unbalanced. Only 3 % of workers are employed in the primary sector. This is because farming, mining and forestry have become so mechanised that they need very few workers. And the USA is rich enough to import any extra food and raw materials that it needs. The number of workers in manufacturing industries is shrinking because of mechanisation – and because countries like China and Mexico can make goods more cheaply than US firms. The USA is therefore said to be 'de-industrialising'.

In the USA, millions are employed in the tertiary sector – in offices, schools, hospitals and banks and in tourism and leisure. Some experts argue that too many people now depend on these service industries, which in future may shrink as they become more mechanised and computerised.

The impact of changes in employment in towns

The case of Mexico

Mexico is an example of the many developing countries where the rapid expansion of manufacturing industry has brought problems to the growing towns and cities.

Mexico used to be a country where most people worked on the land as farmers. But farming gradually became more modernised – so fewer workers were needed on the land. And when Mexico's manufacturing industry began to grow, people left the countryside in large numbers and moved to the towns, where the new factories were being built.

186

Describe and explain the difference between the formal and informal sectors of employment. (4 marks)

■ **Basic answers** (0–1 marks)
Offer ideas only about the sorts of jobs to be found in the formal and informal sectors, but do not explain the differences.

● **Good answers** (2 marks)
Describe the types of jobs in each sector but offer only limited explanation of the differences.

▲ **Excellent answers** (3–4 marks)
Not only describe the types of jobs in each sector, but explain that the differences are related to formal sector jobs giving regular work and income as opposed to the self-employed work with the little security and unreliable income of the informal sector.

The rapid rate of growth was driven by US companies opening branches in its southern neighbour because Mexican wages were so much lower than in the USA. This meant that these firms could produce a whole range of goods such as computers and cameras much more cheaply.

The growth of towns and the growth of manufacturing industries took place together, but the speed of the process caused problems:

◉ The towns could not build enough houses fast enough for the people arriving from the countryside – many people had to live in overcrowded conditions with poor sanitation.

◉ There were not enough jobs in the towns for the huge numbers of people arriving.

The problem in Mexico City – as in many of the towns – is that there are more people than jobs. Some people work in factories (Figure 4), such as the car-making plants of Volkswagen. This is called employment in the **formal sector** – because the workers are formally employed, with permanent jobs, and regular pay (and they pay their taxes). But those people who cannot get a job in the formal sector often find work in the **informal sector** – by setting up their own jobs. These may be jobs like shining shoes, giving hair cuts, selling water, luggage carrying, taking street photographs or selling food they cook on the spot (Figure 5). Nearly all of these 'informal' jobs take place on the streets. The people usually have no shop or office – they just find a spot on a street corner or near a train station or large store.

Figure 4: Workers at a new factory in Mexico City

Figure 5: A person working in the informal economy

The impact of changes in employment in the countryside

People living in the countryside in developed countries have seen big changes in employment. In Scotland, for example, the main changes have been:

- A decline in jobs in farming, fishing and forestry

- A growth in a few areas of new, often high-tech, industries.

The main factors which have led to this growth of new jobs are:

- Intervention by governments to support new firms – grants, loans, new factories, for example – especially in areas like the Highlands and Islands of Scotland

- The growth of new activities which are based on the scenery or natural resources. So jobs in tourism, mountain climbing, canoeing and outdoor activities have grown.

- There has been an influx of 'incomers' – people new to the area. Many are retired people, but there are also younger people seeking a 'better' lifestyle, with clean air and a stronger sense of community.

- The growth of new industries based on new technology, such as broadband Internet access has enabled people to become **telecommuters**. For example in the Highlands and Islands this allows people to work from home and to communicate with other members of their firm via email and other electronic forms of communication (Figure 6). This has allowed more people to live and work in remote parts of countries like Scotland.

Decision-making skills

Draw a table to show the pros and cons of developing tourism on a large scale in the countryside.

Figure 6: Working from home in the highlands of Scotland

Objectives

- Identify some negative and positive environmental effects of changes in jobs.

- Describe some negative and positive environmental effects of changes in jobs.

- Explain some negative and positive environmental effects of changes in jobs.

Activity 2

1. Why does the air not move around freely in Mexico City? Why might it be important to keep some packaging?

2. How is photo-chemical smog formed?

3. Why does the city need so much water?

4. Suggest three things that planners in Mexico City could do to improve air quality.

5. Suggest three things that people in the city could do to reduce the waste problem.

How can the environmental impacts of changes in work be managed more sustainably?

Positive and negative environmental impacts

When industries start to grow in an area, extra jobs are created and major changes to the local environment take place. In the case of Mexico City, the new manufacturing firms which located there from the 1970s onwards attracted millions of people to move to the city from the countryside in search of a better life. Even now, over a thousand people a day arrive in Mexico City from the rural areas. Unfortunately the new firms did not take great care to protect the environment, and the growing population made things worse. The main problems are:

- Pollution of the air by both factories and cars. Because Mexico City is surrounded by mountains the air does not move around freely. This means that pollutants such as nitrogen dioxide build up in the air and are not blown away. As the pollutants build up they react with sunlight to create a photo-chemical smog which attacks the throat and lungs, causing stinging and choking. The pollution is so bad that the city often disappears from view in a brown smog (Figure 7).

- Providing a water supply for the new firms and the 20 million people is a major problem. The city takes water from underground aquifers. Unfortunately a lot more water is pumped out than is replaced by rainfall. So the aquifers are becoming empty. As they empty, the land sinks – the city is sinking at the rate of nine centimetres per year. So buildings become unstable, roads crack and water pipes are broken.

- The water supply is polluted by firms which pump chemical waste into rivers and streams.

- The waste disposal system cannot cope. The authorities collect 10,000 tonnes of rubbish a day. Unfortunately the city generates 11,000 tonnes a day, so the rest is dumped on open ground, or in streets and waterways.

Figure 7: Photo-chemical smog in Mexico City

When industries decline there is unemployment, but there may be benefits to the environment. In Germany the Ruhr area (Figure 8) used to be a major centre for coal mining and for iron and steel making. From the 1980s onwards, both these industries began to disappear. By 2000 there were over 200,000 people out of work – 18% of the workforce. The decline of these industries, however, gave the government the chance to improve the environment. The Ruhr was then an unattractive area of coal tips, slag heaps, polluted water and **derelict land**. There was little open space and there were large numbers of nineteenth-century flats which were in need of replacement. But the environment has now been greatly improved:

- The slag heaps and coal tips were levelled and planted with trees.

- Derelict land has been converted into woodland and open space.

- New green spaces were created as parks for people to enjoy.

- The old flats were demolished and new ones built with more open space between the blocks.

- Air and water quality have both improved greatly – for example, fish can now be found again in the rivers of the Ruhr.

- Five new green belts have been created which run north to south. They separate the main urban areas and give more people access to the countryside.

- The German Methane Capture Project is concerned with capturing methane from the Rhine and Ruhr valleys. Despite the closure of many of the mines, methane – which is 21 times more environmentally damaging than carbon dioxide – continues to escape. The new project will capture the methane before it escapes and use it to generate electricity and heat. Combined Heat and Power (CHP) units feed electricity to the grid and heat to a local district heating system. By burning the methane, its climate impact is reduced by 21 times. Using methane also reduces the area's dependence on the fossil fuels normally used for energy generation.

Figure 8: The Ruhr area in Germany

Activity 3

1. Name three problems for the environment created by the coal mining and iron and steel industries of the Ruhr.

2. Why did the flats in the Ruhr need replacing?

3. How do we know that the water in the rivers of the Ruhr has improved?

4. What are the two main purposes of the green belts?

5. Summarise the main gains to the environment in the Ruhr since the 1980s.

Skills Builder 2

Look at Figure 8.

(a) How far is it across the Ruhr – from Marl-Hüls to Solingen and from Duisburg to Dortmund?

(b) Which river runs from south to north through the area?

(c) Which river runs from east to west across the area?

(d) Describe the position of the 'green' areas on the map. Why are they in these places?

Managing impacts more sustainably – a case study of Sandwell

Employment change can bring both benefits and challenges to the environment. Sandwell is an example of a place which had a huge set of environmental problems. Sandwell lies in the heart of the West Midlands, in an area known as the 'Black Country'. Six main towns make up the Sandwell conurbation – Oldbury, Rowley Regis, Smethwick, Tipton, Wednesbury and West Bromwich – with a total population of 287,500.

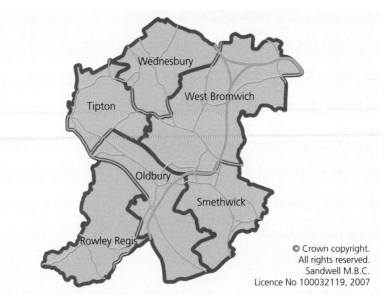

Figure 9: Sandwell in the West Midlands

This is an area which had numerous manufacturing industries, many of which have declined or disappeared over the last twenty years. This has created a lot of environmental problems:

- In the 1990s there were large areas of land which were contaminated by poisonous chemicals, such as mercury and arsenic.

- There were large areas of derelict land where factories and houses had been demolished.

- The air was badly polluted.

- There was a lack of green space.

- In 1995 over 23 per cent of council houses were unfit for habitation.

- In 1997 Sandwell was the seventh most deprived area in the whole of England.

- Traffic jams were a problem on the old, often narrow roads.

- In 1997 the people of Sandwell had the third highest rate of heart disease in the UK.

ResultsPlus
Build Better Answers

Using examples, explain how the growth of industries in developing countries can bring both benefits and problems. (4 marks)

■ **Basic answers** (0–1 marks)
Offer a basic explanation of either benefits or problems.

● **Good answers** (2 marks)
Offer a basic explanation of the main benefits and problems of industrial growth in developing countries.

▲ **Excellent answers** (3–4 marks)
Include detailed explanations of both benefits (greater employment, higher incomes and increased services such as electricity or water) and problems (pollution from factories, the overseas company possibly not paying taxes and profits from overseas companies leaving the country).

How can the environmental impacts of changes in work be managed more sustainably?

191

From the late 1990s onwards, however, a number of projects were developed to overcome the worst of these environmental problems. Sandwell is now building on two earlier **regeneration** projects – the 'Black Country Development Corporation' and the 'Tipton Challenge'. These two schemes were paid for by central and local governments, with some help from private investment. In total they were able to put £1.72 million into the area between 1987 and 1998. Since then, Sandwell has worked hard to continue and expand the progress made in improving the environment for all its people.

Sandwell has one of the 21 Urban Regeneration Companies in existence nationally, operating in its area. RegenCo was established in 2003 to undertake physical regeneration and attract new investment into West Bromwich, Smethwick and Hill Top.

Sandwell also has a New Deal for Communities programme in its area. The Greets Green Partnership delivers this programme, which ends in March 2010, a key theme of which is creating sustainable communities.

One of the UK's nine Housing Market Renewal Pathfinder Areas covers a substantial part of Sandwell. This is a 15-year regeneration programme which aims to improve neighbourhoods with low demand housing.

Sandwell Council is committed to tackling climate change through the Sandwell Declaration. This commits the Council to take a lead on tackling the causes and effects of climate change. In setting an example, Sandwell Council has so far reduced its own emissions by 27% and by 2018 should have reduced them to 45%.

One of the main ways this is being achieved is by upgrading and redeveloping all of Sandwell's 19 high schools between 2010 and 2015. They are being designed to need 60% less energy from gas and electricity. In addition, at least 10% of the energy that they use will come from clean, renewable sources, such as burning woodchips, solar power, wind power and from extracting heat out of the ground.

The comments on the right show the achievements in Sandwell since 1987.

Alternative proposals for a brownfield site

Once employment has declined in an area there are often factory closures. The factory may be demolished, creating a **brownfield site**. People often have different opinions as to how the site should be used in the future. One example of this comes from the USA. In the town of Berlin in New Hampshire there used to be a large pulp and paper mill which employed over 1,500 people. Before the paper mill was built there had been heavy industry on the site, and the land was polluted with mercury and other poisons. In 2005 the paper mill closed and local people began to think about how best to redevelop the site. There were two main proposals for redevelopment (page 192).

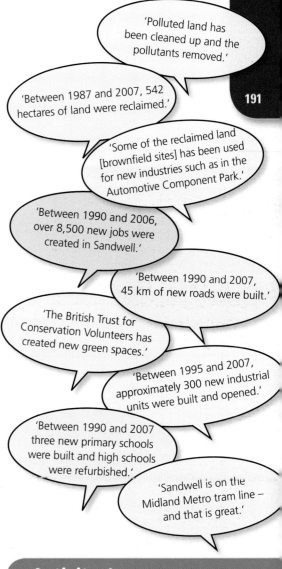

'Polluted land has been cleaned up and the pollutants removed.'

'Between 1987 and 2007, 542 hectares of land were reclaimed.'

'Some of the reclaimed land [brownfield sites] has been used for new industries such as in the Automotive Component Park.'

'Between 1990 and 2006, over 8,500 new jobs were created in Sandwell.'

'Between 1990 and 2007, 45 km of new roads were built.'

'The British Trust for Conservation Volunteers has created new green spaces.'

'Between 1995 and 2007, approximately 300 new industrial units were built and opened.'

'Between 1990 and 2007 three new primary schools were built and high schools were refurbished.'

'Sandwell is on the Midland Metro tram line – and that is great.'

Activity 4

1. Name three environmental problems in Sandwell that were the result of changes in employment.

2. Give two examples of how Sandwell has overcome an environmental problem.

3. How successful do you think has Sandwell been in overcoming its environmental problems?

4. What else do you think people in Sandwell might do to improve their environment even more?

To learn more about Sandwell visit www.sandwell.gov.uk

Proposal 1 – Leave the site as open green space. Clean up any pollution, plant some trees and create parkland. The green space would benefit the local people and help produce cleaner air. The problem for some local people was that this plan would not help to create new jobs – and the were still over 1,000 people looking for work.

Proposal 2 – Build a new biomass-energy project (Figure 10). This would create a new power station which would burn local wood to generate over 60 megawatts of renewable electricity. It would also use 700,000 tonnes of clean wood biomass chips per year. There would be an investment of $80 million and it would employ 40 local people and indirectly create another 500 jobs. When complete, the plant would be one of the largest and most environmentally balanced biomass-energy projects in North America. Some local people did not like the project because they argued it would create few jobs and would not add any open space.

Figure 10: The biomass-energy project in Berlin, New Hampshire

Decision-making skills

Complete a copy of the table below to summarise the advantages and disadvantages of the two projects.

Project	Advantages	Disadvantages
1. Green space project		
2. Biomass-energy project		

ResultsPlus

Exam Tip

⚠ When writing about new jobs created by green technology, be careful to mention a range of jobs from those in water and waste management and green transport, as well as those in renewable energy.

Employment in the 'green' sector

One of the areas of employment that experts predict is likely to grow in the future is the **'green' sector** – the jobs connected with making a **sustainable** environmental future. But just how many jobs will be created by the new green technologies?

The theory

Europe has pledged to cut carbon dioxide emissions by 20% by 2020, and to obtain 20% of its energy from renewable sources, such as wind or water. There are also important developments to reduce the carbon footprint of all transport, construction and electricity firms. This should create new jobs and has been called the 'third industrial revolution'. So where will the new jobs be created?

How can the environmental impacts of changes in work be managed more sustainably?

193

- Most of the new jobs will come from the generation of renewable energy. For example, huge new offshore wind farms are being built or are planned for the coasts around the UK (Figure 11). Firms making new wind turbines and solar power generators as well as hydro-electric power stations could create 2.5 million new jobs by 2020, according to an EU-funded report in 2003. One problem with this is that no one really knows if these new jobs will materialise. Another problem is that in some places such as the coalfields of Germany and Belgium even more coal mines will have to close – and this means that there will be job losses there.

- There may be new jobs in water management. As people become aware of the need to use water carefully – and to reuse it where possible – more workers may be needed.

- There may be new jobs in waste management. For example, as recycling gains momentum more firms will need extra workers to deal with recycling. But there may be only a few jobs because the work will rely heavily on machines.

- Green transport may also create new jobs. As cars and lorries become more environmentally friendly there may be a demand for more skilled workers to design better types of transport.

Figure 11: An offshore wind farm

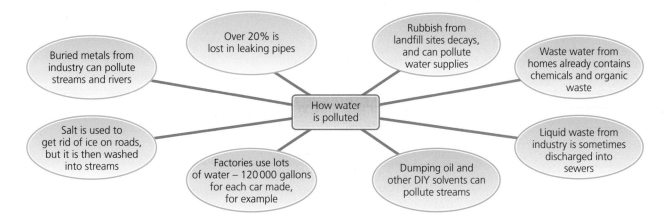

Figure 12: How water is polluted

There is a huge variety in the way we make a living. This not only varies spatially, but also changes over time.

You should know...

- [ ] What the primary, secondary, tertiary and quaternary employment sectors are
- [ ] Which type of jobs belong to each sector
- [ ] How the percentage of employment in each sector changes as a country develops
- [ ] How you fit a country's employment structure into the Clark Fisher model
- [ ] How and why employment varies from place to place
- [ ] What the post-industrial stage is and what type of jobs will people have
- [ ] How employment varies between rural and urban areas
- [ ] How changing employment has both good and bad effects on the environment
- [ ] What is meant by the formal and informal sectors
- [ ] How people manage the problems of industrial pollution to make a more sustainable environment
- [ ] What brownfield and greenfield sites are
- [ ] What green industries are and what new types of green jobs will be available in the future

Key terms

Brownfield site	Green sector	Quaternary employment
Clark Fisher Model	Industrial stage	Regeneration
Deprived area	Informal sector	Secondary employment
Derelict land	Post-industrial stage	Sustainable
Employment structure	Pre-industrial stage	Telecommuter
Formal sector	Primary employment	Tertiary employment

Which key terms match the following definitions?

A An area in which there is a damaging lack of the material benefits that are considered to be basic necessities – employment, housing, etc.

B The proportions of people who work in primary, secondary, tertiary or quaternary jobs

C Work where the people are formally employed, with permanent jobs and regular pay (and they pay their taxes)

D A piece of land that has been used and abandoned, and is now awaiting some new use

E Land on which factories or houses have been demolished

F A generalised description of how societies' employment structures change as they develop

G The period in the development of a society when manufacturing industry has yet to develop

H Forms of employment that are not officially recognised, e.g. people working for themselves on the streets of developing cities

To check your answers, look at the glossary on page 313

Foundation Question: Describe the type of jobs you would find in the 'green sector'. (3 marks)

Student answer ● (achieving 2 marks)	Examiner comments	Build a better answer △ (achieving 3 marks)
Green jobs are those to do with the environment, such as working on farms. They also include people working in rubbish dumps and recycling. Other green jobs include people who work on wind farms in renewable energy.	• *Green jobs are...* This example is incorrect so does not score any marks. Agriculture is not necessarily a green job unless it is organic agriculture. • *They also include...* This is correct and the student gains 1 mark for recycling. • *Other green jobs...* This is correct and the student scores 1 mark for listing renewable energy as an example of green energy.	Green jobs are those that help us to use the environment more sustainably, such as those in green transport. They also include people working in rubbish dumps and recycling. Other green jobs include people who work on wind farms in renewable energy.

Overall comment: Make sure that you describe terms accurately so you do not lose marks.

- - - - - - - - - -

Higher Question: Explain why the 'green sector' is seen as a source of providing new jobs in the future. (3 marks)

Student answer ■ (achieving 1 mark)	Examiner comments	Build a better answer △ (achieving 3 marks)
Green jobs are linked with sustainable development. The main jobs will be in renewable energy developments. And some other jobs will be in waste management.	• *Green jobs are...* This is an excellent start to the answer and scores 1 mark. • *The main jobs...* This is true, but it is not linked to the first statement and is a *description* rather than an *explanation* so does not score any marks. • *And some other...* As with the previous statement, this is a *description* rather than an *explanation*.	Green jobs are linked with sustainable development. Details of the need to reduce carbon footprints or not to damage the environment are needed. Waste management is an example of the third industrial revolution, where the need to damage the environment by landfill is avoided.

Overall comment: The student showed a clear understanding of the green sector, but failed to provide a coherent, linked answer.

Chapter 13 Changing cities

196

Objectives

- Identify the main elements of an ecological footprint.

- Describe some factors which influence the size of 'eco-footprints'.

- Understand that cities creating wealth can also create environmental problems.

Activity 1

Study Figure 1.

1. Do you think this employs many people?

2. Why are sites like this used to get rid of rubbish?

Exam Tip

⚠ Remember that eco-footprints measure the amount of land that would be needed to support people's lifestyles and dispose of the pollution and waste created. They are a useful way of calculating the impact of people's lifestyles on the environment.

What are the environmental issues facing cities?

Urban regions and their eco-footprints

As cities grow and develop they start to have a bigger and bigger impact on the surrounding area. In their early stages, towns and cities use the surrounding countryside to produce food and raw materials for their people and industries. The surrounding area also supplies water, as well as being a place where waste is dumped (Figure 1). But as cities become even larger, the area that they affect also becomes bigger. For example, cities are able to bring in food and raw materials from countries a long way away. They also have to search further for places to get their clean water from – and for places where they can dump their waste.

Recently, people – all over the world – have become worried about the impact that cities are having on the environment. In particular, they are worried about the pollution of air, land and water by the things that we do in our homes, offices, factories and cars. One way of measuring the impact we have on the environment is an **ecological footprint** (often shortened to '**eco-footprint**'). The eco-footprint for any city looks at how much land is needed to provide it with all the energy, water and materials it uses. The footprint also calculates how much pollution is created by burning oil, coal and gas, and it works out how much land is needed to absorb the pollution and waste created by the people of the city. The reason for doing all this is to work out how sustainable any city is and especially what changes we need to make to improve the **quality of life** for people now and in the future. There are a surprising number of changes we can all make, many of which are quite simply a matter of changing our habits. If everyone acts together we can make a big difference.

Figure 1: Landfill sites are used to get rid of many cities' rubbish

The eco-footprint of York

Every day we all contribute to the eco-footprint of cities. We do this by:

● the clothes we wear

● the water we drink

● the food we eat

● how we travel around

● how we heat and light our homes

● the electrical appliances we use

● the waste we create

● the pollution we create.

Scientists can calculate the eco-footprint of individual cities in terms of how much land is needed to support the lifestyle of the people and to deal with all the pollution and waste created. York is a typical British city, and one that is keen to reduce its eco-footprint, which has been calculated as 1,010,585 hectares. This averages out at 5.4 ha of land per person. This is slightly above the UK average, which is 5.3 ha per person. Figure 2 shows what things make up York's eco-footprint. As you can see, **energy consumption** is responsible for 24%, equivalent to 1.3 ha per person. Domestic gas consumption makes up the largest part of this energy, followed by electricity consumption.

Some people, however, consume more energy – and therefore have bigger eco-footprints – than others. Jessica, for example, keeps her central heating high because she likes to be warm. The walls in her rented flat are not insulated. The windows are draughty, and so is the front door which does not fit properly. She has high energy bills and an energy eco-footprint of 1.5 ha. In contrast, David lives in a flat nearby which is newer and better insulated. He keeps his heating tuned down and puts on a jumper before turning up the heating. David has low energy bills and a smaller energy eco-footprint of 0.6 ha. Kate lives in a house which she likes to keep quite warm. She has cavity wall insulation, and some of her household appliances – such as her washing machine – are 'energy-efficient'. She has an energy eco-footprint of 1 ha.

So different lifestyles can generate larger or smaller eco-footprints.

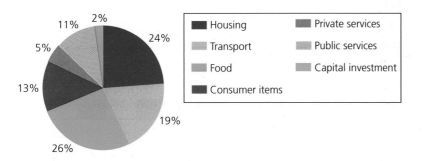

Legend:
- Housing
- Transport
- Food
- Consumer items
- Private services
- Public services
- Capital investment

2%, 11%, 5%, 13%, 26%, 19%, 24%

Figure 2: How York's eco-footprint is made up (Data provided by SEI using REAP baseline 2004)

Eco-footprints vary from place to place

Differences between urban and rural places

People in different parts of a country have different sized eco-footprints. In the UK, for example, people who live in the countryside generally have a lower eco-footprint than those who live in towns. This is because they may produce more of their own food (and in some cases their own power) and they may generate less waste.

Differences between developed and developing countries

The contrasts are even greater between different countries, especially between developed countries like the USA and the UK, and developing countries like Kenya and India, as Figure 3 shows. People in developed countries create much larger eco-footprints than people in developing countries. This is because the people in developed countries use more energy in the form of gas or electricity. More people in developed countries own cars, so again there is much more pollution of air and water than in developing countries where many fewer people own cars.

Developing countries like Kenya have very much smaller eco-footprints than developed countries like the UK and the USA, as Figure 3 shows. You can also see that cities like York and Liverpool also have much bigger eco-footprints than developing countries like Kenya and India. Kenya has an average eco-footprint of 1.5 ha per person, and in India the figure is 1.9 ha per person. This compares to the 5.4 ha for York. The Fair Earthshare target suggests that to be fair to everyone in the world cities should only have an eco-footprint of 2 ha per person. And 90% of the world's population has an eco-footprint smaller than York's 5.4 ha.

People who live in the countryside in developing countries like Kenya and Egypt have the smallest eco-footprint. This is because they grow much of their own food and they are often poor. They also use local wood for cooking, lighting and heating, and get around by walking or on bicycles rather than cars. So they do not create much pollution of the environment.

Skills Builder 2

Study Figure 3.

(a) Which country has the largest eco-footprint?

(b) Which country has the smallest eco-footprint?

(c) What is the size of Liverpool's eco-footprint?

(d) What is the size of Germany's eco-footprint?

(e) What is the size of the world's eco-footprint?

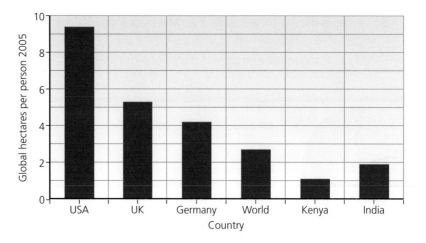

Figure 3: The eco-footprints of different countries and the world in 2005

The environment under pressure

Our environment can be divided into three main elements – land, air and water. The table below shows how we manage to pollute these three elements by our daily activities, such as travelling to work, eating a meal or turning on a light.

Producing electricity	Producing electricity through the burning of coal, oil or gas pollutes the air.
Nuclear power	Using nuclear power to generate electricity threatens land, air and water by the possible escape of radioactivity.
Industry	Industries such as the chemicals industry or car production can also add to **environmental pollution**. Water is often polluted by industries when chemicals escape.
Household waste	Burying **household waste** in **landfill** sites is not a good way to dispose of waste material. The buried material may contain chemicals which can escape and pollute local water supplies.
Oil	The main source of energy for transport (cars, lorries, buses, trains and planes) is oil. Oil production has grown hugely over the Past thirty years and burning all this oil releases chemicals which pollute the air.
Agricultural chemicals	The food most people in developed countries eat is made from crops which are grown with the help of agricultural chemicals. These chemicals can end up in our food and water.
Contaminated water	Many people get their water supply from underground sources. Over the past twenty years many of these underground sources have been contaminated by chemicals washed into the ground from houses, factories, farms and fields.

The challenge now is to cut down this damage to the environment.

Decision-making skills

Study the table. Summarise the severity of the pressures shown on the environment. Rank them in order of what you consider to be the most severe and the urgency of finding solutions.

200

Activity 2

Study Figure 6.

1. Name two ways in which cities pollute the environment.

2. How could one of these sources of pollution be reduced?

3. Write a short report, as if for a newspaper, to explain how cities are polluting the environment.

4. Look at Figures 4 and 5, and explain how wealth is being produced here.

Can cities generate wealth and be eco-friendly?

Cities have become more important with time, and are vital parts of our planet. This is because cities are:

● Centres where decisions are made

● Centres which create wealth from a wide range of activities – from factories to financial offices (Figures 4 and 5)

● Centres of transport – by land, air and water

● Centres of population – millions of people live in them.

Cities have grown up as places which create wealth, but they depend on lots of other places for their power, their raw materials and their resources. They also depend on other places for the disposal of their waste, as Figure 6 shows.

Figure 4: A modern factory in Scotland

Figure 5: Financial offices

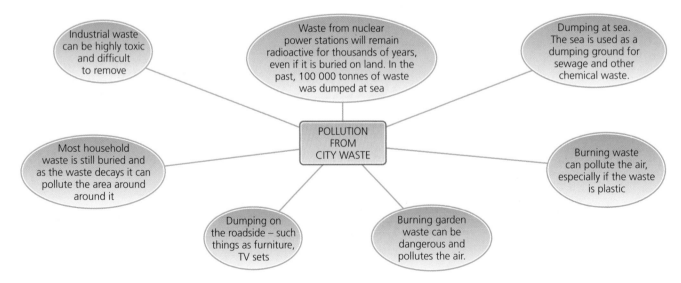

Figure 6: Pollution from city waste

How can we make things better?

We are all responsible for the pollution of our planet. In cities in developed countries, every household is responsible – as Figure 7 shows. For example, the car is the most polluting thing that most people own. Throughout its life each car uses up valuable raw materials. It also pollutes the atmosphere, and finally may be dumped and pollute the land. Houses also contribute to the demand for fossil fuels – for heating and lighting. Part of the problem with houses is that much of the energy used for heating escapes through walls, windows and roofs. Even in the garden, many gardeners rely on chemicals to control weeds and pests. These poisons can damage valuable garden wildlife, and sometimes they find their way into water supplies. People buy too much processed food. The things large corporations put into food, such as a lot of salt and sugar as well as chemicals for colouring, are not as healthy as they might be. Some meat which comes from **factory farming** (such as intensive rearing units for hens or cattle) may contain antibiotics or other chemicals used to keep the animals healthy. Water from many households is polluted by the chemicals used in the home, such as cosmetics, pesticides and some washing powders. The UK is a **throw-away society** – each year people in the UK throw away many times their own weight in rubbish. Mixed rubbish is difficult to recycle and is often dumped as landfill, using up valuable land.

Activity 3

Study Figure 7.

1. Make a list of three things people can do to cut down pollution.

2. Name two things people can do to cut down heat loss in their homes.

3. Why might using fewer chemicals in the garden create problems?

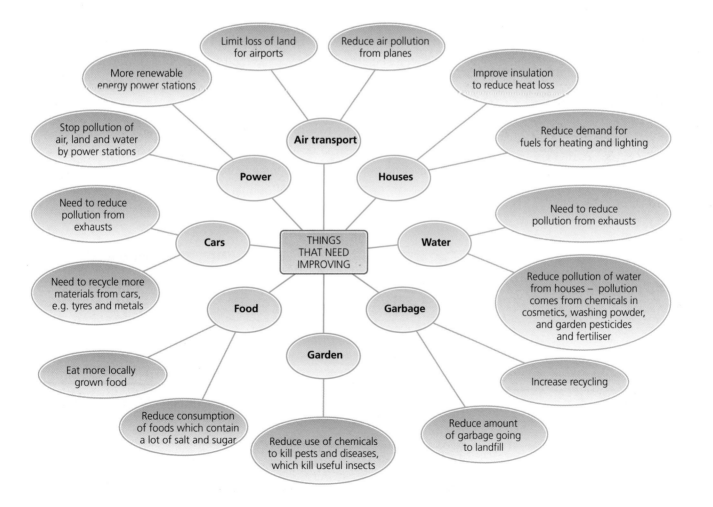

Figure 7: Things that need improving

Objectives

- Identify three ways in which one city is trying to reduce its eco-footprint.

- Describe some factors which could make transport more sustainable.

- Understand that green consumer behaviour has both advantages and disadvantages.

How far can these environmental issues be resolved sustainably?

Cities have the potential for reducing their eco-footprints – the case of York

The people of York are trying to reduce the eco-footprint of their city. The main challenges are:

- **Energy**: 20% of York's eco-footprint comes from energy. Home heating accounts for 60% of the carbon dioxide emissions from household use. An area nine times the size of York would need to be planted with trees to absorb all the carbon dioxide produced from energy use in York.

- **Food**: 33% of York's eco-footprint comes from food. This is largely because so much energy is used at each stage of food processing and because much of it has to travel a very long way (**food miles**) to the tables of York. Over 20% of all food bought by people in York is never eaten and ends up in landfill. For every tonne of food eaten in York, a quarter of a tonne of packaging is produced.

- **Waste**: In 2006 nearly 80,000 tonnes of waste was produced in York. The manufacture of the products and the disposal of the waste needs energy, which produces greenhouse gases. Worse still, as it decomposes waste also produces greenhouse gases such as methane. In York, however, the local waste company – Yorwaste – is now recovering and converting 70 per cent of this methane into electricity.

- **Transport**: The main type of transport used by the people of York is the car. This causes local air pollution and adds to global warming. The impact of travelling one mile by car is three times higher than travelling the same distance by train. The impact of using a bicycle is forty times less than the impact of using a car. Walking has no eco-footprint. High car usage also creates congestion on the roads (Figure 8).

Figure 8: Congestion on the roads in York

Reducing York's eco-footprint – energy use

In order to cut down the amount of energy used by people in York and to tackle global warming and avoid more flooding (Figure 9), the local council have produced a series of tips for saving energy in the home:

- If you are too warm at home, turn down the thermostat on the heating. A cut of 1 °C can cut 10% off your fuel bills.

- Make sure you do not leave the fridge door open longer than necessary. For every minute the door is open it will take three minutes to get back to its cool temperature – and this will use electricity.

- Try to use the shower rather than the bath. It uses only 40% of the water used in a bath.

- Use a 40 °C washing machine cycle rather than the 60 °C cycle – this uses 33% less electricity.

- Don't put really wet clothes in a tumble dryer. Wring them out or spin dry them first.

- Don't fill the kettle if you do not need to heat all that water. Try boiling it half full. This could save you enough electricity every day to run a TV for four hours.

- Wait for a full load when using the washing machine or the dishwasher.

- Fix any taps that are dripping. In just one day a dripping hot water tap can waste enough hot water to fill a bath.

- Try not to leave appliances on standby. Around 80% of the energy used by a VCR is while it is in the standby mode.

- Turn off lights when you leave a room. Lighting accounts for 10–15% of electricity bills.

- Close curtains at dusk to stop heat escaping through the windows.

- Choose the right size pan for the food and cooker you have, and keep lids on when cooking. This reduces cooking time and the energy used.

Figure 9: Flooding in York.

Activity 4

1. Why is using a shower rather than a bath a good idea?

2. Why is it a good idea to wring or spin dry wet clothes before putting them in the tumble dryer?

3. Why is it a good idea to wait for a full load before using the dishwasher?

4. Why should we not leave TVs on standby?

5. Why do we need to turn off lights when leaving a room?

6. Carry out a survey of the class. How many people do half of the things on the list to save water and electricity? How many people do one-third of the things on the list? How many people do none of the things on the list?

7. Some homes do not have a dishwasher, washing machine or central heating. Some homes do not have a shower. How could people in these homes reduce their eco-footprint?

Reducing York's eco-footprint – waste

As well as trying to reduce the size of its eco-footprint in relation to energy, York is trying to reduce the waste it generates. It is doing this in a variety of ways:

- Recycling is very important in York (Figure 10), as in most UK cities. People have different coloured bins and are asked to sort their waste into groups – paper, textiles, plastic, glass and cans. This has led to a big reduction in the amount of waste that has to be sent to landfill sites, but there is still room for improvement.

- Shops and businesses are also asked to sort their waste, for example to separate cardboard which can be disposed of separately.

- On average, only 16% of commercial waste is recycled, and most of this is office paper. So offices are being asked to cut back on the amount of paper they use and to make more use of electronic communication, such as email and text. There is also a drive to increase the percentage of waste that is recycled by improving the sorting of the waste before it is collected.

- The city is recovering more and more (70% now) of the methane gas produced for its landfill sites and this is a big part of cutting down the eco-footprint.

- The amount of waste that each household produces is likely to increase by 30% in the next ten years. One important way to reduce this would be to encourage people to buy fewer goods that are pre-packaged, because packaging (mostly plastic and paper) uses a lot of resources in its production, and then it is simply thrown away.

Figure 10: Yorwaste deals with waste in York

Activity 5

1. Suggest some ways in which packaging could be cut down, especially in supermarkets. Think about all the things that are packaged, such as meat, bread, flowers, fruit, vegetables.

2. Why might it be important to keep some packaging?

How can we make transport more sustainable ?

People in York – and in towns and cities all over the UK – are looking at ways to make transport more sustainable. The number of cars is the main problem – 72% of UK households own at least one car, meaning that there are over 22 million cars on the roads. Cars pollute the air with their exhausts, clog up roads with traffic jams and use valuable raw materials and energy in their construction (Figure 11). Solutions to the pollution problem include:

● Converting cars to Liquid Propane Gas (LPG) instead of petrol or diesel. LPG is cleaner and cheaper than petrol, and grants are available to help with the cost of conversion.

● Using **hybrid cars** which are powered by electric batteries as well as petrol. They are more energy-efficient and cheaper to run.

But changing car fuels will not by itself solve the problem of making transport in cities more sustainable. So other solutions are needed, and the simplest is to improve and develop the public transport systems in cities. Public transport is much more energy-efficient than private cars and causes less pollution. There have been various recent developments in public transport:

● Supertrams have been introduced in cities like Manchester and Sheffield. They can reduce congestion and improve air quality.

● **Bus lanes** help speed journeys to and from city centres. This encourages people to use buses, reduces pollution and is energy-efficient.

● **Congestion charging** encourages some motorists to use public transport – and reduces congestion.

● Underground systems such as the London Underground are very efficient and create little pollution, but they are expensive to build.

● **Park-and-ride schemes** help to reduce car use and congestion in some cities.

Activity 6

1. Carry out a survey in your class of how people travel to school. How many come by car, on foot, by bicycle, by bus? How could this change to create less pollution?

2. Make a list of the ways in which transport can be made more sustainable in your area.

Figure 11: Congestion on London's roads

Activity 7

1. What is the main cause of Curitiba's success?

2. Why do buses stop for such a short time at stops?

3. What are the benefits of using alternative bus fuels?

4. Why is Curitiba a model for other countries?

Curitiba – a role model for a greener urban future

An urban plan guided the growth of Curitiba in southern Brazil, which now has a population of 2.2 million. The plan was based around several axes crossing the city which operate as high-speed, one-way, three-lane roads, with the central lane reserved for express buses. The success of Curitiba is based around its public transport policy, centred on buses. There is an 'Integrated Transportation Network ' in which bus lanes are interconnected through twenty terminals. The system is fast, efficient and cheap. The buses run very frequently, some as often as every 90 seconds. The buses stop at cylindrical clear-walled stations with turnstiles, steps and wheelchair lifts. Instead of steps the buses have extra wide doors and ramps that extend out to the station platform when the doors open (Figure 12). This system of same-level bus boarding results in a typical average time at the bus stop for the bus of only 15 to 19 seconds.

The system transports 2.6 million people every day, and is used by 70% of the city's people. One fare allows passengers to travel anywhere on the system. Buses now use **alternative fuels**, especially natural gas, which creates less pollution. Curitiba uses 30% less fuel per person than the eight other Brazilian cities of the same size. So now this pattern of public transport and of land use is a model for other countries, because it has greatly reduced greenhouse gas production and prevented traffic jams (and their pollution) within the city.

Figure 12: A bus stop in Curitiba

Sustainable city living

Putting all the elements together to make a **sustainable city** would be a major challenge. But what would a sustainable city be like? Figure 13 shows one view of what a future sustainable city might look like. Some of the key points are:

- All waste would be recycled for reuse.

- Organic waste from urban sewage systems would be used as fertiliser.

- There would be more parks and green space.

- Houses would have a dual water supply, one pure for drinking and one less pure for other uses.

- Solar panels would heat water and generate some electricity.

- Public transport would be powered by electricity generated from renewable sources.

- Wind farms and wave generators would collect energy.

- Industry would be required to clean its waste to prevent pollution.

Figure 13 shows an idealised version of a city, but some of its elements are already happening in lots of towns. One of the key parts to improving the environment in cities of the future is the actions of individual people. We cannot leave it to the planners. For future cities to be sustainable we all need to take action. The impact of individual actions – such as buying locally grown **organic produce** or visiting **farmers' markets** – is becoming more and more important. Most supermarkets now stock organic produce, which was not the case ten years ago. The reason for this is that people decided they wanted to eat food that was free from chemicals and were prepared to pay slightly more for it. This 'power of the purse' is one way that individuals can take part in **green consumerism**.

Decision-making skills

Which of the five Rs do you consider to be the most important in our drive towards a sustainable society? Explain why.

- **R**educe – e.g. by using the car less (cutting down on consumption)
- **R**ecycle – waste and compost
- **R**euse – e.g. charity shops
- **R**enew – switch to alternative energy
- **R**espect – the environment and cultures

Which of these affect your life at present? Which are the most easy to get people to do?

Figure 13: A sustainable house of the future with solar panels

Using examples, explain how people can become green consumers. (4 marks)

■ **Basic answers** (0–1 marks)
Give only a limited identification of attempts to become green consumers, referring mainly to looking for environmentally friendly labels and comparing labels on energy-consuming products.

● **Good answers** (2 marks)
Offer ideas of looking for items with less packaging, as well as using labels to identify and compare products.

▲ **Excellent answers** (3–4 marks)
Offer a full range of ideas, including not buying more than is needed, buying organic products, using public transport and the 3 Rs, as well as labels to compare and identify products.

Decision-making skills

Choose four of the schemes from the list of bullets that you think would be the most environmentally sustainable. Provide evidence to say why.

Green consumerism – food purchasing

Green consumerism concentrates on the things people buy, encouraging us to:

● Reduce our consumption of all products, including energy, water and food.

● Buy products that are environmentally friendly. This includes a wide range of things from washing powder to organic food and clothing.

Consumers have a lot of power when it comes to green issues. For example, many supermarkets only began to stock organic foods and clothing when consumers began to demand it – and when it became clear that people were prepared to pay for it. By voting with their money, consumers are telling the large companies what they want. So, for example, there are now many more washing machines available that use a lot less water.

There are many ways to become a 'green consumer', including:

● Don't buy more than you need.

● Go for long-lasting products rather than cheap products with a short life.

● Buy organically grown fruit and vegetables. (There are now organic cotton T-shirts, too.)

● Look for an environmentally friendly label, especially in paper and forest products.

● Compare labels on energy-consuming products – often there is a star rating.

● Look for products with less packaging.

● Try to use public transport.

● Practise the three Rs – reduce, recycle and reuse.

Figure 14: Organic produce in a supermarket

Green consumerism – farmers' markets

Farmers' markets were pioneered in the USA. The first British farmers' market opened in 1997 in Bath, and there are now over 500 in the UK. People spend £120 million a year at these farmers' markets.

What is a farmers' market?

A farmers' market is a set of stalls, usually in a town or city, run by farmers and food growers from the local area (Figure 15). They sell their produce to local people, who get the chance to talk about the food to the people who have grown or produced it. This increases the trust between buyer and seller. The idea is to encourage small-scale, environmentally aware methods of farming and food production, such as organic methods. It is also more profitable for the farmer, who does not have to pay a shop.

What are the advantages of farmers' markets?

For customers, the advantages are:

● Very fresh produce

● Seasonal vegetables and fruit

● Free-range and organic meat and eggs

● Specialities, such as cheeses

● Try before you buy

● Possible bargains

● Face-to-face buying, with the chance to ask questions.

For farmers, the advantages are:

● Immediate payment

● No high costs of transport

● No costs of retailers

● Face-to-face selling, with the chance to explain the product

● Opportunity to find out, first-hand, what the customer wants.

Figure 15: A farmers' market

What future for farmers' markets?

Farmers' markets have grown quickly and they are now an important part of people's view of how to live a green lifestyle in a city. The future looks good in that more and more people are using them and buying from them. One of the issues for the farmers is how much time they can spare to run the stalls and how much time they need to spend producing the food. This is why permanent farm shops have been opened on some farms and are growing in popularity.

examzone

Know Zone
Changing cities

Traditionally cities have been very large users of resources and emitters of pollution, but all this is changing as they develop sustainable strategies to improve their environment (and the peoples' quality of life).

You should know...

☐ What an eco-footprint is and how we can calculate it

☐ Why many cities have very large eco-footprints

☐ Why eco-footprints vary in size from place to place

☐ What are the main causes of environmental pollution

☐ How we can manage the environment more sustainably and reduce our eco-footprint

☐ How we can reduce waste

☐ How we can make transport more sustainable

☐ How sustainable city living can improve the quality of life

☐ What green consumerism is

☐ How farmers markets are encouraging people to buy high-quality local produce and cut down on food miles

Key terms

Alternative fuels	Factory farming	Park-and-ride schemes
Bus lanes	Farmers' market	Quality of life
Congestion charging	Food miles	Sustainable city
Eco-footprint	Green consumerism	Throw-away society
Energy consumption	Household waste	
Environmental pollution	Hybrid cars	
	Landfill	
	Organic produce	

Which key terms match the following definitions?

A Cities that have a number of policies that attempt to reduce their impact on the environment (including the surrounding regions)

B Cars which use electric batteries as well as petrol engines

C Choosing to buy environmentally friendly products

D Disposal of rubbish by burying it and covering it over with soil

E A set of stalls run by farmers and food growers from the local area

F Food grown or produced without the use of chemicals

G Energy sources that provide an alternative to fossil fuels

H Material produced by households that needs to be disposed of

To check your answers, look at the glossary on page 313

ResultsPlus
Maximise your marks

Foundation Question: Describe how the way people living in cities leads to a large eco-footprint. (4 marks)

Student answer △ (achieving 3 marks)	Examiner comments	Build a better answer △ (achieving 4 marks)
In cities people use a lot of resources as they do not grow their own food. They also use a lot of energy commuting to work. They also make a lot of waste and pollution. So this is why they have a large input and output.	• *In cities people...* This is a good start to the question and includes a fair example. This scores 1 mark. • *They also use...* This is partly correct but is not very precise. Commuting could be energy efficient if using a bus or train. • *They also make...* This is correct and scores 1 mark. • *So this is why...* This is correct and scores 1 mark but could be linked to eco-footprints.	In cities people use a lot of resources as they do not grow their own food. The heavy levels of car use contribute to the very high use of energy, using fossil fuels. They also make a lot of waste and pollution. The high inputs and outputs mean that there is a large eco-footprint.

Overall comment: The student could have improved their answer by being more precise. Make sure that you develop your answers to score more marks.

Higher Question: Explain why cities such as London have a large eco-footprint. (4 marks)

Student answer ● (achieving 2 marks)	Examiner comments	Build a better answer △ (achieving 3 marks)
London is an unsustainable city so it has a very large eco-footprint. It uses vast quantities of food, energy, water and building materials. The 8 million people in London have very low levels of recycling (around 9%). In places like Freiburg and Copenhagen they recycle up to 55%. So they are more sustainable.	• *London is an...* This is an introductory sentence that does not contain enough detail to score any marks. • *It uses vast...* This is a valid point on inputs with a good range of examples, scoring 1 mark. • *The 8 million...* This scores 1 mark because it is correct and includes an example. • *In places like...* This part of the answer is not relevant as it is explaining why some cities have a low eco-footprint.	Large eco-footprints occur when cities use huge amounts of resources, recycle almost nothing and create waste and pollution. It uses vast quantities of food, energy, water and building materials. The 8 million people in London have very low levels of recycling (around 9%). Low levels of recycling mean that outputs of waste going to landfills are huge and, combined with the pollution from cars and factories, leads to very high outputs.

Overall comment: The answer is unbalanced, straying from the point at the end. However, the correct vocabulary is used, which shows that the student understands the subject.

Chapter 14 Changing countryside

Objectives

- Learn the main changes taking place in rural areas.

- Recognise that the rural changes in developing countries are very different from those in developed countries.

- Understand that the reasons for the changes often lie outside rural areas themselves.

Activity 1

What would you suggest as good indicators of rural poverty?

ResultsPlus
Build Better Answers

Using named examples, explain the challenges faced by rural areas in the developing world. (6 marks)

■ **Basic answers** (Level 1)
Mention the problems caused by population growth in general terms, with very few specific examples.

● **Good answers** (Level 2)
Include some detail of location and a good description of the challenges posed, although there is less linkage with the reasons for why these are challenges.

▲ **Excellent answers** (Level 3)
Clearly identify and explain challenges and links to climate change and changes in cropping.

What are the issues facing rural areas?

A rural area is usually defined as one that is relatively sparsely populated and either largely given over to farming or left as wilderness. Rural areas are often described as 'countryside'. However, in many developed countries, the traditional focus on farming is becoming less, as other activities move into the countryside. At the same time, the distinction between rural and urban is becoming less clear-cut. In short, the countryside of the developed world today is quite different from what it was, say, 200 years ago. It also contrasts with the countryside typical of much of today's developing world, which is also changing. These contrasts are reflected in the rural issues that confront both types of countryside (Figure 1).

Figure 1: Rural areas in the developing world and the developed world

The challenges faced by rural areas in the developing world

Over most of the developing world, rural areas are important for three main reasons:

● They are lived in by the majority of the population – because levels of urbanisation are still quite low. In many developing countries, less than a quarter of the population lives in towns and cities.

● They provide the food necessary for the survival of both the rural inhabitants and many of the urban dwellers.

● They are an important part of the traditions and heritage of many different tribes.

Figure 2: The spiral of decline in rural areas of the developing world

The rural areas of the developing world now find that they are being increasingly changed and challenged by a number of physical and human processes:

Environmental degradation	This results from desertification and deforestation. Clearance of vegetation, overuse of the soil, soil erosion and an increasing scarcity of water are causing many rural areas to dry out. And, of course, global warming is making these problems worse. As all this happens, so the amount of food that can be grown steadily declines.
Population change	High rates of natural increase are increasing the pressure on dwindling amounts of food. The outcome is malnutrition and even starvation. It is partly this pressure that encourages increasing numbers of people to migrate from the countryside into towns and cities.
Urbanisation	As towns and cities grow, so more people are leaving the countryside. Rural people are attracted by urban jobs and wages that will allow them to buy their food. They are also attracted by better education and healthcare. This rural–urban migration is resulting in a mainly elderly and female population being left behind to do the farming and produce the food.
Human hazards	Epidemics of diseases (such as cholera, malaria, TB and HIV/AIDS) together with wars have two major impacts on rural areas in developing countries. Raised death rates lower the population which, in turn, means fewer people to attend to the important task of food production.
Globalisation	Some rural areas are discovering that they contain valuable resources that are of interest to the global economy. These resources might be minerals. Equally, they might be qualities, such as wildlife and spectacular scenery, that attract foreign tourists. Most recently, areas of fertile soil and tropical climate are being exploited by agribusinesses growing exotic or out-of-season crops for the shelves of the developed world's supermarkets. Land that should be growing food for local people is now feeding people thousands of miles away.

Rural poverty in Kenya

Over the past thirty years, poverty has been on the rise in Kenya. More than half the country's 31 million people are poor. About 75% of the poor population live in rural areas. Ironically, most of the rural poor live in areas that have a high agricultural potential. Population densities in such areas are more than six times the national average of 55 persons per km².

The trebling of Kenya's population over the past thirty years is one of the basic causes of the rising **rural poverty**. Other causes of the spiral of decline (Figure 2) include:

Figure 3: Drought can turn farmland into useless soil and sand

- Low agricultural productivity, made worse by land degradation, insecure land tenure and poor farming methods (Figure 3).

- Shortage of work other than subsistence farming.

- Difficulty in finding money to finance farm improvements.

- HIV/AIDS, which is particularly prevalent among the population of working age.

- Rural–urban migration, mainly of young males, leaving women behind and responsible for growing crops and herding livestock.

- Some of the best remaining farmland has been taken over by commercial farming and is now producing food for export and for the foreign tourists drawn here by Kenya's wildlife.

Quick notes (Rural poverty in Kenya):

- Rural–urban migration and HIV/AIDS deaths have done little or nothing to reduce the pressure of population on Kenya's agricultural land. Rural poverty and food insecurity continue to rise.

- The challenges facing rural areas in the developing world appear to be due in part to the external processes of urbanisation and globalisation.

- But there are also significant internal factors at work, such as overpopulation, environmental degradation and human hazards (war and disease).

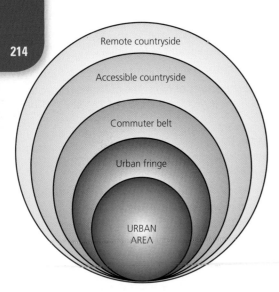

Figure 4: The four types of countryside outside the
urban area

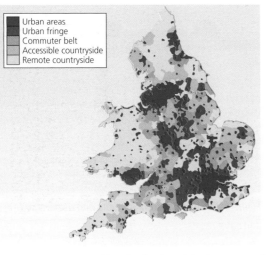

Figure 5: The distribution of the four types of
countryside outside the urban area in England
and Wales

Skills Builder 1

Study Figure 5 and refer to an atlas.

(a) Where are the main areas of remote
countryside?

(b) What are the physical features of
these remote areas?

The challenges faced by rural areas in the developed world

The countryside challenges in the developed world are rather different. As in the developing world, the countryside is also changing, but mainly due to urban and demographic processes. These challenges faced in the developed world will be the focus for the remainder of this section.

Change is everywhere and that includes the UK's countryside. 'Change' is a big umbrella under which many things happen. The nature of countryside change varies from place to place. The most important factor is distance from a major city. According to this, it is possible to recognise four different types of countryside outside the urban area, as shown in Figure 4:

● The **urban fringe** – This type of countryside is being quickly lost to urban growth, particularly where there are no planning controls such as green belts.

● The **commuter belt** – This is countryside, but the settlements within it are used as **dormitory towns** by urban-based workers and their families.

● The **accessible countryside** – This is beyond the commuter belt, but within day-trip reach. Still very much a rural area.

● The **remote countryside** – This takes the best part of a day to reach from a city. Almost totally rural.

Figure 5 shows the distribution of these four types of countryside in England and Wales.

These four types of countryside are being affected by a number of processes of change. As shown in Figure 6, some of the processes span from the city to the remote countryside, whilst others involve just two or three of the countryside types. Remember that these processes are working in two directions – a few towards the city, the majority outwards from the city.

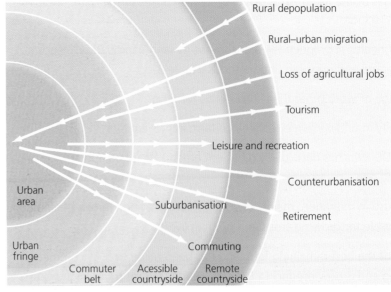

Figure 6: The processes of change

Decline in agricultural employment

The most important activity in the countryside was, and still is, agriculture. But Figure 7 shows that it has declined as an employer. Mechanisation has cut the number of jobs. Although the amount of land being farmed has decreased, what it produces has increased. But the growth of the global economy is resulting in cheaper food being imported from overseas. The UK is now only grows around 60% of its food. The cutback in farming jobs has been marked in the remote countryside. Remoteness means higher transport costs. Slightly higher costs can easily tip the balance of whether or not a farm remains profitable and stays in business. Farming has also declined in the commuter belt and the urban fringe, as more and more of the countryside is converted from fields into building plots.

Skills Builder 2

Study Figure 7.

(a) How many people were employed in agriculture in 1978 and in 2005?

(b) What was the percentage change in agricultural employment between 1978 and 2005?

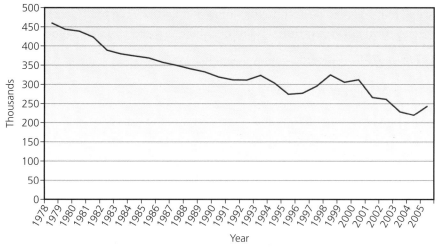

Figure 7: Employment in agriculture in the UK, 1978–2005

Farm diversification

Farming today can no longer support the large number of families it once did. Many farmers are finding that they can scarcely make a profit from their traditional food production alone. Supermarkets are now paying UK farmers very low prices, and even cheaper food is being imported from overseas. So, if they want to stay where they are, hard-pressed farmers have no choice but to diversify – by doing one of two things:

● Find other ways of making money out of the farm – while continuing to farm.

● Turn their farms into completely different businesses.

The first option is by far the more common form of **diversification** (Figure 8). With the second option – changing to a different business – a move into leisure, recreation and tourism is a common choice. In other instances, the land has been sold off and farm buildings turned to micro-businesses or cottage industries, such as making greetings cards, knitwear and beauty preparations. With the extension of broadband services into the remote countryside, some farmhouses have become telecentres or premises for **telecottaging**. Others have been renovated and turned into second homes. The table on page 216 gives some idea of the different ways of diversifying a farm.

Figure 8: Examples of farm diversification – pick your own (PYO) and converting barns into housing

New products	New outlets	Tourism	Leisure & recreation	Development	Energy
Organic crops	PYO	B&B	Shooting	Converting barns into housing	Wind turbines
Herbs, cheese, bottled water	Farm shop	Caravan or camping site	Off-road driving	Industrial units	
Different animals, e.g. bees, goats, ducks, ostriches	Farmers' market	Café or restaurant	Mountain biking	Telecentres	

Quick notes (farm diversification): Many farms in the UK are surviving only by taking on new activities. This is necessary because of low product prices and competition from imported food.

Activity 2

Investigate a farm near you. Has it diversified? If so, in what ways?

Decision-making skills

Work in pairs to provide a recommended location for the following types of farm diversification:

- PYO
- A mountain-biking centre
- Local leaders
- A wind farm (several turbines)
- A riding stables

Think about physical and human factors.

Changes in population, employment and services

For well over a hundred years, with the number of jobs in the countryside falling, many people have chosen to leave and find work in towns and cities causing **rural depopulation**. This rural–urban migration was encouraged by:

- Higher wages and more job opportunities in urban areas

- The availability of more and better services

- The perception that towns and cities offer a better quality of life.

But rural–urban migration has now largely ceased. It is only the remoter countryside that continues to lose population, particularly young people of working age. The problem is that when such people and their families leave, the demand for local services (shops and schools) falls – and they close. The loss of services makes the area even less attractive. More people choose to leave. As Figure 9 shows, there is a chain reaction which leads the rural area into even faster decline. Fortunately, the downward spiral is being checked by the process of **counterurbanisation**. The loss of population from many remote rural areas is now being compensated by an inward movement of newcomers. Thanks to the broadband revolution, people disenchanted with town or city living and enchanted by the 'rural idyll' are able to telework from home. However, while the population is increasing in all rural areas, except the most remote, the services continue to decline.

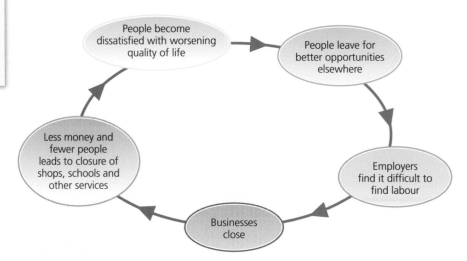

Figure 9: The spiral of decline in rural areas

Village shop closures

A survey in 2000 found that 3,000 British villages, with a combined population of 4.5 million, had no shop and no Post Office. One-third of British villages were dependent on neighbouring towns for even basic goods. At that time, village post offices were closing at the rate of 400 a year. The Post Office announced in 2008 the closure of a further 2,500 of their branches – many of them in rural areas. About half of these double as local shops.

It is now extremely difficult for more than one specialist shop, such as a butcher or baker, to survive. Shops must sell a wide range of goods if they are to compete with the supermarkets and larger chains of stores in neighbouring towns.

The closure of village shops is an important factor in persuading village residents to move home. This is particularly true for those – particularly the elderly and not so well off – who do not own a car and have to rely on poor public transport. In a number of villages, residents have stepped in and taken over the threatened Post Office (Figure 10). In most cases, they have received grants from the Department for the Environment, Food and Rural Affairs (DEFRA).

Figure 10: A campaign against village post office closures

> Quick notes (village shop closures):
> - The closure of village shops and Post Offices is happening at a fast rate.
> - The closures help persuade people to move out of villages altogether.

ResultsPlus
Watch out!

■ It is widely thought that most retirement moves are from cities and towns to the countryside. In fact, most end up elsewhere (see Figure 11).

Also helping to stem the outward flow of people from the remote countryside have been changes in two other aspects of life – retirement and leisure.

Retirement migration

UK residents are living longer. The average life expectancy for women is now 81, and for men 76. Most people can expect to enjoy ten or more years of retirement. With this prospect, more and more people are moving once they have retired. They are doing this for a number of reasons:

- It is no longer necessary for them to live close to what was their place of work

- To downsize into a smaller home

- To sell their home for something cheaper and use the difference in price as some sort of pension

- To move into a quieter, calmer and more attractive environment.

The dream for many is to retire to a small cottage in the countryside or to settle in some picturesque village. In reality, that often does not work out. Figure 11 shows three impressive retirement flows in England and Wales. All three are out of Greater London – to the Outer South East, to the South West and to East Anglia.

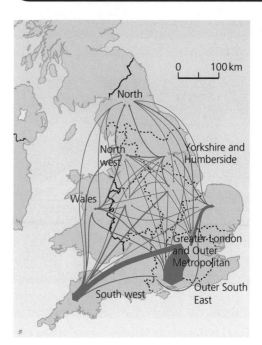

Figure 11: Retirement migration flows in England and Wales

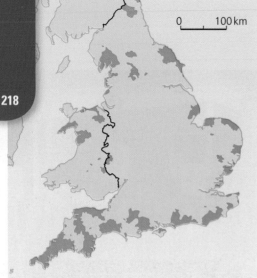

Retirement areas (in orange) are those with a significantly higher than average proportion of people of retirement age and which are experiencing a growth in the proportion of elderly people.

Figure 12: Popular retirement areas in England and Wales

Figure 12 shows that many make for coastal areas where the attractions are milder winters, pleasant scenery and a slower pace of life. Coastal resorts such as Great Yarmouth (on the East coast), Worthing (on the South coast) and Lytham St Anne's (on the West coast) are also popular for another reason. They are already well geared up to meet the needs of a 'greying' population. They are sometimes referred to as the 'grey resorts'.

Recreation, leisure and tourism

One of the features of twenty-first-century living is that many people in the UK have both spare time and disposable income – and both are being 'spent' on leisure, recreation and tourism. Leisure and recreational needs are being met in the urban fringe, commuter belt and accessible countryside. In the urban fringe and commuter belt, you will find sports centres, playing fields, golf courses and country parks (Figure 13). In the accessible countryside, the facilities may be somewhat different. In order to reach them, city people will have to think in terms of a day-trip rather than just a short drive in the car. Maybe it will be a farm visit, a fun day at a theme park, some birdwatching around a nature reserve or pony trekking.

Tourism is a boom business nowadays, thanks to shorter working hours, holidays with pay and modern transport. This is where the remote countryside comes into its own.

Figure 13: Recreation in the urban fringe, and in the accessible countryside

Figure 14: The annual May Day fair at Finchingfield – a honeypot village

Honeypot villages

The growth of tourism in the countryside is not evenly spread. It tends to be focused on what are called **honeypots** – places which offer something that attracts large numbers of tourists. It might be because of:

- The immediate surroundings – e.g. Castleton in Derbyshire, with its limestone caves.

- The picturesqueness of the place – e.g. Finchingfield in Essex, often described as a **'chocolate-box' village** (see Figure 14).

- Its historic associations – e.g. Tintagel in Cornwall, with its legendary King Arthur's castle.

- Its use as the setting for some TV series – e.g. Holmfirth in West Yorkshire, the setting for the long-running *Last of the Summer Wine* series.

Becoming a honeypot village is fine for those making a living out of tourism, but there is a downside – constant crowds and traffic congestion. It is not so much fun if you are a resident and place a value on peace and quiet. It is not much fun either if all the food shops have become tea-rooms and souvenir shops.

Suburbanisation of the countryside

The obvious outcome of the rural–urban movement of people is the building of new homes. They are most likely to be built in three locations: the urban fringe (provided there is no restricting green belt), the commuter belt and in parts of the accessible countryside. In the first, housing is mainly added to the outer edge of the city's built-up area in the form of suburbs. In the commuter belt, existing towns and villages become encircled by new housing estates. The same pattern occurs, on a smaller scale, in parts of the accessible countryside. But the occupiers of all these new homes will need more than just a roof over their heads. They will require services, such as shops, schools and medical centres. So the building of new homes leads to a further round of building to provide these and other services.

In the urban fringe, the demand for both housing and services is increased by people moving out from the city. As Figure 15 shows, the urban fringe offers a number of benefits that attract people, businesses, shopping centres and recreational uses.

Around London and other major cities, suburbanisation in the urban fringe has been discouraged by the creation of green belts. Within this belt, planners have tried to keep suburbanisation to a minimum. Protection of the countryside has been the priority. Instead, population movements have been encouraged to new towns and expanded towns located out in the commuter belt.

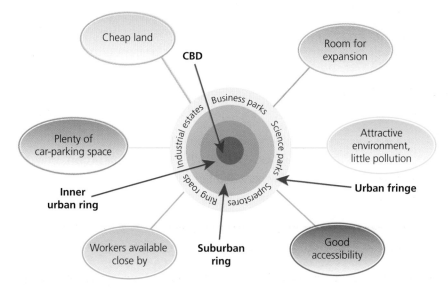

Figure 15: The benefits of the urban fringe

Skills Builder 3

Calculate the straight-line distances from London to the furthest commuting points shown in Figure 16.

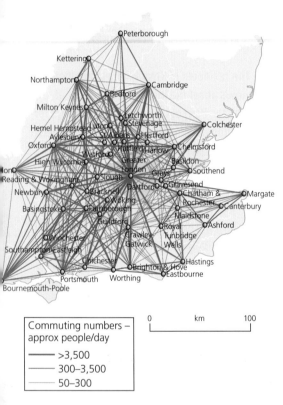

Figure 16: Commuting in Southeast England

The spread of commuting

For a long time now, many people have been moving from the countryside into urban areas. But now there are also movements in the opposite direction. People are leaving the cities and towns and buying homes in the countryside. But they continue to work in the city or town they have just left – they become commuters. This is happening in the urban fringe and in the commuter belt. The reasons for this outward movement include:

● The attraction of cheaper and more spacious housing.

● The availability of fast transport to work. Most jobs are either in the city centre or in new industrial estates and business parks in the urban fringe.

● Fears about personal security in large urban areas.

● The perception that the urban fringe and commuter belt offer a better quality of life.

In short, people feel that the time and money spent on commuting are worth it. Figure 16 shows the complex pattern of commuting in Southeast England. From it, you can gain a clear impression of the extent of London's commuter belt.

The challenges for the four types of countryside

The following may be identified as the main challenges resulting from the processes discussed so far and their impacts on the countryside:

Urban fringe	Preventing the complete disappearance of real countryside, yet providing space for new suburbs as well as shopping complexes, industrial estates and business parks.
Commuter belt	Maintaining good transport links. Ensuring that villages retain some of their original character rather than become one-class dormitories.
Accessible countryside	Farm diversification. Coping with the loss of young adults and the growing numbers of newcomers, particularly retired people and those seeking the rural idyll.
Remote countryside	Halting depopulation and the decline in services. Coping with the recreation and tourism boom. Finding new means of livelihood.

Most of these changes and challenges are to do with urbanisation, counterurbanisation and population change rather than **globalisation**. Only the decline of agricultural employment and the appearance of industrial estates and business parks, mainly in the urban fringe, can be linked to globalisation.

How might the issues facing rural areas be resolved sustainably?

Boosting the rural economy in the developing world

In developing countries, improving the rural economy requires action to be taken in two main directions:

- Raising food production – but in a way that does not cause more **environmental degradation**. In other words, farming needs to be made both more productive and more sustainable – food security needs to be improved.

- Reducing rural–urban migration – by making rural areas more attractive places in which to live and work. This will involve a whole range of initiatives, from providing healthcare, education and new types of work to improving housing conditions and the rural infrastructure.

Crops and a shop in Ethiopia

Worke lives in the Afar region of Ethiopia (Figure 17), and she has five children. Ethiopia is one of the poorest and least developed countries in the world, ranking 169th out of the 177 countries. The Afar region is hot and arid. It is a difficult place to live in at the best of times, but when drought hits, it can be a killer. Worke used to live in fear of drought (Figure 18). She knew that if her animals died and her crops failed, food aid would be the only option for her and her children. So when FARM-Africa, a non-governmental organisation, came to her village several years ago, she jumped at the chance to become involved – and she formed an irrigation group with her neighbours. FARM-Africa works at a grassroots level with rural communities in eastern Africa, providing simple but long-term solutions to poverty. According to FARM-Africa, it is the millions of smallholder farmers of sub-Saharan Africa who hold the key to reversing the downward spiral of poverty and starvation now gripping large parts of the continent.

With FARM-Africa's help, Worke and her neighbours built a canal to channel water from the local river to the fields and they installed two pumps to draw river water into the canal. She also received a starter kit of seeds and went on to harvest 4,000 kg of maize, worth almost £400. She now grows onions, tomatoes and peppers, which she sells locally or takes to market. But Worke's achievements have not stopped there. She has taken out a £130 loan from FARM-Africa and opened a small shop selling everyday items like soap, salt and razor blades. It has been so successful that she has already repaid the loan and its interest.

Thanks to the income from her shop and vegetables, four of Worke's children now attend high school and her eldest son has completed school and become a journalist in a local town. Best of all, Worke finally feels secure, particularly in terms of food. She knows that if the droughts strike again, her family will not have to rely on hand-outs. She stands proudly in her shop window: 'Before, we staggered along on one foot. Since FARM-Africa, we can stand on our own two feet'.

Objectives

- Learn why rural areas in developed countries have been declining.

- Recognise that there are different ways of protecting and conserving rural areas.

- Understand that rural areas need to be managed in a more sustainable way.

Figure 17: The Afar region of Ethiopia

Figure 18: The hot and arid landscape of the Afar region

Activity 7

Which of the designations in the table can be found near to your school?

The table below shows a range of protected and managed areas in England. (Some of these designations – AONBs, ESAs and NNRs, for example – are not made in other parts of the UK.) The areas that are designated are thought to be precious because of one or more of the following factors – their landscape and scenery, their wildlife, their scientific interest or their cultural heritage. Each designation involves a slightly different way of managing the area. The table omits a huge number of small nature or wildlife reserves. Some of these are owned and run by county councils and local authorities. Others belong to county and local wildlife trusts. They are mainly to protect wildlife and natural habitats.

Types of protected areas in England

Designation	Number	Purpose	Examples
World Heritage Site	24	To conserve natural and cultural sites of outstanding importance in terms of global heritage.	Jurassic coast Hadrian's Wall Bath Stonehenge
National Park	9 (8% of England)	To protect areas of landscape beauty, their wildlife and places of interest. To encourage open-air enjoyment of them.	Lake District Norfolk Broads New Forest
National Nature Reserve	222 (<1% of England)	To protect the most important areas of wildlife habitat and geological formations as places for scientific research. Places where wildlife comes first.	Dungeness, Kent The Lizard, Cornwall Malham Tarn, N. Yorks
Country Park	270+	To provide the perfect setting for people of all ages to experience nature, heritage and the great outdoors.	Box Hill, Surrey Lickey Hills, Birmingham Rutland Water
Area of Outstanding Natural Beauty (AONB)	36 (15% of England)	To protect areas that are outstanding in terms of their flora, fauna, historical and cultural associations as well as scenic views.	Cotswolds Lincolnshire Wolds Malvern Hills
Environmentally Sensitive Area (ESA) now replaced by Environmental Stewardship Scheme (ESS)	22 (10% of agricultural land in England)	To work in cooperation with farmers to enhance the conservation, landscape and historical value of the key environmental features of an area. Where possible, to improve public access.	Brecklands, East Anglia North Kent marshes Somerset Levels
Site of Special Scientific Interest (SSSI)	4,000+ (7% of England)	To preserve the country's very best wildlife and geological sites.	Bamburgh Dunes, Northumberland Epping Forest, Essex Ribble estuary, Lancs.
Local nature reserve	1,280	To ensure the conservation of nature and/or the maintenance of special opportunities for study, research or enjoyment of nature.	Little Paxton pits, Cambridge Kirtlington Quarry, Oxon.

Since the middle of the last century, Britain's national parks have been the flagships of countryside conservation. Here we will focus on one of them – the New Forest National Park. Each of the fifteen national parks is unique in terms of its heritage and landscape. But, despite this, all of them are faced with the same general pressures and conflicts.

New Forest National Park

This is one of the most recent parts of Britain to be made a national park. It was designated in 2005 and covers an area of 570 km² to the west of Southampton and immediately north of the Solent (Figure 22). It was once a royal hunting ground. Today it is a mix of heaths, bogs and ancient woodlands. It is renowned for its wildlife, including rare and endangered species, as well as for its ponies, cattle and pigs – which roam freely through much of the Forest. (These are not wild – they are owned by local farmers known as 'Commoners'.)

Like all national parks, the New Forest is managed by a National Park Authority (NPA). The main pressures and conflicts that it has to deal with are:

Pressures

- The Forest is accessible to millions of people living in London, Southeast England and the Midlands. Too many visitors mean traffic congestion, over-full car parks, disturbance of wildlife, footpath erosion, etc.

- Local people need to make a living, and that is not easy when there are all sorts of restrictions on what can and cannot be done inside a national park.

Conflicts

- Between the Forest's function as a recreational area for the enjoyment of people and the need to protect its wildlife and habitats.

- Between incomers and local people. Housing is being bought by well-paid London-bound commuters and by wealthy second-homers. It is very difficult for local young people to find affordable housing.

The task of the NPA is not an easy one. It needs to steer a middle course between rival claims and different interest groups. It is very easy to upset everyone. Successful management lies in the 'three Cs' – consultation (with the public, special interest groups, local government), compromise (getting people to give ground) and control (ensuring that the main aims are achieved). The more the conflicts are resolved, the greater the likelihood that the New Forest (its precious habitats, its people and its settlements) can look to a more sustainable future.

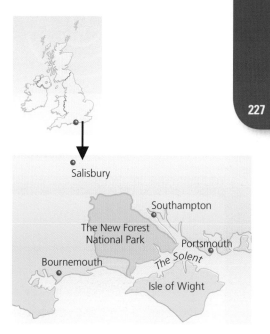

Figure 22: The location of the New Forest

Figure 23: A New Forest scene

Activity 8

Discuss the bullet points in the New Forest National Park section. Explain why there are these pressures and conflicts.

Decision-making skills

Draw a conflict matrix to show how the views of various people differ over how the New Forest National Park should be managed.

227

Know Zone
Changing countryside

The countryside is changing. In developing countries, it can be an area of poverty as many people migrate to the towns. In developed countries, although the population is growing in the countryside, it is no longer a place just for farming, but also for recreation and tourism.

You should know...

- ☐ What the characteristics of a rural area are
- ☐ How environmental degradation is affecting many rural areas in developing countries
- ☐ How many rural areas are changing in developing countries as a result of outside influences such as globalisation
- ☐ How there is extreme poverty and hardship in rural areas such as in Kenya
- ☐ How the different types of rural areas in the UK – both accessible and remote – are experiencing many challenges
- ☐ What changes are taking place on farms, with many examples of diversification
- ☐ How, in general, the countryside population is growing from counterurbanisation, but services are declining
- ☐ How recreation and tourism are now a major source of income for the countryside
- ☐ How accessible rural areas are becoming suburbanised as they are 'taken over' by commuters
- ☐ How sustainable grassroots schemes run by NGOs are improving the life of rural people in developing countries
- ☐ How the countryside in developed countries is being managed sustainably for a variety of uses

Key terms

Accessible countryside	Grassroots scheme
Chocolate box village	Honeypot
Commuter belt	Remote countryside
Counterurbanisation	Rural depopulation
Diversification	Rural idyll
Dormitory town	Rural poverty
Environmental degradation	SSSI
	Telecottaging
Globalisation	Urban fringe

Which key words match the following definitions?

A A small area that has been officially designated for protection because of its wildlife or geology

B The process, led by transnational companies, whereby the world's countries are all becoming part of one vast global economy

C A place of special interest or appeal that attracts large numbers of visitors and tends to become overcrowded at peak times

D The movement of people and employment from major cities to smaller settlements and rural areas located just beyond the city, or to more distant smaller cities and towns

E Working from a home in the country, using computer communication

F Countryside within easy reach of urban areas

G The decline of population in rural areas and regions

To check your answers, look at the glossary on page 313

Foundation Question: Explain, with an example, why some National Parks are under pressure from a range of activities. (4 marks)

Student answer ⚠️ (achieving 3 marks)	Examiner comments	Build a better answer ⚠️ (achieving 4 marks)
Parks such as the Lake District and Peak District are under huge pressure for recreation, with many visitors.	• *Parks such as...* scores 1 mark. The student uses named examples and begins to explain about pressures.	Parks such as the Lake District and Peak District are under huge pressure for recreation, with many visitors.
They are very easily reached by motorways and are near large urban areas.	• *They are very easily...* also scores 1 mark. This is a developed point that adds to the explanation given above.	They are very easily reached by motorways and are near large urban areas.
These visitors carry out damaging actions such as footpath erosion.	• *These visitors carry...* This is another valid point and gains 1 mark.	These visitors carry out damaging actions such as footpath erosion.
These visitors all go to honeypot sites.	• *These visitors all...* Although this part of the answer is not incorrect, it does not answer the question or offer an explanation.	Another pressure comes from quarrying, which is very dusty and can damage the National Park.

Overall comment: This is a very good answer that correctly identifies pressures, the lack of detail about a range of activities means that it does not score full marks.

- -

Higher Question: Explain why the aims of National Parks conflict. (4 marks)

Student answer ⚠️ (achieving 3 marks)	Examiner comments	Build a better answer ⚠️ (achieving 4 marks)
National Parks are owned by the State to conserve the landscape.	• *National Parks are...* This statement is only correct for countries such as the US.	National Parks were set up to conserve the landscape and the wildlife.
They are victims of their own success, with millions visiting every year.	• *They are victims...* This part of the answer is a development of the next point and scores 1 mark.	They are victims of their own success, with millions visiting every year.
This means that they are fulfilling their other aim of allowing public access to the scenery.	• *This means that...* scores 1 mark as it is a valid point.	This means that they are fulfilling their other aim of allowing public access to the scenery.
The public come to the Lake District National Park largely by car as they use the M6 motorway.	• *The public come...* This part of the answer is irrelevant and does not focus on the conflicting aims of National Parks.	Allowing public access almost always conflicts with the aim of conserving the landscape, as the public can erode footpaths and drop litter.

Overall comment: The answer loses focus towards the end. Make sure that you respond to the question being asked.

Chapter 15 Development dilemmas

Objectives

- Recognise that countries have core and peripheral areas, and that there are different approaches to development.

- Explain how economic development can lead to urban cores and rural peripheries, and understand the problems that result from a disparity within a country.

- Explain the impacts of a top-down development scheme, and show that top-down schemes have advantages and disadvantages.

How and why do countries develop in different ways?

Regional disparity

Development within a country does not take place evenly across all the regions at the same time. It occurs at different speeds, and so a spatial difference in the economic pattern emerges. This creates a **disparity** – a great difference – between a **core region** and a **periphery**. The rich core is based on the urban areas, which have the majority of the people, services, businesses, industry and the government headquarters. The periphery is rural – often remote countryside – and involved in the production of raw materials, and so is poor.

Disparities in Brazil

In Brazil, the south-east of the country is the core (Figure 1), based around the cities of São Paulo, Rio de Janeiro and Belo Horizonte. This region has a hospitable temperate climate, access to ports for trading (e.g. Santos), road and rail networks, and fertile soils for farming (e.g. São Paulo state). Over time these advantages have attracted people (rural to urban migration) and investment in business. So the south-east core has grown. The periphery of Brazil can be found in the north, and in the north-east. Acre and Roraima are regions a long way (3,000 km) from the core, with a wet tropical climate, poor access by land due to the extensive tropical rainforest, sparse population and limited access to the sea for trading.

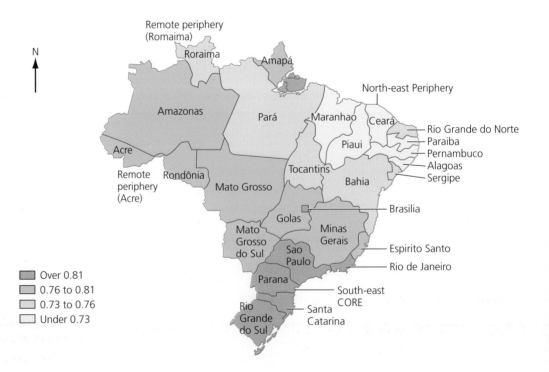

Figure 1: Regional disparities in Brazil, according to the Human Development Index

Maranhão and Piauí are sparsely populated, lack resources, are found in a semi-arid climate zone with frequent droughts, and situated a long way (2,200 km) from the core. The table below, which shows the GDP per capita data for different regions, confirms the disparity – the rich/poor divide – that exists in Brazil.

The large disparity within their country encouraged the Brazilian government to build a new capital city – Brasília – inland in 1960 to help spread wealth and economic development. Figure 1 and the table below suggest that this had some success. Several large schemes were also implemented in the periphery, such as the Carajás Project (based on the world's largest iron ore deposits), and several **Hydro-electric power** (**HEP**) dams (such as Santo Antonio and Jirau on the Madeira River in Rondônia). Schemes like these partly support the core region (e.g. power supplies), but they can also spread the benefits of the core to the periphery.

Regions of Brazil: GDP per capita in R$
(Brazilian reals), 2005 (Source: IBGE, 2007)

Region	GDP per capita
Piauí (NE periphery)	3,700
Maranhão (NE periphery)	4,150
Acre (N periphery)	6,792
Roraima (N periphery)	8,123
Rio de Janeiro (SE core)	16,052
São Paulo (SE core)	17,977
Brasília Federal District (capital)	37,510

Governments all over the world usually wish to spread development throughout their countries because:

● People in the periphery could be left with a lower quality of life than those in the core.

● Conflicts could arise between the people of different regions (e.g. between rich and poor, or between local people and development companies).

● Lots of poorer people from the periphery could move into the core region.

● Overcrowding and lack of jobs could arise in the core.

● Resources in the periphery could be left undeveloped so slowing the development of the whole country.

● If the periphery remains undeveloped, the government receives less tax money to help improve the whole country.

ResultsPlus
Build Better Answers

Study Figure 1. Describe the regional differences that have resulted from economic development. (4 marks)

■ **Basic answers** (0–1 marks)
Fail to include geographical terms such as 'south-east', or the names of states such as 'Roraima' consistently throughout their answer. Statements such as 'the core at bottom of map and periphery at top right' are used.

● **Good answers** (2 marks)
Identify the south-east core based around São Paulo and the north-east periphery from Alagoas to Maranhão.

▲ **Excellent answers** (3–4 marks)
Also mention the spread of development from the core towards Amazonas via Brasilia.

Activity 1

1. Define the terms 'core' and 'periphery'.

2. Why do urban areas form the core and rural areas the periphery?

3. Identify and discuss three problems caused by a disparity within a country.

4. What are the benefits of involving local people and communities in the decision-making process?

Figure 2: Tanzania tax districts

Skills Builder 1

Study Figure 2 and the table below.

(a) Choose one of the columns of data from the table. Consider the range of values and put the data into four categories (e.g. 0–1,500 or 0–0.5). One way to do this accurately is to plot the data on a single-axis scatter graph or another way is to work out the median and quartiles.

(b) Produce a choropleth map on a copy of Figure 2. Your colour or shading scheme should have a logical sequence (e.g. purple, red, orange, yellow). Figure 1 is an example of a finished choropleth map.

(c) Describe the pattern shown by your map.

Tax income from the tax districts of Tanzania, 2005–06

Tax district	Tax revenue per person (Tanzanian Shilingi)	% contribution to total tax revenue
Arusha	49,816	3.16
Dar Es Salaam	678,928	83.19
Dodoma	2,927	0.24
Iringa	4,993	0.37
Kagera	3,261	0.32
Kigoma	1,417	0.12
Kilimanjaro	36,823	2.49
Lindi	909	0.04
Mara	25,217	1.69
Mbeya	13,553	1.37
Morogoro	17,177	1.48
Mtwara	3,236	0.18
Mwanza	13,842	1.99
Pwani	2,897	0.16
Rukwa	1,276	0.07
Ruvuma	1,520	0.08
Tanga	32,092	2.58
Shinyanga	1,740	0.24
Singida	874	0.05
Tabora	2,093	0.18

Source: Tanzania Revenue Authority

Types of development

A country can try different ways to prompt development. Some use large-scale technology, often advanced, and these schemes are usually expensive. In the past, countries wishing to develop often borrowed money from the World Bank, or involved companies from more developed countries to help them start up a big scheme. The decisions about these big schemes were made by the national government and, where necessary, new regulations or laws were passed to enable these schemes to go ahead. Local people – who lived near to where the scheme was to take place were often not involved in the process and had great difficulty influencing the decisions being made. Such an approach is known as '**top-down**' (Figure 3).

Activity 2

Study Figure 3.

(a) Describe the roles of different groups of people in the top-down decision-making process.

(b) How could debt for the country be a result of this process?

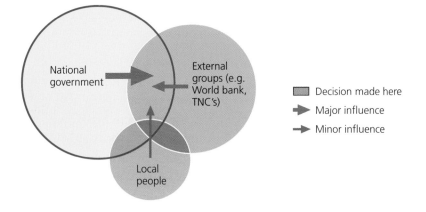

National government

External groups (e.g. World bank, TNC's)

Local people

▢ Decision made here
➡ Major influence
➡ Minor influence

Figure 3: Top-down decision-making

This approach was adopted by Brazil, which has made considerable progress since the 1960s. It is one of the 'BRIC' countries (Brazil, Russia, India and China) which have been recognised as having made significant economic improvements. The table below shows the variety of 'top-down' schemes, but what are their costs and benefits?

Activity 3

1. Compare Figures 3 and 8 (on page 239). What are the main differences between the decision-making processes for a top-down and a bottom-up approach?

Examples of large schemes around the world

Type of scheme	Example
Dam (hydro-electric power)	Bakun, Malaysia
Mining operation	Carajas Project, Brazil
Port facilities	Yantian Container Port, Pearl River Delta, China
Airport	Bangalore International Airport, India
Motorways	Mallam–Tetteh Quarshie road, Accra, Ghana
Power stations and transmission lines	Power Holding Company, Nigeria
New capital cities	Brasília, Brazil
Industrial areas	Songkhla Province, Thailand
High-tech communications	Multimedia Super Corridor, Malaysia
Tourism	Société Burkinabé de Promotion Hôtelière, Burkina Faso

Common problems with large schemes

- Getting the country into debt, because of the large amounts of money borrowed from the World Bank or International Monetary Fund to pay for the expensive schemes.

- Conditions may be attached to the loans, leading to some external control over the economy or other aspects of development (such as spending on health and education).

- Using machinery and technology, rather than providing unskilled jobs for local people.

- Being energy-intensive and expensive to operate after construction.

- Relying on external links and technology rather than internal links and **appropriate technology**.

- Becoming 'growth poles' which take resources away from peripheral areas.

Many people and organisations believe that development schemes that use intermediate technology are more appropriate and more beneficial to peripheral areas of a country. This is sometimes called a '**bottom-up**' approach to development, because local people are fully involved in the process and decision-making. Many contemporary aid programmes use such an approach, raising the level of technology for local people so that their lives get better.

Dams are common large schemes in developing countries, e.g. the High Aswan Dam on the Nile in Egypt, Akosombo in Ghana on the Volta river, or the Three Gorges Dam on the Yangtze in China. This is because as a country begins to develop economically, energy is needed for businesses, industry and housing. A developing country cannot afford to import expensive fossil fuels, and will use cheaper alternatives. Those countries with major rivers have the opportunity to build hydro-electric power (HEP) schemes, in which the force of the water turns turbines in a dam built across the river. As the turbines turn they generate electricity. This is a cheap source of energy because the water is free, so, although the initial cost of building a dam can be high, the electricity costs can be kept low for businesses and industries (so helping them to make greater profits and expand) and low for people (so helping them improve the quality of their lives).

Brazil has several major rivers, including the largest in the world, the Amazon. Several of the Amazon's tributaries, as well as other rivers, provide HEP opportunities. The Santo Antonio dam is the first top-down HEP scheme for fourteen years – the last one was the Xingo Dam on the São Francisco river, in 1994.

Activity 4

In which ways is water an important resource for a low-income country?

Decision-making skills

Working in groups, choose one large scheme and use Google to find out a bit more information on it. Then, draw a diagram to show the environmental impacts of your chosen scheme, e.g. impact on land, air and water.

Top-down development and bottom-up development

The priority of top-down schemes is to help the whole country, especially urban cores. But without careful management, the impacts on small areas of the rural periphery can be severe. An alternative approach is bottom-up schemes which use a smaller, more appropriate technology, and maximise benefits for the rural periphery. But such schemes do not consider the urban cores directly. This is the 'development dilemma' that faces developing countries – which type of development is better for them?

The Santo Antonio dam

The Santo Antonio dam is part of the Madeira River Project – the largest project in the Amazon region's history (Figure 4). It is one of the 'Integrated Regional Infrastructure for South America' projects and includes four dams, a navigation channel, three highways (e.g. BR364) and electricity transmission lines. The Madeira River basin covers over 1.5 million km² in three countries – Peru, Bolivia and Brazil. The main river is formed by the confluence of the Guaporé, Mamoré and Beni tributaries, creating an average discharge of 23,000 m³ per second. The Santo Antonio dam is 5 km upstream from Pôrto Velho, the capital of Rondônia. The dam is designed to produce 3,150 MW of electricity and is costing $5.3 billion to build. It is being constructed by Consorcio Madeira Energetica, which includes Brazilian, Spanish and Portuguese banks, the Brazilian construction company Odebrecht, and Furnas (Brazil's electricity company).

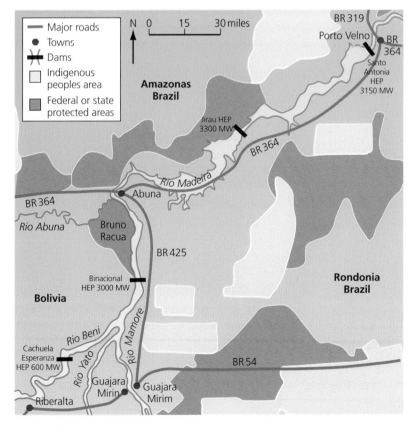

Figure 4: The Madeira River scheme

235

Activity 5

(a) What is an Environmental Impact Assessment (EIA)?

(b) Why is it important for an EIA to be completed before giving permission for a development scheme to start?

(c) How successful was the EIA for the Santo Antonio dam?

Skills Builder 2

Using Figure 4 and an atlas, describe the location and characteristics of the Madeira scheme.

Activity 6

Study the tables on pages 236 and 237

Explain how the Santo Antonio dam, a top-down scheme, has met the needs of Brazil, and assess how successful it has been.

Decision-making skills

Write a report/PowerPoint in which you investigate the environmental, social and economic impacts of the Madeira scheme. Do you think that the Brazilian government should have said yes? Why/why not?

Figure 5: Protests against building the Santo Antonio dam

ResultsPlus
Build Better Answers

For a named example of a top-down development strategy that you have studied, explain how successful it has been in meeting the needs of the developing country. (6 marks)

■ Basic answers (Level 1)
Do not include the name of the scheme and only make general points, such as 'created wealth and jobs in Brazil'.

● Good answers (Level 2)
Name an appropriate scheme and country, and make detailed points, such as specifying the number of jobs created, the amount of electricity created by a dam and how this was used by industries in a specified place.

▲ Excellent answers (Level 3)
Also provide a wider range of information.

Social advantages

- 20,000 jobs created and 100,000 people attracted to the region.
- Only 300 families moved and relocated to nearby villages.
- Social support programmes put in place to help rural communities.
- Health, education, leisure, safety and sanitation infrastructure improved.
- Two Indian reservations paid for by the consortium.
- $30 million given to improve the sewerage system in Pôrto Velho.
- Residents closely monitored for mercury poisoning.
- An education and training centre for immigrants and job seekers established.

Economic advantages

- The scheme will produce the cheapest electricity in Brazil.
- A waterway for barges will be created, by-passing rapids on the river, making it easier to transport soy, timber and minerals from the region. Other rapids also flooded by the small reservoir.
- Better infrastructure (roads and waterways) in the centre of South America (Brazil–Peru–Bolivia hub) provided, so products can be taken to the Pacific or the Atlantic.
- The expansion of agriculture in the area (e.g. soy crops to increase by 13%) will increase the agribusiness in the region (e.g. the Grupo Andre business), creating new jobs.
- An energy corridor created from Pôrto Velho to Araraquara in São Paulo state.
- Essential energy supplies for the development of Brazil and Rondônia created (e.g. the Madeira scheme will supply 8% of Brazil's electricity), especially supplying south-east Brazil (the area with the largest consumption).
- In 2007 a Brazil–Bolivia working group was established to create cooperation.

Natural environment advantages

- HEP is a renewable form of energy and avoids the need for oil or nuclear power.
- The flooded area will be relatively small (529 km^2) considering the power output.
- In the first three years of construction the river will keep its natural bed.
- The area flooded is minimised by employing 'run of the river' technology (using the natural flow rate of the river rather than the release of water from a reservoir).
- Fish channels in the dam created to allow fish to migrate.
- Monitoring of sediment and mercury levels in the reservoir will take place.
- Suggestions from the public meetings have been incorporated into the basic environmental planning.
- Two forest reserves will be paid for by the consortium, with $6 million made available for environmental police.

Social disadvantages

- The Brazilian government forced a yes decision on the dam, despite local and conservation opposition.
- The leadership of the Brazilian Institute of Environment and Renewable Natural Resources was sacked when it appeared that they might not support the building of the dam.
- Political conflict between Brazil and Bolivia has arisen because part of Bolivia will be flooded by the dams in Brazil. For example, a Bolivian wildlife reserve (Bruno Racua) will be affected.
- The public hearings took place over a very short time period.
- An estimated 3,000 people forced to leave their homes, and some towns disappeared (e.g. Mutumparaná).
- Indigenous peoples' lands are at risk from flooding and erosion.
- The riverbank traditional way of life may be lost (e.g. subsistence fishing) affecting 5,000 fishermen.
- More migrant workers may upset the delicate relationship with indigenous peoples (e.g. the Oro Bom), including contact with remote tribes in voluntary isolation (e.g. the Katawixi).
- Too many immigrants (up to 100,000) may enter the area, causing a strain on services and living space, especially in Pôrto Velho. Illegal land grabbing and selling is likely.
- Accumulation of mercury in the food chain may affect people when they eat fish (this is a result of gold mining, which still continues today).
- An increase in malaria is likely, because of the increase in water area.

Economic disadvantages

- The Madeira scheme is very costly ($22 billion).
- Commercial fishing (worth $1 billion) is at risk from the dam, because the preferred fish catch will be disturbed. 2,400 fishermen may lose their jobs.
- Investor banks may be in breach of the 'Equator Principles' (which say that only socially responsible and sound environmental management schemes should be supported).
- There is too much reliance on HEP in Brazil already (76% of electricity).
- High sediment loads will shorten the life span of the dam (to 100 years). 12% of sediments will be held back by the dam.
- High sediment loads may block turbines, reducing their effectiveness. Sediment loads have increased over time as deforestation has caused soil erosion.
- Irrigation water will be lost and fisheries affected downstream of the dam.
- Because the dam is a 'run of the river' design, occasional droughts may prevent it generating electricity.
- Climate change effects are as yet unknown, but the Andean glaciers are melting (30% reduction since the 1960s).

Natural environment disadvantages

- Soy expansion in the twenty-first century is a major cause of deforestation in Brazil, and the dam is opening up even more areas. 80,000 km² could end up being converted to farmland, in an area rich in biodiversity. There are, for example, 750 fish species and 800 bird species along the Madeira River.
- The **Environmental Impact Assessment** was based on insufficient data and so was superficial and lacked thoroughness. There were no measures to help solve a third of the impacts listed in the EIA.
- Even a 'run of the river' design disrupts river processes.
- River food webs will be affected. Some fish will gain from the changes to habitats (e.g. piranha-caju) but others will not (e.g. migratory carnivores). River-bottom feeders (e.g. bagres) and scaly fish (e.g. pescadas) are affected, as well as freshwater shrimp, which form the base of many food chains.
- Fish migrations up river will be impeded by the dam, e.g. the dourada (top of the food chain) migrates 3,000 km each year from the Amazon mouth to the Andean foothills, and it may become extinct. Genetic variation may also be lost because the variety of habitats has been reduced.
- The final flooded area may be twice as large as the official size (i.e. over 1,000 km²).
- The constortium were fined us $3.3 million for killing 11 tonnes of fish at the start of construction.

Objectives

- Recognise and describe the typical features of a bottom-up approach.

- Understand what is meant by sustainable development and appropriate technology.

- Explain the advantages and disadvantages of bottom-up schemes, and evaluate them through a comparison with top-down schemes.

How might countries develop more sustainably in the future?

Bottom-up schemes and their effects at a local scale

Development is usually associated with economic growth, such as secondary and tertiary industries, but it is also associated with the well-being of people – their health, freedom and food supplies. The **Human Development Index** (**HDI**) considers several factors, including income, education and life expectancy (Figure 6). In 2007/08, out of 177 countries, the HDI ranked the UK as 16th, Brazil 70th and Peru 87th.

Micro-hydro schemes

In contrast to large HEP schemes such as Santo Antonio, some low-income countries have decided that their remote rural peripheries would benefit from **'micro-hydro' schemes** – a 'bottom-up' approach. Micro-hydro schemes are defined as those with a generating capacity under 100 KW. Most use a 'run of the river' method like the Santo Antonio dam, but on a much smaller scale. Water is diverted from a stream to a high point on the valley side and then dropped through a pipe to turn a turbine (Figure 7). Much of the equipment can be made in developing countries.

Micro-hydro projects are considered to be appropriate to the local skills level, and they help spread technology to the rural periphery. They are low-cost, and involve local people in all stages of the scheme, from planning to daily operation (Figure 8).

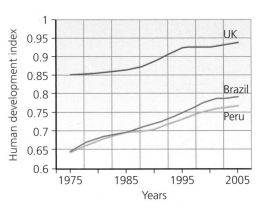

Figure 6: The Human Development Index for the UK, Brazil and Peru, 1975–2005

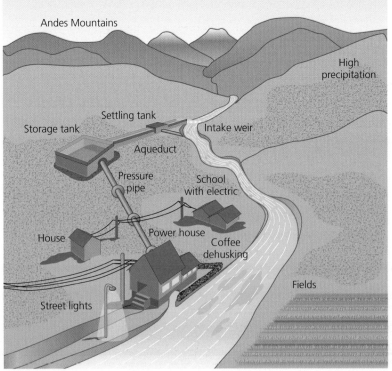

Figure 7: Peruvian valley with features of micro-hydro

Skills Builder 3

Study Figure 7. Use evidence from the sketch to describe the impacts that this micro-hydro scheme has had on the river and the valley.

Bottom-up Decision Making

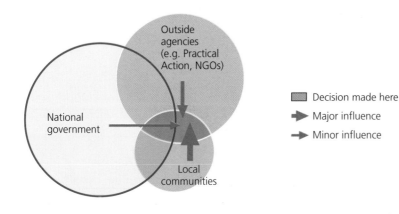

Figure 8: Influences on decision-making in a bottom-up approach

Micro-hydro schemes in Peru

On the eastern side of the Andes are some of the poorest areas of Peru – 44 per cent of the population, for example, live on less than $2 a day. This rural periphery has steep slopes, poor road networks and a scattered agrarian population. The low population density makes it uneconomic to provide grid electricity. However, rainfall is high and there are plenty of streams and rivers. A charity called 'Practical Action' has ten full-time staff working on micro-hydro in Peru, and it has helped install nearly fifty schemes, providing electricity to 30,000 people. Part of the costs are paid for by the local community, who are also responsible for management and maintenance. There is also a government ten-year plan to provide fifty-eight more micro-hydro schemes, with possible funding from Japan.

One community to benefit is Chambamontera, with sixty families (see Figures 9 and 10 on page 240). The village is more than two hours, drive from the nearest town and 1,700 metres above sea level. The farming in the area is well organised and some services are available, but electricity is needed to improve the social and economic environment. This micro-hydro scheme started in September 2008 and when it is finished, it will generate 15 KW from a water flow of 0.035 m³/s. Meetings were held with villagers to share information, and an agreement was drawn up, which included the village commitment of labour, materials and some money. From this the legal community group was formed to make decisions and to obtain loans if necessary. A socio-economic survey of the community was completed and this helped to establish the needs of the local people, and confirm their ability to maintain the electricity service. The total cost of the scheme is £34,000 (55% coming from the Matthiesen/Orbis Pictus Foundation, 22% from other donations, 17% from Practical Action, and 6% from the village community).

Skills Builder 4

Using Figure 9, describe what this area of Peru is like.

Activity 9

(a) Explain how the Chambamontera scheme could be seen to have positive and negative impacts on the social, economic and natural environments of the local area.

(b) What are the benefits of involving local people and communities in the decision-making process?

(c) Are there any ways in which this scheme is helping the development of Peru as a whole?

ResultsPlus
Build Better Answers

Using a named example, explain how bottom-up schemes meet the needs of developing countries, and how successful these have been. (6 marks)

■ **Basic answers** (Level 1)
Include only limited detail and do not name an example.

● **Good answers** (Level 2)
Are structured in relation to a named example, include some factual detail and partially evaluate its success.

▲ **Excellent answers** (Level 3)
Are detailed throughout, use geographical terms, and provide at least two points about its success.

Figure 9: Satellite image of North-western Peru

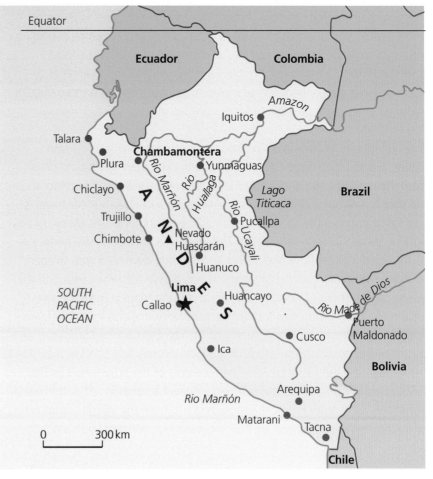

Figure 10: Location of Chambamontera village in Peru

Social benefits

- Local people are involved at all stages of the scheme, including decision-making.
- Reliable electricity supply provides refrigeration (e.g. milk coolers), lighting, entertainment, communication, computer access for families, and street lighting.
- Health care is improved as electricity allows the storage of medicines (e.g. vaccines) in refrigerated conditions.
- Electricity for schools (Figure 12 on page 242) means that they can use equipment (e.g. photocopiers, audio-visual).
- Young adults and teachers are encouraged to stay or return from the large urban areas because opportunities are provided by electricity.
- Training of local people to operate the technology gives them skills.
- Electricity means that the kerosene lamps can be replaced. The lamps produced unhealthy fumes which people (especially women and children) were breathing in, and they created a fire risk.

Economic benefits

- Cheaper electricity than a large HEP scheme (only $0.14 per KW) and with low running costs, the costs are covered by the charges for electricity.
- The scheme uses local expertise and skills, and cheap technology.
- Electricity powers machinery for small businesses, e.g. carpentry workshops, and more small businesses start up, which pay for the micro-hydro installation. 26% of households have started or expanded a business, e.g. battery charging.
- The piped water can drive machinery directly, e.g. for coffee de-husking and processing (Figure 7, page 238).
- 60% of the people in a survey said that their income had increased.

Natural environment benefits

- Avoids flooding a large area of land which would destroy ecosystems and take away farmland.
- Avoids the need to burn wood from local trees, so reducing deforestation and soil erosion.
- Replaces fossil fuels (e.g. kerosene).

> 'We are so happy to have light, we feel very thankful and proud. Our neighbouring communities envy us. At first they thought our work was in vain but now they are asking us what they can do to obtain light.'
>
> **Alfredo Sarango, Pampa Verde, Peru**

Activity 10

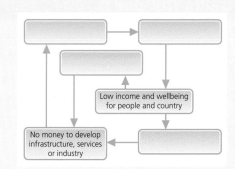

Figure 11: The Poverty Cycle

1. Make a copy of Figure 11 and add the following four labels in the correct boxes: 'Little investment in industry or services', 'Low level of consumer demand for services', 'Low economic output and no community amenities', 'No savings or investments by people'.

2. Make a list of the ways in which the cycle can be broken so that the social and economic environment can be improved.

3. How would a top-down scheme and a bottom-up scheme help to break this cycle?

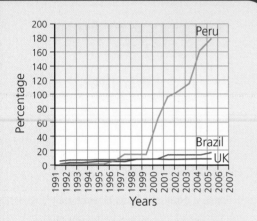

Figure 13: Cumulative change in protected areas (national parks, nature reserves, etc.)

Study Figure 13.

(a) Describe and compare the changes in protected areas between 1991 and 2006.

(b) What else may need to be considered to make a judgement on the significance of the trends?

1. Using the sustainability criteria shown in Figure 14, assess the sustainability of the Santo Antonio dam scheme, and the Chambamontera scheme.

2. What are the advantages of using a bottom-up scheme rather than a top-down scheme from the sustainability point of view?

Figure 12: Electricity for a Peruvian school from micro-hydro

Social problems

● Poor people still have to pay for the electricity, which is metered.
● Some villages have doubled in size, increasing population pressures.

Economic problems

● Demand for electricity is variable, so extra supplies need to be diverted (e.g. to heat water).
● Initial capital cost is high for a poor village, e.g. £500 per household, so loans may be needed such as from the Inter-American Development Bank and these have to be repaid. (10% of households in eastern Peru have problems with this.)
● Some specialised equipment has to be imported (e.g. load controllers from Canada or UK).
● The systems installed have only a 20–25 year lifespan.

Natural environment problems

● Small storage dam is required, along with construction in the valley, which alters the flow of the river and spoils the scenery. The diversion channel and pipe can be long (see Figure 7 on page 238).

Bottom-up schemes and sustainability

Many rural peripheries of developing countries are stuck in a **poverty cycle**, which is extremely difficult for them to break out of. Achieving sustainable development (Figure 14) therefore becomes a problem for these areas. Large schemes that focus on national needs (often the economic needs) may overlook the rural poor and do little to improve the lives of people living a long way from the core. Small-scale schemes, on the other hand, may be more appropriate – and more sustainable.

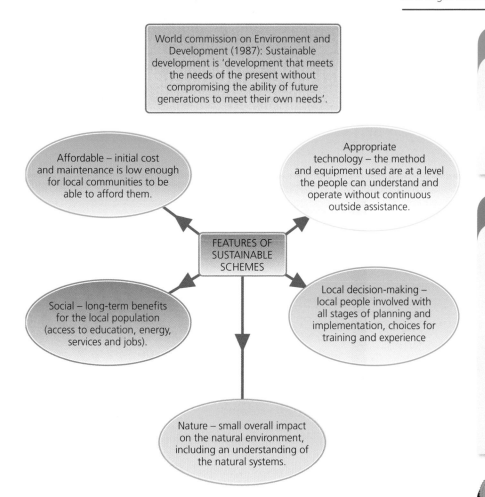

Figure 14: Sustainability criteria

Activity 12

1. Why would a developing country wish to get rid of disparities within the country?

2. Name three ways in which a country may reduce the disparity between urban core and rural periphery.

3. Why may both top-down and bottom-up schemes be needed to bring about long-term economic development in a developing country?

ResultsPlus
Build Better Answers

Identify three characteristics of a bottom-up rural development scheme that makes it sustainable. (6 marks)

■ **Basic answers** (Level 1)
Only mention that natural, economic and social features have to be balanced, and do not link to an example.

● **Good answers** (Level 2)
Mention the importance of access to jobs, energy and protecting food webs.

▲ **Excellent answers** (Level 3)
Also link to a named case study, with some facts relating to jobs, energy or conservation.

Measuring and monitoring progress in the different aspects of development is important so that negative impacts can be reduced and progress be made towards sustainability, as well as then continuing it. The **Millennium Development Goals** provided targets for countries to achieve by 2015. For example, Target 9 for Goal 7 is 'Integrate the principles of sustainable development into country policies and programmes and reverse the loss of environmental resources'.

Target 1 for Goal 1 is 'Halve [between 1990 and 2015] the proportion of people whose income is less than $1 per day'. By 1998, both Peru and Brazil had reduced extreme **poverty** by only 8%, leaving much to be done by 2015. One of the development dilemmas facing developing countries is how to improve the well-being of the population while at the same time conserving natural resources and ecosystems. This balance is not easy to achieve, especially over a long period of time, but it is important that the benefits remain for future generations. HEP is an attractive technology because it can be used while respecting nature. Both the Santo Antonio dam and the Chambamontera scheme have relatively low impacts on nature, but they do vary in terms of their social and economic impacts.

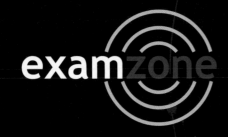

examzone

Know Zone
Development dilemmas

Areas develop at different rates, leading to disparities within and between countries. There is a major dilemma as to how development is defined and whether bottom-up methods are better than top-down practices.

You should know...

- ☐ That the rate that areas develop varies spatially both in countries and in general
- ☐ How this uneven development leads to disparities with a developed core region and a more rural backward periphery in many countries and regions
- ☐ How, at a world scale, disparities can lead to different rates of development
- ☐ How there are different ways development can take place
- ☐ The advantages and disadvantages of named top-down developments
- ☐ What the features of bottom-up developments are and how they compare to top down strategies
- ☐ What the advantages of sustainable development are
- ☐ How appropriate technology can play a vital role in development.

Key terms

Appropriate technology
Bottom-up approach
Core region
Disparity
Environment Impact
Assessment
Hydro-electric power
Human Development
Index
Micro-hydro schemes

Millennium Development
Goals (MDGs)
Non-Governmental
organisations
Periphery
Poverty
Poverty cycle
Top-down approach

Which key terms match the following definitions?

A The use of fast flowing water to turn turbines which produce electricity

B The outer limits or edge of an area, often remote or isolated from its core

C A great difference – e.g. between parts of a country in terms of wealth

D Equipment that the local community is able to use relatively easily and without much cost

E Development projects that originate in local communities rather than in central government or external agencies

F A method of evaluating the effect of plans and policies on the environment

G A set of processes that maintain a group or society in poverty

To check your answers, look at the glossary on page 313

Foundation Question: Describe, using an example, how a mega-dam could help a country develop. (4 marks)

Student answer ● (achieving 2 marks)	Examiner comments	Build a better answer △ (achieving 4 marks)
The Aswan dam on the River Nile in Egypt is vital to Egypt. It has improved fishing in the Nile delta region. It has provided irrigation for cotton and food crops. This is very important as Egypt is a desert country.	• *The Aswan dam...* is a well located example and scores 1 mark. • *It has improved...* is incorrect. The silt starvation has caused all the fish to migrate away from the waters near the delta. • *It has provided...* This is a correct point that has been exemplified so scores 1 mark. • *This is very important...* This part of the answer is not developed enough. It begins to explain but the question asks the student to describe.	The Aswan dam on the River Nile in Egypt is vital to the country. It has provided a new fishing area in Lake Nasser, which is also a tourist site. It has provided irrigation for cotton and food crops. It has provided Egypt with 90% of its electricity for industry and people.

Overall comment: The answer provides an example and one very good statement, but needs a second to score more marks.

- -

Higher Question: Explain the positive impacts, **other** than providing HEP, of building mega-dams. (4 marks)

Student answer ● (achieving 2 marks)	Examiner comments	Build a better answer △ (achieving 4 marks)
Mega-dams are built for many reasons, but providing irrigation water is very important. These mega-dams also provide lakes for extensive water sports, tourism and transport. These lakes can also be stocked with fish, which provide additional food and income for local people. The electricity from the dams is used to develop new industry.	• *Mega-dams are built...* This answer is not fully developed and the point needs to be explained to achieve a mark. • *These mega-dams also...* scores 1 mark as it is a clear point. • *These lakes can...* This is a further clear point that is well developed as an explanation. The scores 1 mark. • *The electricity from...* This part of the answer is related to HEP so is not relevant to the question.	The irrigated water can be used for cash cropping of cotton, citrus fruits and other goods. These mega-dams also provide lakes for extensive water sports, tourism and transport. These lakes can also be stocked with fish, which provide additional food and income for local people. The land around the dam can be afforested, which makes the area less prone to soil erosion and more attractive to look at.

Overall comment: In this question the student needed to **explain** the positive impacts, but not include HEP.

Chapter 16 World of work

Objectives

- Learn what the global economy is.

- Recognise that countries vary in terms of what they contribute to the global economy.

- Understand why the benefits of the new economy are not equally shared.

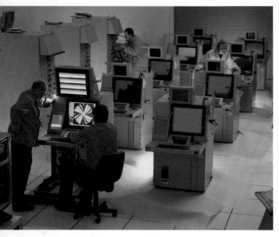

Figure 1: A typical activity in the quaternary sector

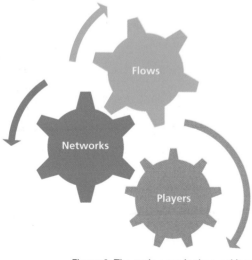

Figure 2: The main cogs in the workings of the global economy

How does the 'new economy' function?

The 'new economy'

The process of **globalisation** is gradually drawing the countries of the world together into a single economic system. This system is known as the global economy. This is the '**new economy**' – and it has winners and losers.

The global economy is divided into four sectors. Each sector involves a different type of activity:

- **The primary sector** – working natural resources. The main activities are agriculture, fishing, forestry, mining and quarrying.

- **The secondary sector** – making things, either by manufacturing (a TV or a car) or construction (a house, road or new airport).

- **The tertiary sector** – providing services. These include services that are commercial (shops and banks), professional (solicitors and dentists), social (schools and hospitals), entertainment (restaurants and cinemas) and personal (hairdressers and fitness trainers).

- **The quaternary sector** – a new sector that is mainly found in developed countries. It is about research, information and communications. Sometimes known as the 'knowledge sector' or 'high-tech industry', it also includes nanotechnology and biotechnology, producing computer hardware and software, as well as the development of new military equipment (Figure 1).

Figure 2 shows three critical parts in the working of the global economy. They provide the links between the four sectors and between the different parts of the world:

- **Networks** are the 'spider's webs' linking countries together – for example, transport networks, the telephone, the Internet and trade blocs.

- **Flows** are the things that move through those networks – for example, raw materials, money, migrant workers, information and aid.

- **Players** are the organisations that have a great influence on the workings of the global economy. They include the huge business empires known as **transnational companies** (**TNCs**, see page 249) and global organisations such as the World Bank, the United Nations, the International Monetary Fund and the World Trade Organization. Other powerful players are the USA, the European Union, Japan and the oil-producing countries.

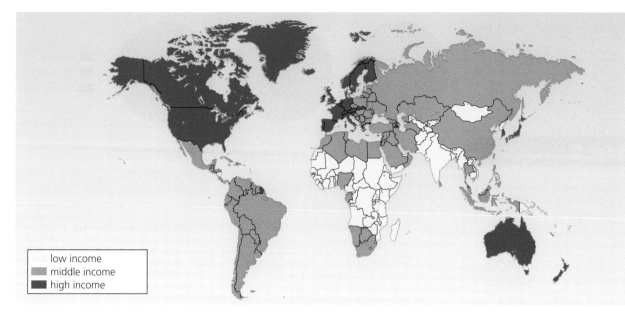

Figure 3: Global distribution of three income groupings, based on Gross National Income per capita, 2003

A particular feature of the global economy is the **dependence** that grows between countries. Country A depends on raw materials supplied by countries B and C to manufacture a particular product. Country A also depends on countries D and E as markets for that manufactured product. That makes B and C indirectly dependent on D and E, and vice versa. In short, countries are drawn into a complex network of economic flows.

What each country contributes to the global economy depends very much on their level of development. Figure 3 divides countries into three groups, based on a measure of development known as '**GNI per capita**':

● Low-income countries (LICs)

● Middle-income countries (MICs)

● High-income countries (HICs).

It is still commonplace to divide the countries of the world into developing and developed countries. The problem with this simple scheme is that it ignores the middle-income countries that are making the transition from the developing to the developed world. Since these countries are clearly developing, they tend to be seen as members of the developing world.

In each of the three economic groupings shown in Figure 3, the balance of the three economic sectors is different, as shown in Figure 4 (on page 248). The primary sector is the most important one in most LICs (e.g. Ethiopia). The secondary sector is quite strong in most MICs (e.g. China). The tertiary sector is top in all HICs (but is supported by the quaternary sector in some of them). The UK is an example.

Activity 1

In which sector do the following occur: a warehouse, a public park, an Internet café, an oil refinery, a plantation, a university?

Skills Builder 1

Study Figure 3.

Make notes that summarise the distributions of the low-income, middle-income and high-income countries. For example, your notes about the distribution of low-income countries might read: 'Occur largely in Central Africa and in South and Southeast Asia'.

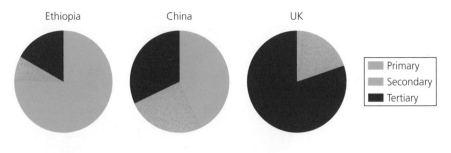

Ethiopia   China   UK

Primary
Secondary
Tertiary

Figure 4: Balance of economic sectors in a typical LIC, MIC and HIC (2005)

Skills Builder 2

Study Figure 4.

In Russia, the primary sector accounts for 11% of all jobs, the secondary sector 29% and the tertiary sector 60%. Draw a pie chart to show these values. How does this pie chart compare with those in Figure 4 on page 248? Do you think Russia is ahead of or behind China along the development pathway? Give your reasons.

For LICs, being part of the global economy usually means growing food or selling minerals and buying cheap manufactured goods. For MICs, it means manufacturing and selling goods, and buying raw materials, perhaps food and some services. For HICS, it means buying food and manufactured goods and selling services and new technologies.

Finally, remember that the global economy is changing all the time. Besides growing larger by the day, there are other important changes which are shown in Figure 5.

● New services are appearing within the tertiary and quaternary sectors.

● The locations of particular activities are changing. For example, there is a **global shift** of manufacturing from many HICs to cheaper, more profitable locations in MICs and LICs (see page 252).

● Modern communications are also changing the location of employment (see teleworking and outsourcing on pages 259–260).

● As each country develops, new activities are added to its own economy. For example, as an LIC begins to develop into an MIC, new jobs in manufacturing begin to replace traditional ones in agriculture. As an MIC grows into an HIC, a whole new range of jobs within the tertiary sector will appear, possibly replacing factory jobs. A quaternary sector will gradually develop.

● More and more resources are being consumed.

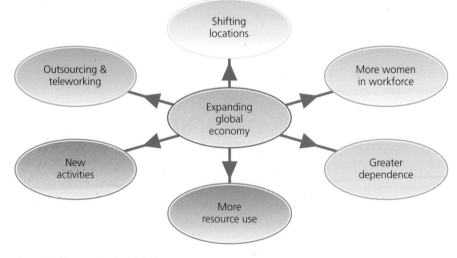

Figure 5: Changes in the global economy

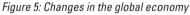

The growing economy is affecting people, particularly workers, throughout the world, but in different ways:

● Workers in the developed world have had to re-skill, as jobs in agriculture and manufacturing have given way to jobs in the tertiary and quaternary sectors.

● Workers in the developing world have had to learn new skills, especially those of commercial farming, manufacturing, tourism and outsourcing.

● More women are becoming involved in the global labour force – and this applies to all three groups of country. More mothers at work may be good news for the family budget, but we are beginning to realise that this may be creating serious social problems. In some Islamic countries, however, the home is still regarded as the place for women.

● The use of child labour is traditional in the some countries. Such labour is cheap and it also allows the poorest families to earn a little more income. In the new economy, however, the use of child labour is banned. No one would object to the ban if it meant that those children would go instead to school. However, outlawing child labour is condemning more families to persistent poverty.

We will return to most of these changes in the section on 'Changes in the workplace'.

In theory, playing a part in the global economy should benefit every nation in the world. But that is true only to a limited extent. Some countries are more 'connected' to the global economy than others and therefore stand to gain more. Because they are among the most powerful players in the global economy, the HICs are doing particularly well. Sadly, the LICs are benefiting least, because they are still being exploited by other countries. This vital issue will be discussed and illustrated in the second half of this chapter.

Transnational companies

The transnational companies (TNCs) are thought by many to be the most powerful players in the global economy. They might be seen as the 'builders' of the global economy. In particular, they build the 'bridges' that link together the national economies of the world because they produce goods for a global market.

The scale and activities of TNCs

The table on page 250 makes it clear that the top five companies are involved in only three different industries. However, other activities appear in the list of the top 25 TNCs, including mining (Rio Tinto – UK), publishing and printing (Thomson Corporation – Canada), food, beverages and confectionery (Nestlé – Switzerland), pharmaceuticals (AstraZeneca – UK) and telecommunication (France Telecom – France). But there are also major TNCs that are much less specialised. This is particularly true of Japan-based companies like Sumitomo and Mitsubishi which have wide business interests that stretch across the economic sectors.

ResultsPlus
Build Better Answers

Explain why the benefits of the 'new economy' are unevenly spread. (4 marks)

■ **Basic answers** (0–1 marks)
Describe the growth of the global economy and some of its changes, but the varying benefits are not directly addressed.

● **Good answers** (2 marks)
Develop some ideas, including the rise of economies such as India and China, but without much distinction between types of work.

▲ **Excellent answers** (3–4 marks)
Identify the attractions of some areas connected with a good infrastructure, while the disadvantages of other areas are also addressed. Locations are identified and some students identify an urban–rural contrast.

Activity 2

In your own words, explain what being 'connected' to the global economy means. Give examples.

The top five TNCs, based on foreign assets, 2006

Rank	Company	HQ	Industry	Revenue ($bn)
1	Exxon Mobile	USA	Oil	377
2	Wal-Mart	USA	Retailing	351
3	BP	UK	Oil	318
4	Shell	UK / Netherlands	Oil	274
5	General Motors	USA	Motor vehicles	207

Skills Builder 3

Study Figure 6. Make notes about the distribution of Rio Tinto's mining ventures.

Case study – Rio Tinto

Rio Tinto is a transnational mining and resources group, founded in 1873. It was in that year that a British company acquired from the Spanish government the name and mines located in the south of the country. Soon the Rio Tinto mine was the world's leading producer of copper. The company has since turned its attention to other minerals. Today, for example, it is the third-largest coal mining company in the world. But, as Figure 6 shows, it is now involved in the working of a whole range of minerals, and its network of mines is truly global in extent.

AFRICA

South Africa	copper, titanium
Namibia	uranium
Zimbabwe	diamonds
Guinea	iron ore
Madagascar	titanium

EUROPE

UK	aluminium
France	talc, borates

ASIA

Indonesia	copper, gold
Mongolia	aluminium, copper

AUSTRALASIA

Australia	aluminium, coal, copper, diamonds, iron ore
New Zealand	aluminium

NORTH AMERICA

Canada	aluminium, diamonds, lead, silver, titanium, zinc
USA	borates, coal, copper, gold, nickel, talc, uranium

SOUTH AMERICA

Argentina	potash
Brazil	aluminium, iron ore
Chile	borates
Peru	copper

Figure 7: A Rio Tinto mining operation

Figure 6: Rio Tinto's mining interests extend all over the world

The extraordinary size of the top TNCs can be shown by comparing their revenues (see the table on page 250) with the GDP of some countries in the same year. Sweden's GDP, for example, was $444 bn, Greece's $360 bn, South Africa's $277 bn and Malaysia's $180 bn. It is TNCs' size that gives them so much economic and political influence.

Production chains

Most TNCs have their HQs in one of the world cities – London, New York or Tokyo – from where they set up and run their **production chains** (also known as supply chains or commodity chains). The production chain consists of a number of stages involved in the making of a particular product. At each stage, value is added to the emerging product. Figure 8 shows the production chain of a pair of trousers.

Figure 8: The production chain of a pair of trousers

Companies set up these 'transnational' production chains for five reasons (Figure 9). The two main reasons are to reduce costs and increase both sales and profits. For example, the Japanese company Nissan produces cars in Sunderland. This means it is close to the large consumer markets within the EU. Better still, it escapes the tariffs and quotas that make it difficult for foreign companies to export to the EU. Many of the jeans marketed under well-known American labels are made in sweatshops in countries as distant as Mexico, Bangladesh and China in order to reduce costs. Labour is less regulated in such countries. This means that a TNC can pay the workers less and make them work longer hours – often in less safe conditions.

Activity 3

What is copper used for?

Activity 4

Research the production chain of one of the following: a BMW Mini, a personal computer or a pair of Nike trainers.

Decision-making skills

Study Figure 9.

Which do you consider to be the most important reason for TNCs to go global? Give reasons for your choice.

Are there any other specific reasons you can think of for TNCs like Rio Tinto and Unilever.

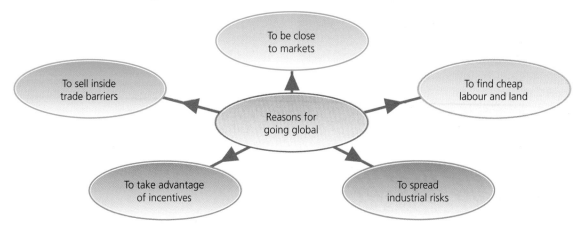

Figure 9: Five reasons why TNCs operate globally

Activity 5

1. Study Figure 10. There are at least two modes of transport missing from this diagram. What are they, and what would be their average speeds?

2. Research and find out whether there is any evidence of deindustrialisation in your home area. If there is, what was being made, and where have the businesses moved to?

ResultsPlus
Exam Question Report

Transnational companies (TNCs) affect the countries where they locate their factories. Choose a TNC that you have studied. Explain the advantages and disadvantages the TNC has for a country or countries where it has factories. (8 marks)

How students answered

Some students failed to identify any specific TNC and the impacts were not clearly defined as either an advantage or a disadvantage.

28% (0–3 marks)

Most students identified a TNC and were able to state at least one advantage and disadvantage but without very much development.

61% (4–6 marks)

A few students could identify and explain the advantages or disadvantage for the countries and some could identify groups who might benefit and groups who might lose. Good case-study evidence was used in these answers.

11% (7–8 marks)

Many developing countries are keen to attract the attention of the TNCs – so they offer grants and other incentives. For example, countries such as China and Malaysia have set up free trade zones along their coasts to lure the TNCs. Cheap building sites and a good infrastructure are provided. Imports and exports are tax-free. By taking advantage of these 'carrots', TNCs are able to reduce their costs and increase their profits.

Another reason for becoming transnational is that it reduces business risk. If a civil war breaks out or a major natural disaster happens in one country, the TNC can simply raise production in another until the situation returns to normal.

Two technological developments have made these drawn-out production chains possible. First there has been the development of modern transport networks. These are now capable of moving commodities and people quickly and relatively cheaply over long distances. Look at the distances involved in making a pair of trousers (Figure 8). Figure 10 illustrates how the increasing speed of transport has 'shrunk' the world. Secondly, there have been major advances in information and communications technologies. The fast transfer of information by the Internet and mobile phones means that factories scattered around the globe can keep in close touch with market trends in the UK, for example. Changes in fashions and styles, as well as new orders, can be instantly relayed to the factory, whether it is in Dhaka or Shanghai.

Perhaps the greatest impact that the TNCs have had is on the global distribution of manufacturing. As a result of their policy of seeking out the cheapest locations for making particular products, factories have been closed down in the UK and other HICs. They have been replaced by new factories (branch plants) set up in these cheap production locations. As a consequence, there has been a global shift in the location of manufacturing. **Deindustrialisation** in the developed world is matched by industrialisation in the developing world.

1500–1840

1850–1930

1950s

1960s

Best average speed of horse drawn coaches and sailing ships was 10 mph

Steam locomotives averaged 65 mph
Steam ships averaged 36 mph

Propeller aircraft 300–400 mph

Jet passenger aircraft 500–700 mph

Figure 10: A shrinking world

For some developing countries, the greatest employment change has been caused by the global shift in manufacturing. Countries to benefit early from the shift were South Korea, Taiwan and Hong Kong (now part of China) – known as newly industrialised countries (NICs). There is now a new generation of beneficiaries, known as the 'recently industrialising countries' (RICs). Examples are China, India and Brazil. Industrialisation is promoting them to become MICs.

Made in China

Take a look at the labels on your clothes and shoes. You will probably find that a number of them say 'Made in China'. The same is likely to be true for your TV set, mobile phone and photocopier. As we have seen, China is currently ranked as the third most important manufacturing country in the world and is famous for its **consumer industries**. For example, China produces:

- Half of all the world's clothes

- Two-thirds of the shoes

- One-third of the mobile phones

- Two-thirds of the photocopiers

- Half the microwave ovens.

But China has many other industries too – from iron and steel to chemicals and fertilisers, from ships to aircraft, and from military equipment to space satellites. 30 years ago, there were far fewer industries in China and 'made in China' products were scarce outside the country. So what has caused this spectacular 'industrial explosion'? Figure 16 identifies six major factors. The most important one has been the political shift from communism towards capitalism. Evidence of this move is the fact that almost half of the world's top 500 TNCs now have some form of investment or presence in China. It is they that are helping the process of industrialisation.

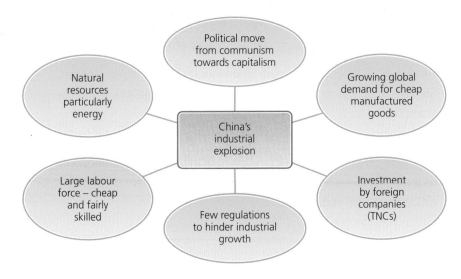

Figure 16: The factors behind China's industrial explosion

Explain why some developing countries have experienced a rise in employment in the service sector in recent years. (6 marks)

■ **Basic answers** (Level 1)
Make one or two generic comments about the rise in tertiary employment but are not specific about developing countries or the type of employment.

● **Good answers** (Level 2)
Identify some locations and mention the idea of global shift. However, the reasons are not developed beyond the observation that they have occurred.

▲ **Excellent answers** (Level 3)
Include the relocation of call centres and the growth of back-office work among the list of service industry growth, and some might identify growth within India and China as a cause. Reasons for re-location include cheaper labour and improved systems of communications. The located evidence is accurate.

Quick notes (Made in China):

- Industrialisation can take place very rapidly, particularly when governments and businesses work together.
- Industrialisation has created many forms of employment that are new to China.

Thanks to its industries, China now has the third-largest economy in the world. Thanks to its industries, a whole range of new jobs has been created. These jobs are offering millions of Chinese people secure employment and the opportunity to raise their standard of living and quality of life.

The three new employments in developing countries that we have just looked at – tourism, commercial agriculture and manufacturing – share a number of characteristics:

- They are creating job opportunities.
- They connect developing countries to the growing global economy.
- They have been made possible by advances in transport and technology.
- They involve exploiting aspects of the natural environment – fine scenery, beaches, climate, soils, energy resources and minerals.

Changes in the workplace

Closely linked to the employment changes illustrated above, there have been changes 'in the workplace' – which means things like working conditions rather than types of job.

In HICs, thanks to government laws and trade union pressure, working conditions have been improved in a number of ways:

- Health and safety regulations now apply to most forms of employment and protect the well-being of workers. The workplace has been made a safer and healthier environment.
- Working days are shorter and regulated.
- National minimum wage scales mean that a particular job is paid at the same rate no matter where it is located.
- Equal opportunities legislation means that there should be no discrimination on the basis of gender, age or ethnicity.
- Flexible working hours and part-time work mean that more mothers are able to go out to work. There are now many more women in the workforce.
- 'Hot desking' is a new working practice in which workers literally share a desk with others. This allows employers with staff who are frequently out of the office (sales people, for example) to make better use of the available office space and equipment.

Decision-making skills

Work in groups and look at the changes that have been made to improve workplace conditions. Decide which you think are the most important and give reasons why. Can you see any disadvantages in 'hot desking' for workers?

New technology has not only produced new jobs – as we have seen – it has also produced new ways of working. Perhaps the most striking example is what is variously called **teleworking**, telecottaging or telecommuting. In the UK today, over 2 million people, who are mainly self-employed, work from home. Such home-based work relies on phones, computers and the Internet. Many office activities in banking, accounting and publishing do not require direct contact with customers and can be done at home. It really does not matter whether you live 10 km or 1,000 km from London (or anywhere else) in order to telecommute. Some of the remote rural areas of the UK have particularly benefited, and this new technology has certainly encouraged the process of counterurbanisation.

Skye makes a comeback

Skye is one of the largest Western Isles. It has an area of 1,656 km² and a population of just over 9,000. Population numbers have been rising for some time as a result of **net in-migration**. The incomers to the island fall into three main categories:

● People involved in tourism, either as hotel staff or setting up their own B&Bs and guest houses.

● People moving to the island to retire.

● People taking advantage of modern ICT and telecommuting.

But why should Skye be different from the rest of the Western Isles which are still suffering from depopulation? The answer is that Skye is part of the Inner Hebrides and lies much closer to the mainland than the other islands. And in 1995 a bridge was opened which gave Skye a direct link to the mainland for the first time (Figure 17). It is this bridge, plus two good ferry links, that have improved the accessibility and the appeal of the island, particularly to tourists. The extension of the broadband network means that those who want to can work in the modern world and yet live in a relatively untouched environment.

Figure 17: The Skye Bridge, which connected Skye to the mainland for the first time

Activity 8

Make sure that you understand these workplace improvements. Are you able to think of any more?

Activity 9

1. Find out about the process of counterurbanisation and explain how new technology is helping it.

2. Can you think of any teleworking businesses that could be run from Skye?

Quick notes (Skye):
Social change and modern technology can turn around the fortunes of remote areas by encouraging new forms of employment. In this instance, the jobs have been in tourism and teleworking.

In LICs and MICs, the same advances in telecommunications have produced a form of teleworking known as **outsourcing**. Most forms of information can now be digitised and then sent in a matter of seconds for processing somewhere else in the world. This means that companies in Western Europe and North America can transfer certain types of service sector jobs to cheaper countries on the other side of the world.

Call centres in India

Since the 1990s, many companies have moved their **call centre** operations to countries where the wages are lower. There are now thousands of call centres dealing with such services as Internet banking and airline booking, help-lines and credit card billing (Figure 18). Many well-known companies are making use of call centres – HSBC, British Airways, Dell, Orange, American Express and Tesco.

But why are so many call centres in India? The main factor is that India produces large numbers of well-educated, English-speaking workers. They are prepared to work for much lower wages than are paid to staff in the UK or North America. This is good for company profits. It also means that these companies do not have to pass on heavy service changes to their customers in developed countries.

Call centres are good news for India's new economy. There are responsible for nearly 5 million jobs and earn something close to $60 million a year. By Indian standards, those jobs are secure and relatively well paid. But for the UK and the USA, outsourcing to call centres means a further loss of jobs, to be added to those already lost through deindustrialisation.

In developing countries, jobs in outsourcing are much sought after and filled by those who have enjoyed a good education. In other sectors of the economy, the workplace still shows many unwanted characteristics. These include:

- Exploitation of workers – women and children, in particular, are paid minimal wages that do not even meet the basic costs of living. There is no trade union protection.

- The working environment is often unsafe. There are many accidents causing injuries or death. Pollution has a bad impact on the workers' health.

- Many jobs are still part of the **informal sector** and therefore lie beyond the reach of any laws that might be introduced to improve working conditions.

But all is not gloom and doom. Many of the new jobs, as for example in tourism, are jobs being taken up by women. With two wage-earners, the chances of families breaking out of the cycle of poverty are slightly improved. Women at work also help their families' chances because they tend to have fewer children.

> **Quick notes (Call centres in India):**
> Call centres bring benefits, but also have their costs.

Figure 18: An Indian call centre

The question of sustainability

The last two sections have clearly shown that the global economy is opening up opportunities for those countries that wish to connect with it. But it is also clear that the global village is now occupied by a two-class society. Most developed countries are guilty of exploiting the developing countries in some way or another. And, surely, this cannot be sustainable on moral grounds?

At the same time, however, both types of country are guilty on another charge – overexploiting the environment for economic gain. Today's global economy is characterised by the following unsustainable features:

- A quickening rate of resource use, particularly to provide energy and raw materials for the global economy. This is most worrying in the case of non-renewable resources which, once used, are lost for ever.

- The mass transport of raw materials, products and people around the globe involves burning fossil fuels – this not only means further use of non-renewable resources, but also higher carbon dioxide emissions. In short, it is contributing to global warming.

- Much damage is being done to the environment, not just by increased carbon emissions. Two obvious examples are the clearance of forests to make more farmland and the clearance of mangroves to provide sites for tourist hotels. Air and water are being badly polluted by cities and industries (Figure 19).

- There is an increasing shortage of water resources as more is used to irrigate crops and meet the needs of industry and a growing population.

All of these unsustainable features are linked, in some way or another, to the new global patterns of employment and perhaps also to the new workplaces.

So, can anything be done to make the future more sustainable? The short answer is perhaps yes, but not very much. In the world of work there are some obvious actions that might help the situation, including:

- More reliance on renewable energy

- More recycling of waste

- More efficient use of resources

- More local sourcing of materials

- Less transporting of commodities and products

- Less water usage.

Putting these actions into effect will involve extra costs. But the global economy is all about minimising costs and maximising profits. So what hope is there for a more sustainable world of work?

Activity 10

(a) Research the informal sector, in terms of the activities involved and what it contributes to a developing economy.

(b) Explain the links between women at work and the reduction of poverty.

Figure 19: The environmental costs of a global economy

Decision-making skills

Rank the six bullet points, left, in terms of their priority for companies to carry out. Justify your order.

Activity 11

Explore ways in which the developing countries are being exploited in the global economy.

Globalisation and the global shift have led to huge changes in the pattern of employment around the world. As countries develop, the type of work available and working conditions change. A key driver of change has been the ICT revolution.

You should know...

- ☐ How the new global economy works
- ☐ How globalisation has led to the inter linkage of the world's economy, and that some countries are winners and some LDCs are losers
- ☐ How the global shift has led to changes in the global economy as production shifts from developed to developing countries
- ☐ How transnational companies drive the processes of globalisation and the global shift
- ☐ How transnational companies operate
- ☐ How changes are taking place in the world of work
- ☐ How these changes are impacting on people and the environment in named locations
- ☐ How there are winners and losers as a result of changing employment
- ☐ How new technology (ICT) is transforming the ways people work
- ☐ How changes have occurred in the workplace, with improved working conditions for many in developed countries
- ☐ Whether work and employment are becoming more sustainable

Key terms

Agribusinesses
Automation
Call centre
Commodity/production chains
Consumer industries
Deindustrialisation
Dependence
Depopulation
Disposable income
Flows
Global shift

Globalisation
GNI per capita
Informal sector
Labour intensive
Net in-migration
Networks
New economy
Outsourcing
Players
Teleworking
TNCs

Which key words match the following definitions?

A Any form of work in which telecommunications replace work-related travel (commuting)

B A large company operating in several countries

C A condition in which something (e.g. a country) is only able to survive by relying on outside support (e.g. from another country).

D An office equipped to handle a large volume of telephone calls (especially for taking orders or serving customers)

E The process, led by transnational companies, whereby the world's countries are all becoming part of one vast global economy

F The movement of manufacturing from developed countries (HICs) to cheaper production locations in developing countries (MICs and LICs).

G The decline in industrial activity in a region or an economy

H A process in which a company subcontracts part of its business to another company

To check your answers, look at the glossary on page 313

Foundation Question: Explain why transnational companies locate their headquarters in developed countries and many other factories in developing nations. (4 marks)

Student answer ● (achieving 2 marks)	Examiner comments	Build a better answer △ (achieving 4 marks)
The headquarters are in developed nations because this is where they started up, for example GM in the US.	• **The headquarters are...** scores 1 mark. This is a clear, correct statement.	The headquarters are in developed nations because this is where they started up, for example GM in the US.
The factories are in developing countries as this is where products such as palm oil or chocolate grow.	• **The factories are...** scores 1 mark. This is one reason for the pattern but there are others.	The factories are in developing countries as this is where products such as palm oil or chocolate grow.
The workforce in these countries works for less.	• **The workforce in...** Although this statement is true, it needs developing to score a mark.	They also took advantage of the lower costs such as wages in developing countries.
The main reason is that they often took over existing factories.	• **The main reason...** is incorrect and does not score any marks.	A further advantage is that environmental and safety laws are slack, so there are lower production costs.

Overall comment: The student could have gained an extra mark, but the third point was not sufficiently developed.

Higher Question: Explain how transnational companies have led to a global shift in industry and services. (4 marks)

Student answer ● (achieving 2 marks)	Examiner comments	Build a better answer △ (achieving 4 marks)
The global shift is the movement of economic activity from developed to developing countries.	• **The global shift...** This is a good, clear definition that scores 1 mark.	The global shift is the movement of economic activity from developed to developing countries.
It has occurred because wages are much lower in developing countries.	• **It has occurred...** This statement is correct and scores 1 mark, but there is no mention of TNCs.	TNCs have been attracted to LDCs because of the low wages/low costs for manufacturing.
There are also very slack environmental laws so again the product is cheaper.	• **There are also...** This does not score any marks as it is not well linked to the idea of TNCs.	TNCs have also outsourced their business functions, such as call-centre operations, which also leads to the shift.
The result is that TNCs have moved all their offices to developing countries.	• **The result is...** This is incorrect as usually it is only branch plants that go transnational.	The headquarters remain in developed countries, but the production activities in branch plants are often run by contractors.

Overall comment: Although the reasons given are correct, the answer does not focus enough on TNCs to be awarded more than 2 marks.

Unit 3 Making geographical decisions

Your course

Unit 3 is about making well thought-out and evidenced decisions – on topics which affect our planet.

As you will have found out when studying Units 1 and 2, the increasing demands of a rising population and economic development are putting pressure on the planet's scarce resources, such as energy, food and water. These pressures may lead to conflicts between nations, or between rival developers, or between those who wish to exploit the resources and those who wish to conserve them.

Your assessment

- You will sit a 1-hour written exam worth a total of 50 marks.

- Your exam will relate to a pack of resource material that you will study before the exam.

- The exam is broken down into Sections A, B, and C. You must answer all questions.

Chapter 17 Making geographical decisions

Planning for making geographical decisions

Long-term preparation

The theme for your decision-making paper is announced two years in advance. This gives your teacher time to select useful case studies for you to study during your course. The sample used in this chapter is on the need to build sustainable cities and towns, and looks at some of the new ideas that planners are having when developing eco-cities and eco-towns. This theme links very well with 'Living spaces', and with 'Changing cities' and 'Changing countryside'. It may be that there is a fieldwork option for your Controlled Assessment (see Chapter 18) which means that you can investigate sustainable living first-hand, and see what people think about developing sustainable lifestyles.

Throughout this book you will find suggestions as to how you can build up your skills, knowledge and understanding of geographical decision making (see Figure 1). In chapters 1 to 16 in this book there are decision-making skills boxes which set mini activities aimed at developing your skills throughout the course. Because the units you are studying cover a variety of scales, from local to global, you will be able to practise your skills using resources that focus on all different sorts of environments and places.

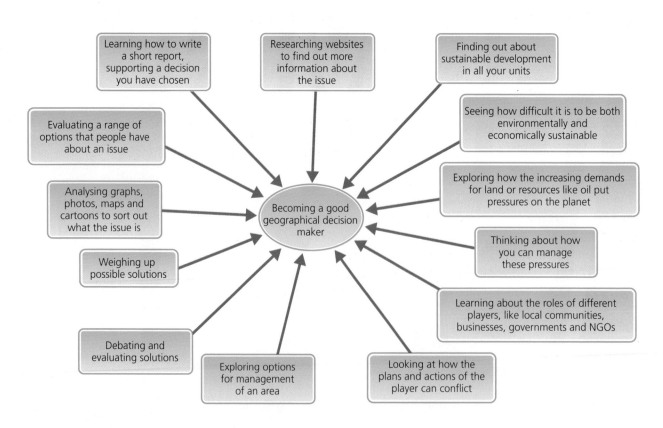

Figure 1: Becoming a good geographical decision maker

When the 'resource booklet' arrives

At least three months before you sit the 1-hour exam, the resource booklet will be made available for you to work on in class time. The first page of the booklet will give you detailed advice on exactly how you can use the resources to prepare for the exam. It will also provide you with a list of other sources – mostly websites – which you can research. Many of the websites will be ones designed for students – and they will have interesting activities to do.

The rest of the booklet will consist of the various resources that you will need when you do the exam. (But the resource booklets must be taken into the exam completely unmarked.) Sometimes the resources are quite long but your teacher will help you to understand them by explaining any difficult concepts. Your teacher may even develop quizzes and games to help you learn them.

Figure 2 shows you the three stages involved in making a geographical decision – and how they relate to the types of question in the exam and to the sample used in this chapter. The stages get harder as you go from one to the next. The sample is a shorter version – with fewer resources – but it does give you a good idea of how the exam paper works. The same style of paper will be set for Foundation and Higher.

The three stages in making a geographical decision	The three sections in the exam	The sample in this chapter
Setting the scene Finding out about the issue and the geographical background.	**Section A question** Short structured tasks based on resources, usually marked by point.	Sustainable cities.
Exploring the issue Exploring a specific example of the issue in greater depth, looking at pressures, players and conflicting opinions.	**Section B question** A mixture of tasks, some earning up to 6 marks.	The controversy about proposals to build a number of eco-towns in the UK.
Making the decision Looking at options – sometimes two, often three or four.	**Section C question** More extended writing tasks, marked by a levels mark scheme.	Two options – proposed sites for eco-towns. Justifying why one was rejected, while the other is being seriously considered.

Figure 2: The three stages, which get progressively harder

The next section of this chapter shows you how the resource booklet will look. This is a shorter version than what you will get in the exam, but provides detailed and useful advice about how to use the resources.

The last section of this chapter provides exam-style questions so you can practise your decision-making skills.

Sample resource booklet

General

You should:

- Begin by reading through the materials to have a good idea what each resource is about.

- List all geographical terms used and make sure you understand their meaning (see the glossary on page 313).

Section A – Sustainable cities

- Look at the models of sustainable and unsustainable cities on page 269 to see how they are different.

- Revise the concept of eco-footprints so that you can see how and why unsustainable cities have such a large footprint.

- Look at the various sustainable plans for existing cities, and try to list the initiatives on Figure 5 in terms of environmental and socio-economic sustainability.

- Use Google to research the plans for the eco-cites in Dongtan (China) and Masdar (United Arab Emirates).

- Think about how they could aspire to be zero-carbon communities and test beds for green technology.

Section B – The controversy about the plans for eco-towns in the UK

- Look at Figure 8 and consider the distribution of these proposed eco-towns. What type of area are they found in – both environmentally and politically?

- Consider the three aims for building eco-towns. Are the they compatible with each other? What does the compatibility matrix on page 275 show you?

- Study Figure 9, which shows you some of the features of eco-towns, and consider how they would be environmentally and socio-economically sustainable. Think about whether you would like to live in an eco-town or not.

- Look at the opinions about eco-towns and the blog on page 274. Think about whether the opinions are justified and whether they are for or against the building of eco-towns. Could it be that there are so many arguments against the idea that only a few will be built? In particular, consider CPRE's concerns about sustainability.

Section C – Evaluating two of the proposed eco-town sites: Imerys (Cornwall) and Weston Otmoor (Oxfordshire)

- Look at the resources on Imerys and carry out an evaluation of this site and proposal. Why do you think that it was considered to be one of the better sites, and one that might be on the final shortlist?

- What do you think the good points are about the plan? What do the local people feel about it?

- Look at the resources on Weston Otmoor and carry out an evaluation of this site and proposal. Why was opposition to it so strong? Why is there little chance of this proposal going ahead?

Sample resources for Section A – Sustainable cities

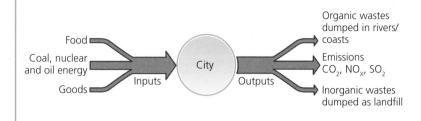

Figure 3: The unsustainable city

Why we need sustainable cities

Many urban areas and settlements are unsustainable in the following ways:

- They 'suck in' enormous quantities of food, water, energy and other resources from the surrounding regions – and even from all over the world – so they have large eco-footprints.

- They produce enormous amounts of waste, the majority of which is dumped, because little waste is recycled.

- The high resource use and the dumped waste results in high levels of land, water and air pollution.

- They lack open space, and are very overcrowded and noisy, which leads to a poor quality of life for their inhabitants.

- They are frequently very congested, with many people having very difficult journeys to work, resulting in stress.

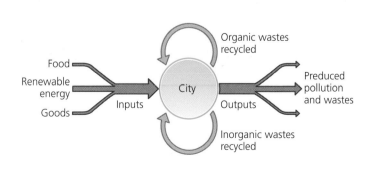

Figure 4: The sustainable city

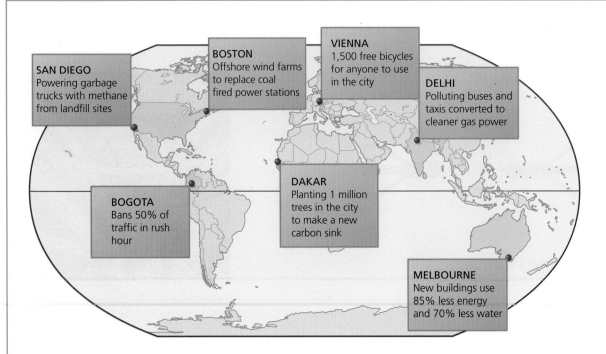

Figure 5: Sustainable initiatives in cities around the world

Two rural ideas for the future

Dongtan (near Shanghai, China)

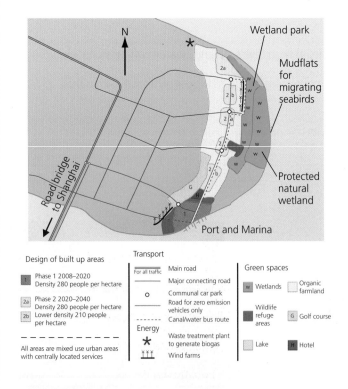

Figure 6: The plans for the eco-town of Dongtan

- Initially a modest-sized community of 5,000.

- By 2020 an eco-town of 80,000.

- By 2050 a series of eco-towns to form an eco-city of 500,000 people.

- Planning problems in China mean the development should start in 2010. The main problem was the impact on the migrating birds who live in the estuary mudflats.

Masdar (United Arab Emirates)

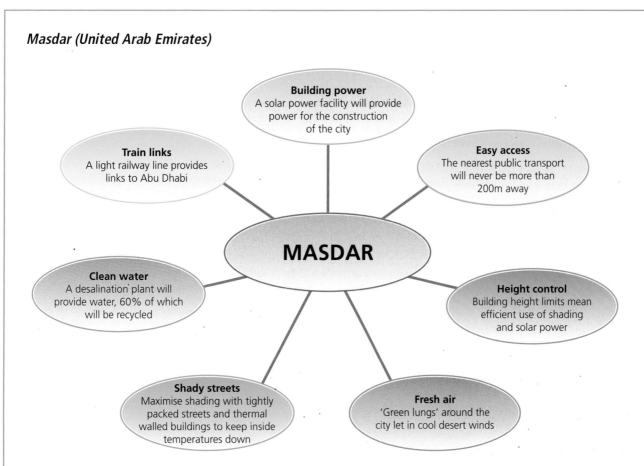

Building power
A solar power facility will provide power for the construction of the city

Train links
A light railway line provides links to Abu Dhabi

Easy access
The nearest public transport will never be more than 200m away

MASDAR

Clean water
A desalination plant will provide water, 60% of which will be recycled

Height control
Building height limits mean efficient use of shading and solar power

Shady streets
Maximise shading with tightly packed streets and thermal walled buildings to keep inside temperatures down

Fresh air
'Green lungs' around the city let in cool desert winds

Figure 7: Features of Masdar

WELCOME TO ECO-CITY – NO FUEL BILLS, NO TRAFFIC JAMS, NO POLLUTION. MOVE IN NOW!

There are several revolutionary plans for brand new eco-cities. Unlike existing cities, which can only develop green ideas to modify their existing infrastructure, these new eco-cities can be built as models for the future – as test beds for sustainable ideas. They can be zero-carbon communities that 'showcase' green technology, and demonstrate to the world what twenty-first-century planning can achieve.

- Building began in 2009.
- Funded by an oil-rich desert state.
- Cars banned entirely.
- Seen as a 'test bed' for new technological development.

The planned land use is:

- 30% housing
- 24% special economic industry
- 19% zone service and transport
- 13% commercial space
- 8% cultural space
- 6% university.

Sample resources for Section B – The controversy about the plans for eco-towns in the UK

Introduction to eco-towns in the UK

The proposed eco-towns are small new towns for the twenty-first century, with between 5,000 and 20,000 houses. They have three aims:

(1) To respond to the challenge of climate change.
(2) To meet the need for more sustainable housing.
(3) To increase housing supply, including affordable housing.

They will have sustainability standards that are significantly above the levels in existing towns and cities. Because they are in the countryside they are not intended to be entirely self-sufficient, but they should be linked to existing settlements by high-quality public transport (*Living a Greater Future*, a government report, 2008).

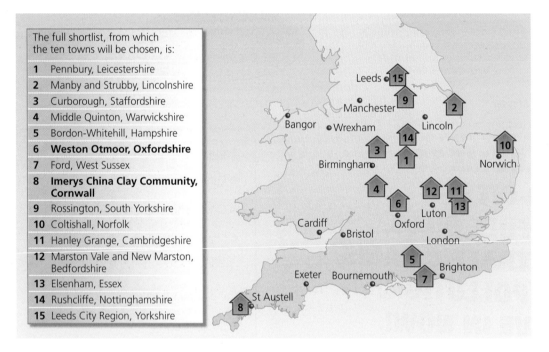

The full shortlist, from which the ten towns will be chosen, is:

1 Pennbury, Leicestershire
2 Manby and Strubby, Lincolnshire
3 Curborough, Staffordshire
4 Middle Quinton, Warwickshire
5 Bordon-Whitehill, Hampshire
6 **Weston Otmoor, Oxfordshire**
7 Ford, West Sussex
8 **Imerys China Clay Community, Cornwall**
9 Rossington, South Yorkshire
10 Coltishall, Norfolk
11 Hanley Grange, Cambridgeshire
12 Marston Vale and New Marston, Bedfordshire
13 Elsenham, Essex
14 Rushcliffe, Nottinghamshire
15 Leeds City Region, Yorkshire

Figure 8: The fifteen proposed eco-town sites. These were the long list proposals in 2008, from which there may be a final shortlist chosen.

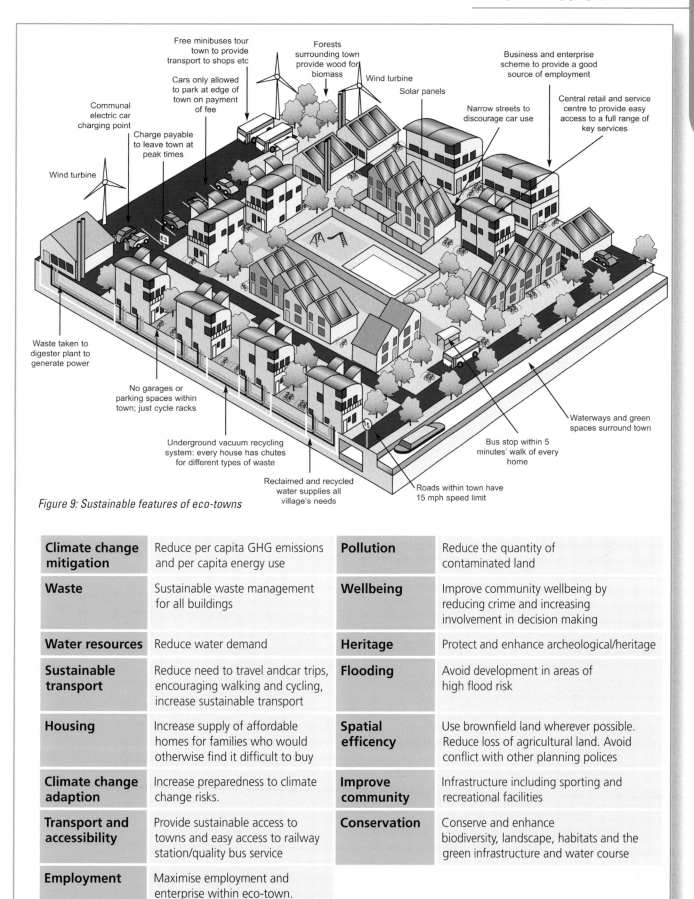

Free minibuses tour town to provide transport to shops etc

Cars only allowed to park at edge of town on payment of fee

Communal electric car charging point

Charge payable to leave town at peak times

Wind turbine

Forests surrounding town provide wood for biomass

Wind turbine

Solar panels

Narrow streets to discourage car use

Business and enterprise scheme to provide a good source of employment

Central retail and service centre to provide easy access to a full range of key services

Waste taken to digester plant to generate power

No garages or parking spaces within town; just cycle racks

Underground vacuum recycling system: every house has chutes for different types of waste

Reclaimed and recycled water supplies all village's needs

Roads within town have 15 mph speed limit

Bus stop within 5 minutes' walk of every home

Waterways and green spaces surround town

Figure 9: Sustainable features of eco-towns

Climate change mitigation	Reduce per capita GHG emissions and per capita energy use	**Pollution**	Reduce the quantity of contaminated land
Waste	Sustainable waste management for all buildings	**Wellbeing**	Improve community wellbeing by reducing crime and increasing involvement in decision making
Water resources	Reduce water demand	**Heritage**	Protect and enhance archeological/heritage
Sustainable transport	Reduce need to travel andcar trips, encouraging walking and cycling, increase sustainable transport	**Flooding**	Avoid development in areas of high flood risk
Housing	Increase supply of affordable homes for families who would otherwise find it difficult to buy	**Spatial efficency**	Use brownfield land wherever possible. Reduce loss of agricultural land. Avoid conflict with other planning polices
Climate change adaption	Increase preparedness to climate change risks.	**Improve community**	Infrastructure including sporting and recreational facilities
Transport and accessibility	Provide sustainable access to towns and easy access to railway station/quality bus service	**Conservation**	Conserve and enhance biodiversity, landscape, habitats and the green infrastructure and water course
Employment	Maximise employment and enterprise within eco-town. Improve diversity of local economy		

Figure 10: The sustainable objectives for eco-towns

Opinions about the idea of eco-towns

These eco-towns will, with few exceptions, fall far short on their green credentials, as the listed sites are unlikely to be truly sustainable. (Campaign to Protect Rural England)

These New Towns could become soulless suburbia or countryside shelters unless they were well linked to existing settlements. If they became too big they could 'kill' existing market towns. (Royal Town Planning Institute)

We already have nearly 300,000 long-term vacant homes on our books. Building new eco-homes emits 4½ times more carbon than refurbishing old homes. (Empty House Agency, a housing NGO)

We will abandon eco-towns. All but two of the fifteen sites are in Conservative seats. We will only sanction the go-ahead if there is no significant local opposition. (Conservative Party spokesman)

These towns could be the eco-slums of the future as we fear that they will be built without due regard to where the residents can get jobs and training. (Local Government Association)

We broadly welcome the project, as it will deliver nearly 200,000 new homes to high environmental standards. (Town and Country Planning Association)

These proposals are doomed by the credit crunch. Builders are short of cash, the banks are afraid to risk finance and the local people hate them. It will be down to a couple of test bed sites only – death of another government big idea. (Political journalist)

Eco-towns will kill our countryside

The politicians in London have come up with a sinsister plan designed to destroy England's rural heritage once and for all: the eco-town. These poorly-sited developments will shatter the fragile communities which already exist, putting an intolerable burden on roads, resources and...

Posted by a leading member of Crede, the Committee for Responsible Ecological Development Elsewhere

Comments

1 08:35 today Another blethering NIMBY protecting his turf. Next.

2 08:41 today This is one of those complex town vs country issues, and as someone who lives in London, I don't care.

3 08:49 today People who live in the country are always complaining about the lack of affordable housing, but the minute somebody suggests building some, a load of down-shifting luvvies start forming committees.

4 08:58 today I am not a NIMBY. I just happen to care passionately about protecting the heritage of rural communities right near my house.

5 09:07 today These homes must be built somewhere, but I believe that many developers are passing off concrete commuterghettos as 'eco-towns' in order to circumvent planning restrictions.

6 09:11 today It's not enough houses, and it won't stop global warming. When will people realise that there is no solution?

7 09:13 today I agree we need to do something about climate change, and we do need more affordable homes in this country. What we don't need is a lot of outsiders swarming into the village and taking over the pub quiz.

The key criteria for eco-towns

- Eco-towns must be new settlements, separate and distinct from existing towns but well linked to them. They need to be additional to existing planse, with a minimum target of 5,000 homes.

- The development as a whole should reach zero carbon standards, and each town should be an exemplar of not only combating climate change but also in at least one other area of environmental sustainability such as energy or transport.

- Eco-town proposals should provide for a good range of facilites within the town – a secondary school, a medium retail centre and leisure facilities (to encourage self sufficiency).

- Affordable housing should make up between 30% and 50% of the total with a particular emphasis on large family homes (this is a very controversial feature for existing rich village dwellers).

- A management body which will help develop the town, provide support for people and businesses moving to the new community, and use simplified planning processes to support innovation.

- Eco-towns Prospectus 2007

274

Compatibility matrix of eco-towns' sustainability objectives

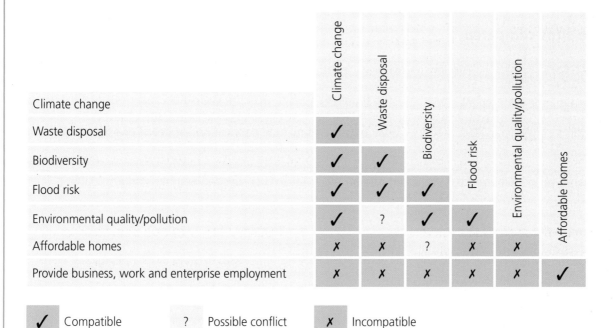

	Climate change	Waste disposal	Biodiversity	Flood risk	Environmental quality/pollution	Affordable homes
Climate change						
Waste disposal	✓					
Biodiversity	✓	✓				
Flood risk	✓	✓	✓			
Environmental quality/pollution	✓	?	✓	✓		
Affordable homes	✗	✗	?	✗	✗	
Provide business, work and enterprise employment	✗	✗	✗	✗	✗	✓

✓ Compatible   ? Possible conflict   ✗ Incompatible

What the Campaign for Rural England (CPRE) looks for in a sustainable site

- Large amounts of brownfield land – limited greenfield

- Low levels of opposition

- Potential for business and employment

- Avoiding high-/medium-quality agricultural land

- Avoiding flood risks

- Avoiding areas of high biodiversity or vital green spaces

- Fitting in with existing local and regional planning strategies

- Availability of sustainable transport links

- Nature of existing site and settlements. Would it improve area?

Sample resources for Section C – Evaluating two of the proposed eco-town sites: Imerys (Cornwall) and Weston Otmoor (Oxfordshire)

The two sites compared

Imerys

- The area is in need of regeneration – it has a poor image, with many derelict china clay sites.

- Cornwall is currently an EU 'Category 1 area' because of its rural deprivation – it needs investment and employment.

- The plan is for genuinely mixed use and mixed tenure, with a much-needed emphasis on affordable housing.

- The proposal to build on six separate sites suggests that development of sustainable transport will be vital.

- The scheme (which is near the highly successful Eden Project) fits in with existing local and regional plans.

- It would be largely built on brownfield land – but still 25% on greenfield sites.

Weston Otmoor

- This is an agricultural area – with medium- and high-quality land.

- This is mainly a greenfield site (16% is technically brownfield but it is currently grassland).

- There is a big risk that the development will be car-based because this district has highest level of commuting in England.

- 25% of the land is in Oxford's greenbelt.

- The plan would have an impact on a nearby wetland SSSI.

- The site is in the River Cherwell basin, which is an area of long-term water stress (although the site itself is prone to flooding).

- The plan might divert resources from the regeneration plans for nearby Bicester.

- The local infrastructure is already stressed because this part of Oxfordshire (just off the M40 – and A34) is a very-fast growing area for wealthy middle-class housing.

- New settlements are consistently rejected by the planning process.

- There is very well organised opposition to the plan.

Source: CPRE Report on Eco-towns, 2008

The Imerys plan

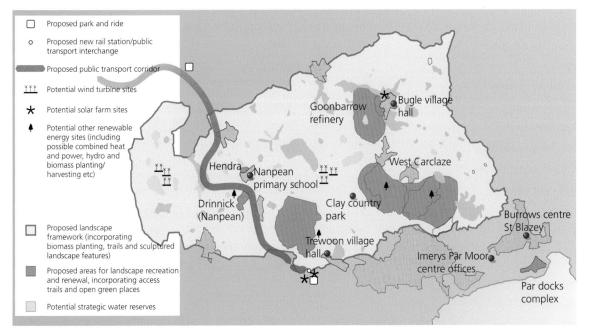

Figure 11: The six sites in the Imerys plan

When the Imerys China Clay company was restructuring its operations in Cornwall it saw an opportunity to link up with the government's eco-town initiative. The plan – known as 'Clay Country Vision' – would involve 700 hectares of the company's land in six separate sites (see Figure 11). The vision is to promote the mid-Cornwall china clay area as a national centre of innovation in sustainable living, employment, education, tourism and recreation. Together with the South West Rural Development Agency and the local councils, the idea would be to provide an environment for job-led growth and to improve transport in the local area. 225 hectares of the brownfield land will be reclaimed and regenerated, with a new link to the A30.

All six sites will be surrounded by open spaces for recreation and to promote biodiversity. The sites are:

1. Goonbarrow Refinery – new employment and 500 houses (including affordable homes) set in a restored and regenerated environment.

2. Drinnick and Nanpean village – a new business park on a landscaped site (possibly powered by a biofuel power station) and 210 affordable homes.

3. West Carclaze – a high-quality technology park to foster businesses, from small start-ups to larger businesses.

4. Baal – 1,000 zero-carbon-emission homes in a beautiful parkland and lakes (in a former quarry), with new community buildings and facilities.

5. Blackpool Pit – a mixed development of employment and 2,000–2,500 homes, plus recreational facilities, a solar farm and a transport interchange.

6. Par Docks – a mixed-use waterfront development with 200–500 homes, marina, retail and leisure .

(www.claycountryvision.imerys.com)

The opinions of the local people about the Imerys plan

As a result of extensive consultation (160 replies received) the proposal received considerable support.

Policy	Supported by
Regeneration strategy	79%
Sustainable housing	82%
The A30 link	77%
Reuse of brownfield sites	81%
Renewable energy proposals	83%
New job opportunities	86%

The responses recorded the need for better transport links, greater job diversity, affordable houses and improved community services. There was some concern over environmental impacts of the developments (which landscape restoration would help to meet). Concern was also expressed that some greenfield land would be used.

There was strong support for all sites, especially Par Docks. Major concerns were over the scale and pace of the scheme (especially the Blackpool Pit site) and the need to build the housing in Cornish traditional style. The worst fears concerned transport issues and congestion – the area already has problems with holiday traffic in the summer and the Eden Project visitors all year round. The biggest plus was the redevelopment of derelict land.

Headlines from local newspapers about the Imerys plan

The Weston Otmoor plan

- Located 12 km north of Oxford

- Site covers 828 hectares, of which 84% is working farmland

- 12,000 jobs

- 15,000 homes (for 35,000 people)

- 30% affordable homes (1.5% social housing)

- Compact urban core covering only a quarter of the site

- Limited private car access, but free trains to Bicester and Oxford

- Combined heat and power station (CHP)

- Water and waste management innovation

- Small communities, local services throughout the town

- Site developer: Parkridge.

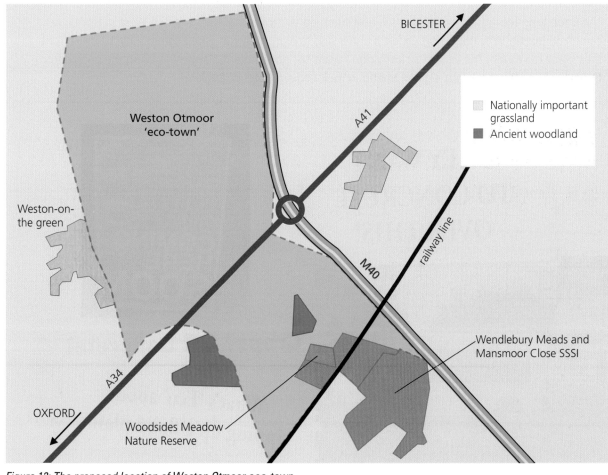

Figure 12: The proposed location of Weston Otmoor eco-town

Objections to the Weston Otmoor plan

The Arup report, commissioned by Oxfordshire County Council and the South East Development Agency, concluded (2009) that the Weston Otmoor development would have a damaging impact on the neighbouring towns of Bicester and Kidlington.

In addition to these regional concerns, there were numerous local objections, including:

- The overwhelming of the historic village of Weston on Green and the destruction of a viable and vital local community.

- With only 12,000 jobs there would be considerable extra car commuting on the already congested A34.

- The limited amounts of truly affordable social housing.

- The transport improvements would not fit in with existing plans.

- The possibility of congestion charging for car use, in and out.

- The eyesore of a huge 6,000-space car park right near the SSSI.

- The very attractive rural landscape would be engulfed by a new town.

- Fast-track planning for eco-towns could lead to a 'land grab'.

The underlying feeling among residents is best summed up by the statement on the Weston Front campaign poster 'Why not export Oxford's underclass to a privately run gated community 10 miles from their place of work?'. The residents clearly don't want the government planners to bring the people from Oxford's problem estates into a gated ghetto in their beautiful part of England.

Figure 14: The Weston Front campaign (www.westonfront.com)

Sample exam questions for Section A

Foundation paper

Question 1

(a) (i) What is meant by an 'eco-footprint'? (2 marks)

(ii) State three ways in which unsustainable cities have a large eco-footprint. (3 marks)

(iii) How is the model for a sustainable city (Figure 4) different from the model for an unsustainable city (Figure 3)? (2 marks)

Inputs _____

Outputs_____

(b) Look at Figure 5. Which of the cities has sustainable initiatives designed to:

(i) Cut down on the use of cars?

(ii) Enable carbon to be stored?

(iii) Cut down on the use of fossil fuels? (3 marks)

(c) State two ways in which the plans for Dongtan help to conserve wildlife. (2 marks)

(d) Explain how the plan for Masdar has been adapted to the local desert environment. (4 marks)

Total for Question 1 = 16 marks

Higher paper

Question 1

(a) (i) What is meant by an 'eco-footprint'? (2 marks)

(ii) Explain why unsustainable cities have a large eco-footprint. (3 marks)

(iii) In what ways do the models for an unsustainable city (Figure 3) and a sustainable city (Figure 4) differ? (2 marks)

(b) Choose three of the city initiatives shown in Figure 5 and suggest, using your own words, how they could be sustainable. (3 marks)

(c) Explain how Dongtan is aiming towards zero-carbon emissions. (3 marks)

(d) Explain how the plan for Masdar is adapted to the needs of the desert area of Abu Dhabi. (3 marks)

Total for Question 1 = 16 marks

Sample exam questions for Section B

Foundation paper

Question 2

Use the resources in Section B to answer the questions.

(a) Describe the general location of the proposed eco-town sites shown in Figure 8. (2 marks)

(b) (i) Which two of the sustainability objectives shown in the compatibility table are likely to cause most conflicts? (2 marks)

(ii) For either of your choices, suggest reasons why. (2 marks)

(c) State how eco-towns will be environmentally friendly in terms of the following:

(i) Energy generation (2 marks)

(ii) Waste disposal (2 marks)

(d) Look at the opinions about the idea of eco-towns and the blog.

(i) Choose one opinion which is supporting the building of eco-towns and explain why. (2 marks)

(ii) Suggest reasons why many people think that very few eco-towns will be built. (4 marks)

Total for Question 2 = 16 marks

Higher paper

Question 2

Use the resources in Section B to answer the questions.

(a) Describe the location of the proposed eco-town sites shown in Figure 8. (2 marks)

(b) Explain how the need to provide affordable homes and to make eco-towns 'provide business, work and enterprise employment' conflicts with the other sustainability objectives shown in the matrix on page 275. (4 marks)

(c) Explain how the design of eco-towns makes them environmentally friendly. (4 marks)

(d) Summarise the arguments raised in the blog about the evils of eco-towns. (6 marks)

Total for Question 2 = 16 marks

Sample exam questions for Section C

Remember – on these longer questions you are assessed for the quality of your written communication.

Foundation paper

Question 3

(a) (i) Examine the advantages of Imerys as a location for an eco-town. (6 marks)

   (ii) State three disadvantages of the Imerys proposal. (3 marks)

(b) Explain why there was opposition to the choice of Weston Otmoor at a local and regional level. (9 marks)

Total for Question 3 = 18 marks

Higher paper

Question 3

Use all the resources in the booklet, but with particular reference to those in Section C.

(a) Explain why the Imerys proposal for an eco-town has received a **generally** favourable reaction and could make the shortlist. (9 marks)

(b) Suggest reasons why the proposal for Weston Otmoor led to such enormous opposition and is therefore likely to be rejected. (9 marks)

Total for Question 3 = 18 marks

Unit 4 Researching geography

Your course

The controlled assessment is an investigation in which you will be asked to complete a fieldwork task, chosen from a list provided by Edexcel. Having completed your fieldwork, you will then be required to compile a report based on your results, produced under examination conditions and supervised by your teacher.

Your assessment

This list shows you how the marks (50 in total) are allocated between the different elements, the suggested time to be spent on each, and what is actually being assessed.

Planning the topic for study (8 marks)
Suggested time: 5 hours

Evidence of your planning, including background research, how you set the context, justify the study, establish the aims and describe the location.

The methods of collecting data (7 marks)
Suggested time: 1 day for collecting your data, plus 2 hours for your write-up

Your description and explanation of the methods of data collection, sharing and selection that you use.

Data presentation and report production (15 marks)
Suggested time: 5 hours

The data presentation techniques you choose and their quality, together with the overall presentation of your report (its structure, use of GIS, use of geographical terminology, grammar, etc.).

Analysis and conclusions (14 marks)
Suggested time: 5 hours

How the findings of your investigation are brought together and conclusions are drawn, including their links back to your original aims and, where relevant, to geographical theory.

Evaluation (6 marks)
Suggested time: 3 hours

Your evaluation of your methods of data collection and presentation, and the analysis and conclusions drawn.

What are the controls in the controlled assessment?

You need to be aware of the level of control which occurs at each stage in the investigation. There are two levels which affect the way you must work:

High level of control

Certain stages of your investigation must be completed individually – by yourself without help from others, in the classroom and under the close supervision of your teachers. This work cannot be taken home – it has to be handed in at the end of each lesson, so that you cannot continue working on it in your own time. These stages are:

- Deciding on the focus of your investigation. The list of possible investigations is created by Edexcel, and your teacher will help you decide the focus of your investigation.

- The completion of your final report, including the writing up of your analysis, conclusions and evaluation.

Limited level of control

For some stages of your investigation, you will be able to work in your own time – at home, in the library or elsewhere. You will also be able to work in groups. These stages are:

- Any secondary research you need to do, such as reading about the processes involved in your work or the location of your sites.

- The writing up of your purpose and methods of data collection.

- The collection of your data. You can work in groups to collect data, but any extra research should be completed individually.

- The development of your data presentation.

Chapter 18 Your fieldwork investigation

Objectives

- Identify aim(s) suitable for small-scale fieldwork.

- Develop clear methods to gain data in response to the aims.

- Be able to present geographical data.

- Understand the geographical processes responsible for patterns seen in the data.

- Be able to analyse and explain geographical processes.

- Be able to evaluate the strengths and weaknesses of the study when it has been completed.

Planning your study

Planning the topic for study (8 marks)

To get top marks you need to:

- Provide a clear, focused statement of your aims, purpose and location of the issue you will study

- Include appropriate maps, including GIS

- Say why you have chosen the study - does it link to any geographical theories?

- Use secondary data and research to inform your study

An important element of geography is its use in the real world. It is a subject which allows us to investigate issues and problems with the intention of making a situation better, or developing a better understanding of a chosen issue. Most geographical investigations cannot be achieved by the use of a laboratory or computer simulation alone – they require us to collect information through the use of fieldwork. This 'Controlled Assessment' is an opportunity to develop the skills which are important in becoming a geographer, skills which are also transferable to other situations.

Planning your study

Identifying the purpose of your investigation

All studies require a purpose. Geographical enquiry is not simply a case of walking around a place and seeing what we can see. In this assessment, you will be asked to react to a question which is deliberately written to allow you to develop a more focused purpose of your own. For example, a typical task question might read: *'Remote rural areas must diversify if they are to survive.' How far does the research you have carried out make you think that this statement is correct?*

This is obviously a very wide-ranging question – which would need a great deal of research and a huge amount of writing to answer properly. You therefore need to choose a smaller focus from such a question – a specific issue which you can investigate. In the case of the rural areas question above, you might aim to record the services available to a rural settlement and compare them to those which existed 30 years ago, in order to understand if diversification or greater diversity has occurred and, if so, in what ways. You might alternatively decide to interview local service providers to see if they feel that diversification is important for the survival of the area.

One of the first things that you must do in developing your work is to give a clear aim for the issue to be studied, such as 'to gain the perceptions of local service providers concerning the need or otherwise to diversify services in X'. You will also need to locate your study clearly.

Locating your investigation

To make your investigation informative for the person reading it, you must first locate it by describing where you carried out your fieldwork. This description will normally include various pieces of information (Figure 1) and maps to show the location graphically. You should also consider why the location you have chosen was appropriate for your study.

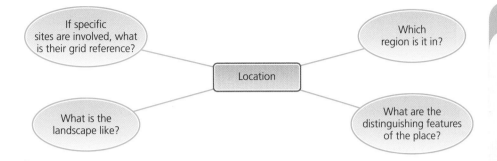

Figure 1: Questions to consider when writing a location description

The location of your study is an ideal opportunity to use websites such as Google Earth or Multimap, as GIS is a required element of the controlled assessment. Figure 2 shows a possible pair of images for showing the location of two villages used to investigate service provision. Annotated and used with a short written description, this would lead to a clear impression for the reader. Figure 3 does the same thing for an investigation of a river. Note the difference in the scale of the images used and how, in both cases, the images have been annotated so that they have some clear information regarding the locations of the investigation.

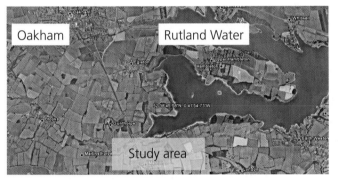

© 2008 Europa Technologies © 2007 Google
© 2008 Tele Atlas
© 2008 Infoterra Ltd & Bluesky

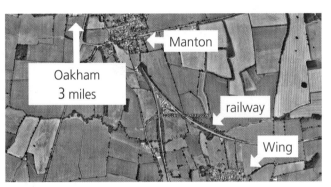

© 2008 Tele Atlas © 2007 Google
© 2008 Infoterra Ltd & Bluesky

Figure 2: Images taken from Google Earth used to locate study villages

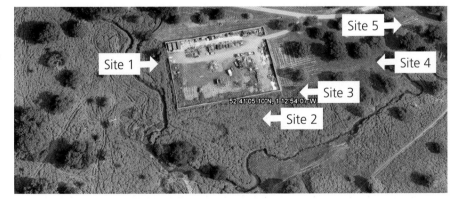

© 2008 Tele Atlas © 2007 Google
©2008 Infoterra Ltd & Bluesky

Figure 3: An image used to locate sites for a river investigation (note the difference in scale from those images used in Figure 2)

ResultsPlus
Exam Tip

Make sure you annotate any maps you include, giving details about the locations which are important, and perhaps the data collection methods you have used at particular locations.

Control

When you collect your data, there is only 'limited control' – so you can work as part of a group, if appropriate.

288

Results**Plus**

Exam Tip

Remember to explain why you have used a particular sampling method.

Methods of collecting data

The methods of collecting data (7 marks)

To get top marks you need to:

- Include a clear description of the methods you used to collect and record data

- Explain the methods you used to collect and record data. Say why you did it that way

- Link your evidence of data collection to the task

- Use GIS, e.g. to select your site

Having stated a clear aim for your study, and having linked it to a clear location, you then need to contextualise your study. To do this you should include:

1. **A justification** – You should justify your particular study. Why is it important or interesting? Is there a particular local issue which makes the study relevant?
2. **Background information** – You should try to find some simple secondary data, which can be used to give a greater sense of the local area. This might include a headline from a local newspaper which relates to your aim, and some data from the National Statistics website which gives some idea of the population of the area, or its level of wealth, if these issues are relevant to your chosen aim.

In other words, the purpose of the contextualisation is to demonstrate some of the main relevant features of the area or location you are studying. You should also consider the format and medium of your report at a very early stage. You may have the opportunity to choose from a number of different formats when developing your final report.

Methods of collecting your data

When you have introduced your study, and described the locations to be used, you will then need to decide which methods you are going to use for collecting your data. This requires careful planning, and includes a consideration of the types of data that you will use – primary data or secondary data, or both.

Primary data	Data which is collected first hand. In the case of a school-based investigation this is the data which is collected by a group of students while undertaking an exercise of fieldwork.
Secondary data	Data which has already been collected by others for a particular purpose, and then 'published' on websites, in books, official reports, etc. It can be included to support primary data in an investigation. A good example is the UK census, which is taken every ten years and provides a large amount of data about the population for academic studies.

Sampling

As well as deciding on the types of data which are to be collected, the sampling method must be considered. Sampling is nearly always necessary because it is just not possible to measure every item or interview every person. A carefully chosen sample will be much easier and quicker to investigate and will still give fairly accurate results. Choosing a representative sample – the when, what, which and where of the data collection – is an important part of the process. If you were using questionnaires, for example, you would need to make sure that the right people were targeted. There is little point conducting a survey to find out which services young people in a town would like to see developed if you only go to a skate park for views, as only a single interest group is likely to be asked. This would give you a very untypical data set and would call into question any eventual analysis of that data. Therefore, you need to decide on the timing of your data collection, as well as being sure of the groups you wish to measure, to make sure that the results gained are not biased. There are a number of ways in which sampling can be carried out, and you should decide which is most appropriate for your investigation.

Sampling methods
There are a number of ways in which samples can be taken, so you should be careful to plan – and identify – how you have carried out your sampling.

Random sampling – the locations for data collection are chosen by chance. An example might be the use of a quadrat to count the number of plant species at a number of sites. As you finish at one site, you would throw the quadrat and count the next site as being where it landed.

Systematic sampling – the locations of the sites used are found at equal intervals from each other. This might be each fourth shop within a shopping area, or at points on a grid if sampling the size of pebbles on a beach.

Line-intercept sampling (also known as a transect) – sites are sampled along a line. This might be used to measure the environmental quality across a city centre, or a river at points along its course.

Having decided on the sampling strategy of the study, and having perhaps gained some secondary data which can be used as background information to support your work, you need to decide on the primary data collection methods you wish to use. In geography there are a whole range of techniques which can be used to collect data, and you should look beyond the few introduced below for further ideas. Remember that you need to consider how your data collection will feed into your presentation and analysis.

Sketching/photos
It is often useful to include a clear picture of your main sites, or close-up details of the ideas which you discuss, e.g. photos of graffiti. Sketches should be annotated to show features, processes, etc., giving a clear description and explanation of important features (see Figure 4 and Figure 5).

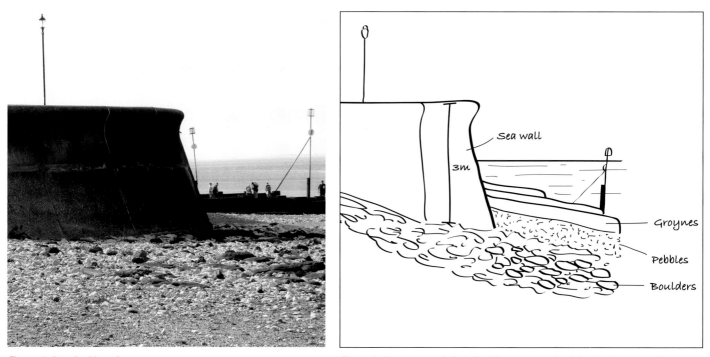

Figure 4: A typical beach

Figure 5: An annotated sketch of Figure 4, emphasising its important features

Mapping

Geography uses mapping a great deal. There are many different types of mapping. You might be looking at parking provision, and need to shade in areas of car parking on a street map. You might be looking at the pattern of plant growth on some sand dunes, and use a map of the dunes to shade in where different plant species are found. A frequently used method is that of land use mapping, where different types of shop or other building use are identified by differently shaded colours. You could look up different types of map in books to see what might be possible, and then consider the type of map you need. At this stage, you may well need to ask for help in finding the correct map. You might use:

- An Ordnance Survey map

- A town street plan

- A sketch map that you have created yourself.

Questionnaires

Questionnaires are a very useful way in which to collect data – but you need to consider carefully what you are trying to find out. There are a number of choices you need to make:

- Which questions will allow you to collect the information you need, without making the questionnaire too long?

- Will the questions be open (allowing respondents to offer opinions or supply information) or closed (requiring respondents to choose from a set of fixed alternatives, such as Yes/No, 1/2/3/more than 3)?

- Can some questions be answered without being asked, such as the sex of the person you are interviewing, and possibly estimating their age rather than asking for it?

- Will you fill in the questionnaire yourself, from someone's answers, or will they fill it in by themselves?

- How will you introduce yourself when you first ask someone to respond to your questionnaire?

Once you have thought about these issues, you should draft a questionnaire and then ask a friend to pilot it by filling it in, to make sure that it works as you want it to.

Measuring physical features

Measurement is central to the study of physical geography. It includes activities such as measuring:

- the width of a river

- the density of species found in a woodland

- wind speed.

ResultsPlus
Exam Tip

⚠ As with sampling, you will need to be able to justify why the methods you have chosen are appropriate.

In each case it is important to make sure that the measurements are taken accurately. By using appropriate sampling strategies, and measuring particular features, a good understanding can be gained as to the processes occurring in a physical environment (see Example 1 below).

Data sheets

You should be confident about developing a clear and efficient data sheet from your work on your local area. Remember that you should leave plenty of space to ensure that results can be clearly accommodated. If this means you need to use more than one sheet of paper, then do so.

Example 1: Basic measurements used to investigate river characteristics

When investigating the changing characteristics of a river as it flows downstream, a number of different measurements can be taken:

- Channel width, by using a tape measure

- Channel depth, taken in several equally spaced places across the stream so that a cross-section can be drawn. This can allow you to calculate the cross-sectional area.

- Water velocity, using a flow meter or float and stop watch

- Sediment size, measuring the size of sediment on the river bed. This might be done using a ruler (where the sediment is large) or sieves (where the sediment is small).

© 2008 Tele Atlas © 2007 Google
©2008 Infoterra Ltd & Bluesky
Figure 6: Location of Site 3

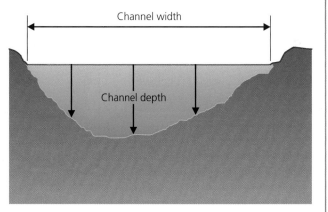

Figure 7: Cross-section of the river channel at Site 3

Whichever investigation you undertake, there is a clear opportunity to collect data as part of a group. If you carry out data collection in this way, it is important that you share that data before starting the subsequent elements of the controlled assessment. This can be done by working within the group to make sure that all data has been copied for each person where necessary, and that everyone knows the exact locations from which data has been collected, so that once the assessment becomes more controlled, there is no chance of someone not understanding the data.

Methods of presenting your data and producing your report

Once you have collected your data, you then need to decide how to present it, so that those reading your investigation can fully understand it. As with data collection, there are a number of ways in which you can present your data.

Tables and graphs

The simplest way of representing data is by using tables with the numeric results in them. Graphs and charts are more visually interesting and they can make a description and analysis of data very clear, but you need to be careful that you use the correct type for the data.

Some people simply work their way along a spreadsheet toolbar to provide some variety in their presentation, but this shows a lack of understanding of the use of graphs. You should use the type of graph that does the job most accurately and efficiently. Some basic types are explained below.

Line graphs

Line graphs (Figure 8) can be used to plot continuous data such as a population size increase over a number of years, or the changes in pedestrian flow over a number of hours. The variable should be plotted on the y-axis, while the time should be plotted on the x-axis. It is also possible to plot more than one set of a particular data type on one graph – for example, the temperature change at three sites – allowing for comparison.

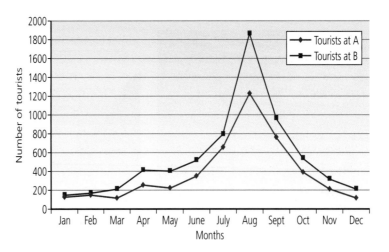

Figure 8: A line graph showing the distribution of monthly tourist numbers at two sites

Pie charts

This graph type is used to present group values such as the number of different transport types observed at a road junction over the course of a day (Figure 9). The values are first converted into percentages, and then into degrees to allow the plotting of the data. If you use a graphics package on a computer, this will normally be done for you automatically. Pie charts tend to be overused, and are best used when there are several categories of data in the group.

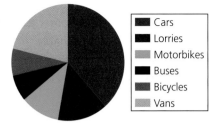

- Cars
- Lorries
- Motorbikes
- Buses
- Bicycles
- Vans

Figure 9: A pie chart showing traffic composition at an urban location

Bar charts

As with pie charts, bar charts are a common form of graph used in investigations. The y-axis is used for a numeric scale, such as the frequency of an event (Figure 10) while the x-axis is used to identify the categories of the data.

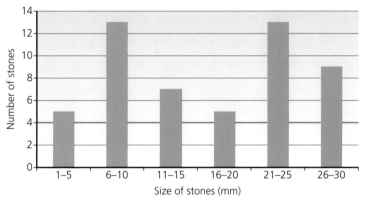

Figure 10: A bar chart showing the number of stones of different sizes found on a stream bed

Scattergraphs

Scattergraphs are more complex than the other types of graphs described here, as they do more than just present data visually. Scattergraphs are used to plot two sets of data to find out if there is a link between them. For example, you might count the number of services or amenities in 12 settlements, and find out what the population of each is. You would then plot the results as a number of points (Figure 11). These show a pattern from bottom left to top right. This is what is called a positive correlation, because as one variable (population size) gets larger, so does the other (number of services).

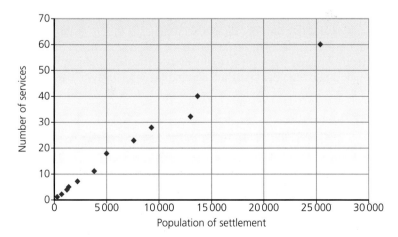

Figure 11: A scattergraph showing the relationship between settlement population and the number of services. (In this case, a positive correlation is shown.)

If a scattergraph shows a clear pattern of points from top left to bottom right (Figure 12), this shows a negative correlation – as one variable gets larger the other gets smaller. Both these patterns would show that there is probably some link between the two variables being plotted. But if the points are random and show no pattern (Figure 13), there is no apparent correlation, so the two variables are probably not linked.

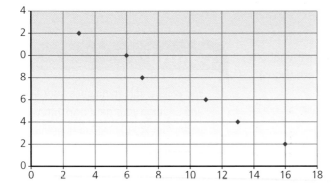

Figure 12: A scattergraph showing a negative correlation

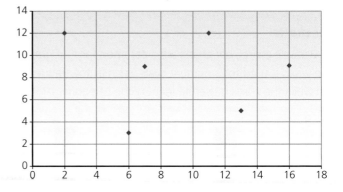

Figure 13: A scattergraph showing no correlation (i.e. random results)

Simple statistics

You may also want to use some simple statistics to interpret your data. The most frequently used statistics are the mean, mode and median. These can help in describing either the most frequently occurring or largest group of data in a data set, and can therefore help quantify and make clearer the patterns in the data you have collected.

Mean	The **mean** is the statistical average of a set of numbers. This is found by adding the values together and dividing the result by the number of values present. For example, the average of the four values 3, 4, 6 and 8 is 3+4+6+8 divided by 4 = 21 divided by 4 = 5.25.
Median	The **median** is the middle value in a set of numbers. It is found by arranging the values in order, and identifying the middle value. If the number of values is even, the two middle values are added together and divided by two. For example, with values 3, 6, 2, 7, 9 and 4, the median is found by first ordering the values: 2, 3, 4, 6, 7, 9, and then – because there are an even number of values – adding the middle two together (4+6) and dividing the result by two. Hence the median value is 10 divided by 2 = 5.
Mode	The **mode** is the most frequently appearing value in a group of numbers. For example, if a set of values is 5, 3, 4, 8, 5, 7, 5, 2, 3, 5 and 1, then the mode value is 5, because it appears more frequently than any other value.

Maps

Maps are another valuable way of presenting your data. You should try to make them clear and colourful. As with graphs, there are a number of different types of map which can be used.

Choropleth maps (Figure 14) use shading to show patterns in data, with shading normally becoming darker with larger numbers in the data. This type of map is used to compare areas in terms of grouped values, such as infant mortality rates, or house prices. Data must be sorted into groups, with clear boundaries, such as county boundaries, or regional boundaries.

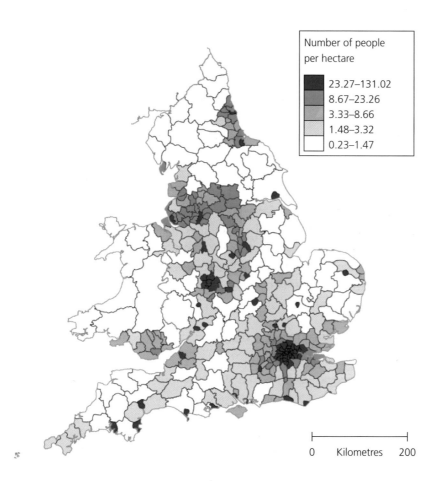

Number of people per hectare

■	23.27–131.02
▨	8.67–23.26
▨	3.33–8.66
▨	1.48–3.32
□	0.23–1.47

0   Kilometres   200

Figure 14: A choropleth map of population density in England and Wales

Flow lines (Figure 15) are used to show data relating to movement. Examples might include the number of pedestrians who walk down a certain street or the amount of traffic on a road. The direction in which the flow is moving is indicated by an arrow head at one end, and the width of the flow line is dependent on the volume of the flow. The scale used to determine the width of the lines must be chosen carefully, to allow the lowest and highest value to be shown clearly on the map.

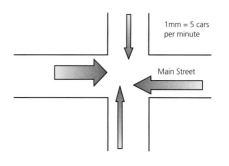

1mm = 5 cars per minute

Main Street

Figure 15: A flow line diagram showing traffic numbers at a crossroads.

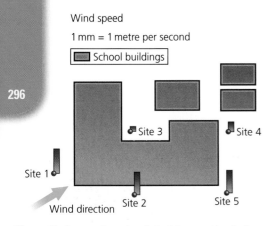

Wind speed

1 mm = 1 metre per second

School buildings

Figure 16: A map of a school's buildings, with wind speeds plotted at five sites

Where data cannot be easily grouped and shown on a choropleth map, you may want to plot separate graphs or charts on to a map (Figure 16). This is a way of showing simple data distributions relative to each other in space, such as environmental quality values at given sites in an area. Again, you need to consider scale carefully so that the highest and lowest values can be clearly shown on the same map.

If you are using maps to present data, this might be a good opportunity to use GIS applications. These packages are designed to allow for the presentation of spatial data, and will also make decisions on issues such as shading and scale far easier to handle.

Using GIS

You will need to demonstrate your ability to use GIS to generate images, especially maps, with data located on them. This can be done in many different ways, and these will in part depend on the GIS packages you have available to you. Some GIS software is available as a program on your computers, and where you have these, you may well have used them before and know how they work. Examples of the types of GIS package you might have used already include those such as Google Earth, and Google Maps, as well as dedicated mapping or GIS software, such as Anquet Maps, Infomapper, Aegis and ArcMapper. If you do not have software installed, there are a number of websites which can be used instead, all of which provide an alternative way of presenting data. You can even complete simple maps with data presented using standard packages such as PowerPoint. Whichever approach you take, your teacher will guide you as to how you can incorporate a GIS component into your study.

Examples of possible GIS outcomes
Presentation of data on a map
A standard GIS style activity is to plot data in a graphic form (normally as a pie chart, bar graph, etc.) on a map or image. This helps locate the data you collected as well as showing the data itself in a graphic form (Figure 17). This is easy to achieve by using either a GIS package, or even just plotting your data carefully, using PowerPoint, as with the example given here.

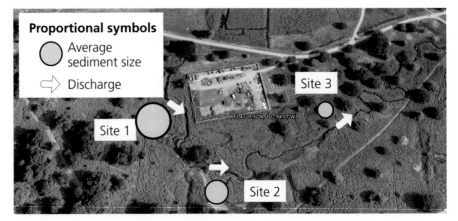

© 2008 Tele Atlas   © 2007 Google   ©2008 Infoterra Ltd & Bluesky

Figure 17: Presentation of data on an image to show changes in average sediment size and discharge

Creating land-use maps

Another GIS application is to create land-use maps, shading areas different colours according to their particular use (Figure 18). Again, this is a simple use of GIS, and can be completed on packages such as PowerPoint as well as dedicated GIS packages. It is also possible to combine both of these suggestions, by adding bar graphs or pie charts to your land-use map. But remember not to allow the maps you create to become confusing – it is better to use two maps and make sure they are clear to read, than one map which has too much information on it.

These two uses are just examples of the types of data presentation you should use to demonstrate that you are beginning to understand the uses of GIS. Your teacher will help you develop other ideas which you can develop for your study.

Producing your report

It is important that you carefully consider how you intend to present your fieldwork investigation. You have the opportunity to present your work in a number of different formats – just as professionals in a commercial setting might use different ways to present their information and ideas.

Figure 18: A simple land-use map using a draw tool in PowerPoint

Report formats

You can use a number of different formats in your report. The most obvious is a written report of approximately 2,000 words. Where this is chosen, it is simply a case of writing the section by hand, or using a word processor to write up your report. However, alternative formats can be used:

- DVDs – used perhaps where interviews are part of the data collection, or when you have filmed yourself explaining some of your results, or collecting data on a stretch of river.

- PowerPoint presentations – which might be written in a similar fashion to a word-processed piece, but completed as a presentation (Figure 19 on page 298).

- Personalised GIS maps – these could be used in data presentation to show results, and possibly in annotating satellite images to develop issues or ideas in a more graphic form.

- Web pages – these can be developed to create a website format rather than a simple word processed project. One advantage of this format might be the ability to link different parts of the report together so that the reader can move backwards and forwards between elements. You might also decide to include hyperlinks to other websites which might help give background information or provide sources of secondary data (Figure 20 on page 298).

You can use more than one format, but your teacher may decide to use particular formats, having considered both the availability of technology within your school and any organisational restrictions which might apply. This might lead to reports which can be handwritten, through to those which use other media.

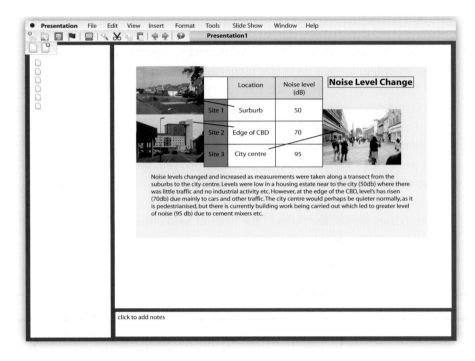

Figure 19: Example of a power point slide

![Website example showing "Noise Pollution in the City"]

Noise Pollution in the City

	Location	Noise level (dB)
Site 1	Surburb	50
Site 2	Edge of CBD	70
Site 3	City centre	95

Noise levels changed and increased as measurements were taken along a transect from the suburbs to the city centre. Levels were low in a housing estate near to the city (50db) where there was little traffic and no industrial activity etc. However, at the edge of the CBD, level's has risen (70db) due mainly to cars and other traffic. The city centre would perhaps be quieter normally, as it is pedestrianised, but there is currently building work being carried out which led to greater level of noise (95 db) due to cement mixers etc.

Figure 20: Example of a website

If you are given a choice of formats, you should consider the following points in planning and developing your ideas:

◉ You need to understand how the elements and formats of your report will fit together. Will the reader understand how to 'read' your work? Are you trying to use too many formats?

◉ Given that you have limited time, you should be confident that you can use the format you have chosen efficiently. A website might sound like an exciting idea, but if you do not know how to write one, you will not have enough time to teach yourself.

● Remember that the geography is what is important. Students can often get carried away in spending time on the design aspect of a PowerPoint presentation or website, and forget that it is the geographical content that they are gaining credit for.

You should decide on a format, if given a choice, at the initial planning stage and discuss it with your teacher so that you are both happy with your decision.

As you complete your fieldwork investigation, you will begin to gain a lot of paper, data, photos and other evidence. During the process it is important that you organise your work so that you do not lose information. And when it comes to writing up your study, it is very important that you organise your work and thoughts into a well-planned and coherent end product.

Putting a study together

When a study is planned, it needs to be clear what the parts of the finished product will be and how they will fit together. If you plan to use more than one medium, you should be clear before you start about how the different elements will come together to make sense. This means you should carefully consider the amount you write for each element of the report, and in each case make sure your writing is focused.

You must remember that it will only be possible for each element to be worked on for a limited period of time. So you should plan your time carefully, to ensure you finish each element – you do not want to run out of time, leaving some elements unfinished. You should always allocate time to check each element – making sure that you have presented your information well, and that the spelling and grammar is accurate (see below), as well as ensuring that everything flows properly from one section to the next.

Spelling, punctuation and geographical terminology

Finally, you need to be careful with your spelling and punctuation. If you are using a word processor, the software may pick up on spelling, grammatical and punctuation 'errors', but do not assume that the computer is always right. You must always read through what you have written and check it carefully. Remember that marks are obtained for good use of language, and that poor spelling and punctuation can lead to a loss of marks.

You should also use geographical terminology where possible. For example, rather than writing 'amount of water flowing through the river', use 'discharge'. You will gain credit for using geographical terminology – if it is used correctly – because it shows a higher level of understanding on your part, and can often lead to briefer, but better explanations.

Analysis and conclusions

Analysis and conclusions (14 marks)

To get top marks you need to:

- Analyse your data in detail using appropriate processing tools

- Identify links and connections between your data

- Include links to geographical theory

- Include clear, relevant and focused conclusions

- Provide evidence to support your conclusions

- Comment on the wider geographical significance of your study

- Provide links to the original aims of your investigation

Developing your analysis and reaching your conclusions

Once you have presented your results, you will need to consider how to make the best use of them in explaining the patterns you see. For students, the analysis of data can be the most challenging part in the process of completing a fieldwork-based project – often because they find it hard to distinguish between describing the data and explaining the data. Analysis always includes explaining what you have found.

Making sense of your data

It is useful to start an analysis by laying out all of the results which you gained during the data collection and data presentation phases. Make a list of the data collected. Next to each set of data, describe what you think it shows (using any graphs, tables or maps you have produced to help). Finally, try to explain the data, highlighting why you think the results appear as they do. Remember that you will need to describe your results – what do they show – before explaining them – why are the results as they are? The description of results should be completed fairly quickly, merely outlining what a group of data shows.

If you look back at Figure 14 on page 295, for example, you might describe the population density data given by highlighting the general pattern, i.e. that in general terms the population density is greater in and immediately surrounding large conurbations such as London, Birmingham and Liverpool, while rural areas have a much lower population density. This should then be supported by some specific examples such as Greater London having the highest population density of between 23.27 and 131.02 people per hectare, while North Devon has a low density of 0.23 to 1.47 people per hectare. Therefore, any description of the data should give a general impression of the patterns and include specific examples such as numbers, or quotes from interviews.

Explaining the patterns in your data

Your description of the data should be brief, while a greater focus should be given to explaining their patterns. This section of your project – where you give reasons for the patterns in your data – is crucial. You have presented and described your data, but can you now suggest why the results and associated patterns look like they do?

Consider the question *'How do channel characteristics vary along your chosen river?'* The results for the size of the sediment in the channel from several sites might appear as they do in the table below.

Site	Average size in millimetres	Sediment shape
1 (upstream)	12	Angular
2	8	Sub-angular
3	7	Sub-angular
4	5	Sub-rounded
5 (downstream)	2	Rounded

We can describe the pattern as one showing decreasing sediment size as we move downstream from an average of 12mm at Site 1 to 2mm at Site 5. We then need to explain this. There might be a number of explanations, but we could argue that the decrease in size is due to erosive processes in the channel, causing pebbles and gravel to constantly hit each other, leading to attrition of the sediment. It is at this point that we might then draw together some of our other results to make our argument stronger. If we have collected data on the shape of sediment at the five sites and can show that the sediment is becoming increasingly rounded, this might give extra evidence that attrition is occurring and wearing down the pebbles/gravel. Your explanations therefore need to give reasons for the data presented and, at the same time, they should also attempt to make links between different elements of your data, rather than simply reading it like a detailed list. Finally, to explain these results successfully, you need to relate your results and explanations back to geographical theory. In other words, are the explanations you are using the same, or similar to those that are suggested in your textbooks and other geographical sources which you have looked at? This link back to geographical theory allows you to consider if your results are those which would be expected when considered against what geographers have found before. For example, if studying service provision in rural areas, theory would suggest that smaller settlements will have fewer services. Do your results agree with theory, or are they different? If they are different, can you explain why they might be different?

Concluding your study

Having explained your results, you should write a conclusion. This should include the following three elements:

1. You should summarise your main findings, in relation to the extent to which they answer your original aim which you have given at the start of the study. Having posed that question, have you been able to answer it to some extent, and what is your answer?
2. You should clearly and concisely summarise the conclusions you have reached, supporting them from the evidence you have developed, and use this summary to answer the question posed in the original task question provided by Edexcel.
3. You should state what you believe the wider significance of your study to be. For example, if you have focused on coastal management in an area, how might your results be useful to others?

Therefore, having presented your findings in graphs, maps, etc., you then need to use these results to describe and explain what you have found, before concluding your study by relating what you have found to your original aim, and how your study might have wider significance.

ResultsPlus
Exam Tip

⚠ Remember that it is important to explain your results, not just describe them. Make sure you understand the difference between the two.

Control

Remember that the analysis and conclusion element of the assessment must be completed individually, in lesson time, under exam-style conditions.

Evaluating your study

Evaluation (6 marks)

To get top marks you need to:

- Include a good review and/or evaluation of the process and your findings

- Link your findings to the original task statement

- Consider the limitations of you data and findings

Evaluating your study

The evaluation of a piece of work is an area which students often find very difficult. This is the part of a study which aims to reflect on the processes of collecting data and analysing it, and how they might have impacted on the quality of the results gained and the conclusions made. It is very important that you accept that no study is perfect – even those carried out by university academics – and it is therefore perfectly reasonable to highlight where you think the shortcomings of your work are.

What is an evaluation?

An evaluation should be based on three basic questions:

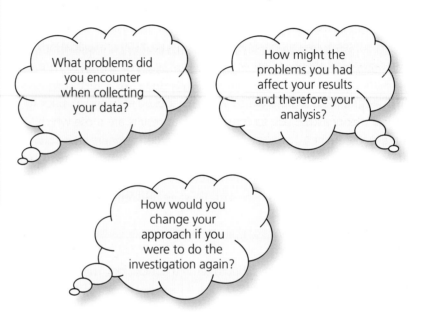

What problems did you encounter when collecting your data?

How might the problems you had affect your results and therefore your analysis?

How would you change your approach if you were to do the investigation again?

Exam Tip

⚠ When deciding on a report format, remember that it is the geography which is being assessed, not the standard of graphic ability.

What problems did you encounter when collecting your data?

When collecting data, you need to be aware of the possible problems involved. If you were measuring the impact of tourism on a local environment, collecting your data on a Bank Holiday Monday might have given unusual pedestrian flow counts. Whilst this is useful in showing the extremes of use that the environment sees, it might not be representative. If you were focusing on traffic volumes at locations around a CBD, you would not have been able to be in all the locations at once. This means that some of the differences in traffic might be due to time differences. For example, you might have visited two locations during the rush hour but, by the time you have reached the third, rush hour might have been over.

Whatever your focus, you should consider the shortcomings of your data collection. Remember that this evaluation does not make your results incorrect – it demonstrates that you have a clear understanding of the difficulties involved in any data collection exercise.

How might the problems you had affect your results and therefore your analysis?

If you have identified any problems experienced in collecting your data, you next need to suggest how they might have affected your results. If we take the example of traffic flows around a CBD, having described the problems in collecting the data, you might then go on to say that, given that the third location was visited after the end of rush hour, its results might have been lower because of the time difference. Hence, your analysis that the third location is much quieter than the other two may be correct, but you have to accept that it may in part be inaccurate and it is possible that the site might be much busier during rush hour, much like the other two sites. Having explained this, you should finally explain how this might impact on your conclusions. Again, you should remember that this should not be seen as showing that you have done a poor piece of work, but that you are aware of the impact of any problems on your results.

How would you change your approach if you were to do the investigation again?

Finally, having identified the problems you had when collecting your data, you should now suggest how you would try to alter your collection methods if you did the study again. Therefore, if you identified the problems with collecting traffic flow data within a CBD, you might suggest that you would alter the locations used to ensure that all of them could be covered in the rush hour, or that you would ask a friend or parent to collect data in one or more locations, so that the timings were as close to each other as possible. This section, therefore, is focused on developing solutions to the problems identified in the first part of the evaluation.

ResultsPlus
Exam Tip

▲ When planning your investigation, remember to refer back to pages 284 and 285 to give you an idea of how long your teacher will ask you to work on particular parts of your work, and the level of control which you will be asked to work under.

ResultsPlus
Exam Tip

▲ Your evaluation should not be descriptive, but should offer explanations to highlight reasons for any problems encountered – and how these might have impacted on your results.

exam zone

Welcome to ExamZone! Revising for your exams can be a daunting prospect. In this section of the book we'll take you through the best way of revising for your exams, step-by-step, to ensure you get the best results that you can achieve.

Zone In!

Have you ever become so absorbed in a task that it suddenly feels entirely natural? This is a feeling familiar to many athletes and performers: it's a feeling of being 'in the zone' that helps you focus and achieve your best. Here are our top tips for getting in the zone with your revision.

UNDERSTAND THE PROCESS

Understand the exam process and what revision you need to do. This will give you confidence but also help you to put things into proportion. These pages are a good place to find some starting pointers for performing well at exams.

BUILD YOUR CONFIDENCE

Use your revision time, not just to revise the information you need to know, but also to practise the skills you need for the examination. Try answering questions in timed conditions so that you're more prepared for writing answers in the exam. The more prepared you are, the more confident you will feel on exam day.

DEAL WITH DISTRACTIONS

Think about the issues in your life that may interfere with revision. Write them all down. Think about how you can deal with each so they don't affect your revision. For example, revise in a room without a television, but plan breaks in your revision so that you can watch your favourite programmes. Be really honest with yourself about this – lots of students confuse time spent in their room with time revising. It's not at all the same thing if you've taken a look at Facebook every few minutes or taken mini-breaks to send that vital text message.

FRIENDS AND FAMILY

Make sure that they know when you want to revise and even share your revision plan with them. Help them to understand that you must not get distracted. Set aside quality time with them, when you aren't revising or worrying about what you should be doing.

GET ORGANISED

If your notes, papers and books are in a mess you will find it difficult to start your revision. It is well worth spending a day organising your file notes with section dividers and ensuring that everything is in the right place. When you have a neat set of papers, turn your attention to organising your revision location. If this is your bedroom, make sure that you have a clean and organised area to revise in.

KEEP HEALTHY

During revision and exam time, make sure you eat well and exercise, and get enough sleep. If your body is not in the right state, your mind won't be either – and staying up late to cram the night before the exam is likely to leave you too tired to do your best.

Planning Zone

The key to success in exams and revision often lies in the right planning. Knowing what you need to do and when you need to do it is your best path to a stress-free experience. Here are some top tips in creating a great personal revision plan:

JUNE

1. Know when your exam is

Find out your exam dates. Go to www.edexcel.com/i-am-a/student/timetables/pages/home.aspx to find all final exam dates, and check with your teacher. This will enable you to start planning your revision with the end date in mind.

2. Know your strengths and weaknesses

At the end of the chapter that you are studying, complete the 'You should know' checklist. Highlight the areas that you feel less confident on and allocate extra time to spend revising them.

3. Personalise your revision

This will help you to plan your personal revision effectively by putting a little more time into your weaker areas. Use your mock examination results and/or any further tests that are available to you as a check on your self-assessment.

4. Set your goals

Once you know your areas of strength and weakness you will be ready to set your daily and weekly goals.

5. Divide up your time and plan ahead

Draw up a calendar, or list all the dates, from when you can start your revision through to your exams.

6. Know what you're doing

Break your revision down into smaller sections. This will make it more manageable and less daunting. You might do this by referring to the Edexcel GCSE Geography B specification, or by the chapter objectives, or by headings within the chapters.

7. Link it together

Also make time for considering how topics interrelate. For example you might have looked at Singapore as an example of a country with a pro-natalist policy for Topic 1 in Unit 2. But it is also an interesting example of a city that takes sustainability quite seriously and it is certainly a 'changing city'. It would be very useful to make a list of examples and case-studies that you have studied and think laterally about how else you might use them in an examination. You know more than you think!

8. Break it up

Revise one small section at a time, but ensure you give more time to topics that you have identified weaknesses in.

9. Be realistic

Be realistic in how much time you can devote to your revision, but also make sure you put in enough time. Give yourself regular breaks or different activities to give your life some variety. Revision need not be a prison sentence!

10. Check your progress

Make sure you allow time for assessing progress against your initial self-assessment. Measuring progress will allow you to see and celebrate your improvement, and these little victories will build your confidence for the final exam

Finally – stick to your plan!

0

7

27

29

30

23

1

Know Zone

Remember that different people learn in different ways – some remember visually and therefore might want to think about using diagrams and other drawings for their revision, whereas others remember better through sound or through writing things out. Think about what works best for you by trying out some of the techniques below.

REVISION TECHNIQUES

Highlighting: work through your notes and highlight the important terms, ideas and explanations so that you start to filter out what you need to revise.

Key terms: look at the key terms highlighted in bold in each chapter. Try to write down a concise definition for this term. Now check your definition against the glossary definition on p313.

Summaries: writing a summary of the information in a chapter can be a useful way of making sure you've understood it. But don't just copy it all out. Try to reduce each paragraph to a couple of sentences. Then try to reduce the couple of sentences to a few words!

Concept maps: if you're a visual learner, you may find it easier to take in information by representing it visually. Draw concept maps or other diagrams. These are particularly good at showing links. For example, you could create a concept map which shows how to learn about sustainability.

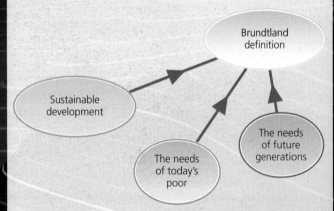

Mnemonics: this is when you take the first letter of a series of words you want to remember and then make a new word or sentence. An example of this is SNAP. This stands for Singapore Needs Additional People.

Index cards: Write important events, definitions and processes on index cards and then test yourself.

Quizzes: Learning facts can be dull. Why not make a quiz out of it? Set a friend 20 questions to answer. Make up multiple-choice questions. You might even make up your own exam questions and see if your friend can answer them!

And then when you are ready:

Practice questions: go back through all the ResultsPlus features with questions to see if you can answer them (without cheating!). Try writing out some of your answers in timed conditions so that you're used to the amount of time you'll have to answer each type of question in the exam. Then, check the guidance for each one and try to mark your answer.

Use the list below to find all the ResultsPlus questions.

Chapter 1: p 17, 25 and 27

Chapter 2: p 30, 34, 37, 38 and 41

Chapter 3: p 43, 44, 49, 52 and 55

Chapter 4: p 60, 62, 64, 66 and 69

Chapter 5: p 73 and 85

Chapter 6: p 88, 93 and 101

Chapter 7: p 102, 105, 111, 115 and 117

Chapter 8: p 124, 125, 128 and 133

Chapter 9: p 139, 141, 143 and 151

Chapter 10: p 157, 163 and 165

Chapter 11: p 172, 174, 179 and 181

Chapter 12: p 186, 190 and 195

Chapter 13: p 204, 208 and 211

Chapter 14: p 212, 219, 222 and 229

Chapter 15: p 231, 236, 240, 243 and 245

Chapter 16: p 249, 252, 257 and 263

Don't Panic Zone

Once you have completed your revision in your plan, you'll be coming closer and closer to the big day. Many students find this the most stressful time and tend to go into panic-mode, either working long hours without really giving their brain a chance to absorb information, or giving up and staring blankly at the wall. Follow these tips to ensure that you don't panic at the last minute.

TOP TIPS

1. Test yourself by relating your knowledge to geography issues that arise in the news – can you explain what is happening in these issues and why?

2. Look over past exam papers and their mark schemes. Look carefully at what the mark schemes are expecting of candidates in relation to the question.

3. Do as many practice questions as you can to improve your technique, help manage your time and build confidence in dealing with different questions.

4. Write down a handful of the most difficult bits of information for each chapter that you have studied. At the last minute focus on learning these.

5. Relax the night before your exam – last-minute revision for several hours rarely has much additional benefit. Your brain needs to be rested and relaxed to perform at its best.

6. Remember the purpose of the exam – it's for you to show the examiner what you have learnt.

LAST MINUTE LEARNING TIPS FOR GEOGRAPHY

● Remember that an intelligent guess is better than nothing – if you can't think of an example of an LIC city then take a guess – you cannot lose marks.

● Know your categories – don't go into the examination unclear about basic definitions: developed and developing, urban and rural, erosion and weathering. Check out the glossary.

● Many examination questions ask you to interpret resources. Make sure that you revise the skills that help you do this effectively.

ASSESSMENT OBJECTIVES

The questions that you will be asked are designed to examine the following aspects of your geography. These are known as Assessment objectives (AO). There are three AOs.

AO1	Recall, select and communicate knowledge and understanding of places, environments and concepts.
AO2	Apply knowledge and understanding in familiar and unfamiliar contexts.
AO3	Select and use a variety of skills, techniques and technologies to investigate, analyse and evaluate questions and issues.

THE TYPES OF QUESTION THAT YOU CAN EXPECT IN YOUR EXAM

The examination papers are designed so that the opening part of each question is the easiest part. The difficulty becomes progressively harder as you move through the question. The level of difficulty is controlled by the command word and content required in your answer. The Foundation tier papers have questions which have more 'scaffolding' (helping you to structure and develop your answer) to make these papers more accessible.

There are four different types of question:

TYPE
Short – single word answers or responses involving a simple phrase or statement.
MCQ – Multiple Choice Question.
Open – free-response questions that involve a limited amount of continuous prose.
Long – free-response questions where candidates have the opportunity for extended writing and allow opportunities for assessing the quality of your written communication.

UNDERSTANDING THE LANGUAGE OF THE EXAM PAPER

It is vital that you know what 'command' words ask you to do. Common errors are:

1. Confusing *describe* with *explain*.

2. Adding *explanation* when you are only asked to *describe*.

Identify...	Name a process or a location
Complete	Finish of a task that has already been partly done
Name	Like 'identify'
Describe...	Give the main characteristics of a topic or issue
Explain..	Give reasons why something is as it is
Examine..	Describe something with some detail
Outline..	Give the main features of something
Define..	Say what something means
Suggest reasons...	Say why something might have happened or occurred
Give the reasons...	Say why something happened or occurred
Comment on...	Give some reasons why or how something is as it is
State...	Like 'name' or 'identify'

Meet the exam paper

This section shows you what the exam paper looks like. Check that you understand each part. Now is a good opportunity to ask your teacher about anything that you are not sure of here.

Print your surname here, and your other names in the next box. This is an additional safeguard to ensure that the exam board awards the marks to the right candidate.

Ensure that you understand exactly how long the examination will last, and plan your time accordingly.

Ensure that you read the instructions carefully and that you understand exactly which questions from which sections you should attempt.

Here you fill in your personal exam number. Take care when writing it down because the number is important to the exam board when writing your score.

Here you fill in your school's centre number. You will be given this by your teacher on the day of your exam.

Note that the quality of your written communication will also be marked. Take particular care to present your thoughts and work at the highest standard you can for maximum marks.

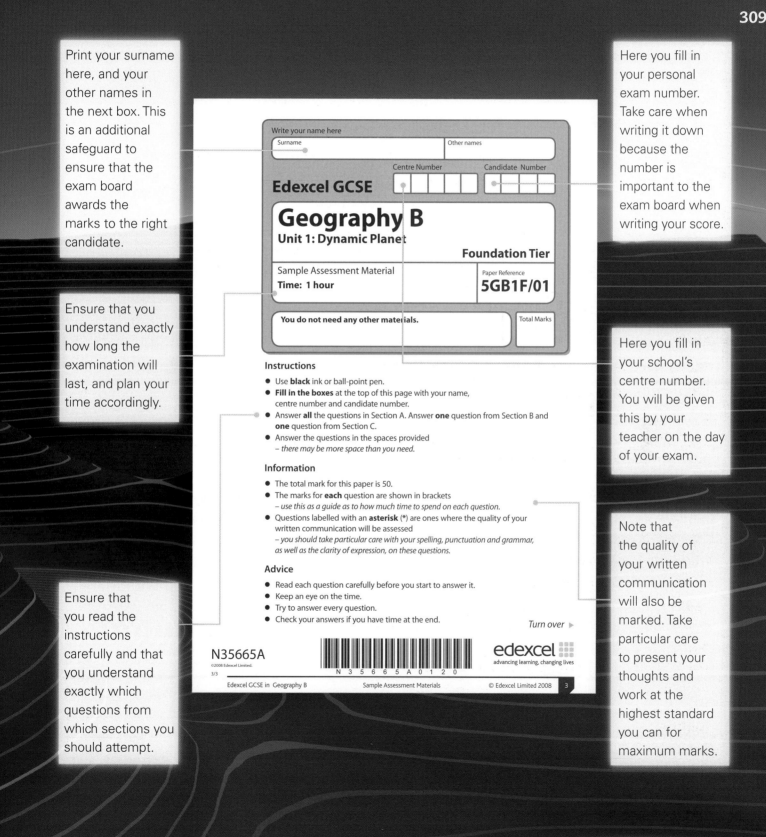

Write your name here

Surname

Other names

Centre Number

Candidate Number

Edexcel GCSE

Geography B

Unit 1: Dynamic Planet

Foundation Tier

Sample Assessment Material
Time: 1 hour

Paper Reference
5GB1F/01

You do not need any other materials.

Total Marks

Instructions

- Use **black** ink or ball-point pen.
- **Fill in the boxes** at the top of this page with your name, centre number and candidate number.
- Answer **all** the questions in Section A. Answer **one** question from Section B and **one** question from Section C.
- Answer the questions in the spaces provided
  – there may be more space than you need.

Information

- The total mark for this paper is 50.
- The marks for **each** question are shown in brackets
  – use this as a guide as to how much time to spend on each question.
- Questions labelled with an **asterisk** (*) are ones where the quality of your written communication will be assessed
  – you should take particular care with your spelling, punctuation and grammar, as well as the clarity of expression, on these questions.

Advice

- Read each question carefully before you start to answer it.
- Keep an eye on the time.
- Try to answer every question.
- Check your answers if you have time at the end.

Turn over ▶

N35665A
©2008 Edexcel Limited.
3/3

Edexcel GCSE in Geography B   Sample Assessment Materials   © Edexcel Limited 2008   3

edexcel
advancing learning, changing lives

If Economic change is the topic that you have studied in class and you wish to answer the question on it, remember to indicate this where you are asked on the paper.

It is not always one examiner who will mark your entire paper. Sometimes, one examiner will mark one question and another will mark a different question. So, you must indicate which question you have answered so that your paper is sent to the correct examiner!

Read the instructions each time – they are there to provide guidance.

SECTION B – SMALL SCALE DYNAMIC PLANET

Answer ONE question in this section.

Topic 5: Coastal Change and Conflict

If you answer Question 5 put a cross in this box ☐.

5   Figure 5 shows coastal management at Hornsea in Yorkshire.

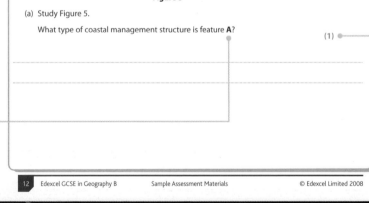

Figure 5

(a) Study Figure 5.

What type of coastal management structure is feature **A**?

(1)

....

....

The marks for each question are shown on the right-hand side of the page. Make sure that you note how many marks a question is worth as this will give you an idea of how long to spend on that question.

Pay attention to any text highlighted in bold. It is highlighted to alert you to important information, so be sure to read it and take note!

This is the 'stem' of a question – it often includes important information that you need to think about in your answers.

This is the resource – be careful, it may not be exactly like resources that you have seen before.

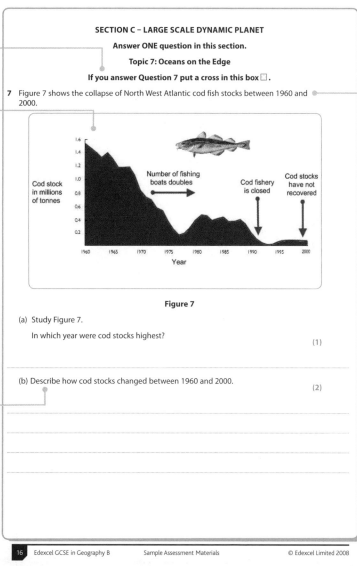

SECTION C – LARGE SCALE DYNAMIC PLANET

Answer ONE question in this section.

Topic 7: Oceans on the Edge

If you answer Question 7 put a cross in this box ☐.

7   Figure 7 shows the collapse of North West Atlantic cod fish stocks between 1960 and 2000.

Cod stock in millions of tonnes

Number of fishing boats doubles

Cod fishery is closed

Cod stocks have not recovered

Year

Figure 7

(a) Study Figure 7.

In which year were cod stocks highest?

(1)

(b) Describe how cod stocks changed between 1960 and 2000.

(2)

Words in bold are highlighted to catch your attention so be sure to note them.

These are the command words that tell you what to do.

Zone Out

Well done, you have finished your exam. So, what now? This section provides answers to the most common questions students have about what happens after they complete their exams.

About your grades

Whether you've done better than, worse than or just as you expected, your grades are the final measure of your performance on your course and in the exams.

When will my results be published?

Results for summer examinations are issued in August, with GCE first and GCSE second. January exam results are issued in March.

Can I get my results online?

Visit www.resultsplusdirect.co.uk, where you will find detailed student results information including the 'Edexcel Gradeometer' which demonstrates how close you were to the nearest grade boundary. Students can only gain their results online if their centre gives them permission to do so.

I haven't done as well as I expected. What can I do now?

First of all, talk to your subject teacher. After all the teaching that you have had, tests and internal examinations, he/she is the person who best knows what grade you are capable of achieving. Take your results slip to your subject teacher, and go through the information on it in detail. If you both think that there is something wrong with the result, the school or college can apply to see your completed examination paper and then, if necessary, ask for a re-mark immediately. The original mark can be confirmed or lowered, as well as raised, as a result of a re-mark.

If I am not happy with my grade, can I re-sit a unit?

Yes, you are able to re-sit each unit once before claiming certification for the qualification. The best available result for each contributing unit will count towards your final grade.

What can I do with a GCSE in Geography?

Geography is well known as a subject that links to all other subjects of the curriculum, so a GCSE in Geography is a stepping stone to a whole range of opportunities. A good grade will help you to move on to AS, Applied A Level or BTEC course. You may want to continue your study of Geography or take a course such as a BTEC National in Travel and Tourism which has a more work-related approach.

The skills that you develop can lead you to employment opportunities in journalism, media, engineering, ICT, travel and tourism, environmental management, marketing, business management and teaching. Geographers are everywhere!

Glossary
A-C

Abrasion: (in rivers) erosion caused by the river picking up stones and rubbing them against the bed and banks of the channel in the flow.

Accessible countryside/rural areas: countryside within easy reach of urban areas.

Adaptation: changes that take place to react to a situation or condition. (They may or may not be successful.)

Ageing population: a population with a rising average age.

Agribusiness: commercial agriculture that is owned and managed by large corporations.

Alternative energy: energy sources that provide an alternative to fossil fuels.

Alternative fuels: fuel sources that provide an alternative to fossil fuels.

Appropriate technology: equipment that the local community is able to use relatively easily and without much cost.

Aquaculture: commercial fish farming, e.g. rearing fish or prawns in ponds or submerged cages.

Aquifer: an underground store of water, formed when water-bearing (permeable) rocks lie on top of impermeable rocks.

Asthenosphere: the upper part of the Earth's mantle, where the rocks are more fluid.

Attrition: (in rivers) gradual wearing down of particles by erosion as they collide with each other, making them smaller and rounder.

Automation: the use of machinery, rather than people, in manufacturing and data processing.

Backwash: water from a breaking wave which flows under gravity down a beach and returns to the sea.

Bay: a feature produced when erosion creates an indent in the coastline, often located between two headlands.

Biodiversity: the number and variety of living species found in a specific area.

Biofuels: fuel sources derived from agricultural crops.

Biome: a plant and animal community covering a large area of the Earth's surface.

Biosphere: the living part – plants and animals – of the Earth.

Birth rate: the number of births per 1,000 people in a year.

Bleaching: degradation of coral reefs under conditions of increased acidity in sea water.

Boserupian theory: the view that when population grows it stimulates technological changes that produce increases in output, ensuring that living standards can be maintained for the growing population.

Bottom-up approach: development projects that originate in local communities rather than in central government or external agencies.

Brownfield site: a piece of land that has been used and abandoned, and is now awaiting some new use.

Bus lane: a marked lane in a road in which only public transport vehicles such as buses and taxis are permitted.

Call centre: an office equipped to handle a large volume of telephone calls (especially for taking orders or serving customers).

Carbon footprint: a measurement of all the greenhouse gases we individually produce, through burning fossil fuels for electricity, transport, etc., expressed as tonnes (or kg) of carbon-dioxide equivalent.

Carrying capacity: the maximum number of people that can be supported by the resources and technology of a given area.

Chocolate box village: a rural settlement that appears to match the picturesque, pretty image sometimes used on boxes of chocolates, etc.

CITES: Convention on International Trade in Endangered Species of Wild Fauna and Flora – an international agreement.

Clark-Fisher model: a generalised description of how societies' employment structures change as they develop.

Climate change: long-term changes in temperature and precipitation.

Coastal flooding: the inundation of low-lying areas in coastal areas and regions.

Coastal management: the processes and plans applied to coastal areas by local authorities and agencies.

Collision plate boundary: a tectonic margin at which two continental plates come together.

Commodity/production chains: the linkages between a product and the sources of its basic materials and/or its components.

Commuter belt: a residential area within relatively easy reach of (and often surrounding) a city, where many of the residents travel to and from the city daily.

Concordant coast: a coastline created when alternating hard and soft rocks occur parallel to the coast, and are eroded at different rates.

Congestion charging: a system of traffic control that charges drivers who enter the congested central area of a city.

Conservation: managing the environment in order to preserve, protect or restore it.

C-E

Conservative plate boundary: where two tectonic plates slide past each other.

Constructive plate boundary: tectonic plate margin where rising magma adds new material to the diverging plates.

Constructive waves: small, weak waves with a low frequency that tend to add sand and other sediment to the coastline because they do not break with much force.

Consumer industries: industries that produce goods for people to use/consume.

Consumption: the using up of something.

Continental crust: the part of the crust dominated by less dense granitic rocks.

Continental shelf: the submerged edge of a continental land-mass.

Convection currents: (in tectonics) circulating movements of magma in the mantle caused by heat from the core.

Coral reef: a hard stony ridge, just above or below the surface of the sea, formed by the external skeletons of millions of tiny creatures called polyps.

Core: (in tectonics) the central part of the Earth, consisting of a solid inner core and a more fluid outer core, and mostly composed of iron and nickel.

Core region: the most important social, political and economic area of a country or global region – the centre of power.

Corrosion: chemical erosion caused by the dissolving of rocks and minerals by water.

Counterurbanisation: the movement of people and employment from major cities to smaller settlements and rural areas located just beyond the city, or to more distant smaller cities and towns.

Cultural background: the origins of an individual's or group's belief system.

Cultural dilution: where a particular culture is changed and weakened, usually by exposure to other competing cultures.

Death rate: the number of deaths per 1,000 people in a year.

Deforestation: the chopping down and removal of trees to clear an area of forest.

Degradation: the social, economic and environmental decline of an area, often through deindustrialisation.

Deindustrialisation: the decline in industrial activity in a region or an economy.

Dependence: a condition in which something (e.g. a country) is only able to survive by relying on outside support (e.g. from another country).

Depopulation: the decline of a population, both by natural processes and, occasionally, by government policy.

Deposition: the dropping of sediment that was being carried by a moving force.

Deprived area: an area in which there is a damaging lack of the material benefits that are considered to be basic necessities – employment, housing, etc.

Derelict land: land on which factories or houses have been demolished.

Destructive plate boundary: tectonic plate margins where oceanic plate is subducted.

Destructive waves: large, powerful waves with a high frequency that tend to take sediment away from the beach, because their backwash is greater than their swash.

Developed countries: countries at a late stage of development. They are generally quite rich, with a high proportion of people working in secondary and, especially, tertiary occupations. Also known as More Economically Developed Countries (MEDCs).

Developing countries: countries at an early stage of development. They are generally poor, with a high proportion of people working in primary occupations. Also known as Less Economically Developed Countries (LEDCs).

Development: economic and social progress that leads to an improvement in the quality of life for an increasing proportion of the population.

Discordant coast: a coastline created when alternating hard and soft rocks occur at right angles to the coast, and are eroded at different rates.

Disparity: a great difference – e.g. between parts of a country in terms of wealth.

Disposable income: the amount of money which a person has available to spend on non-essential items, after they have paid for their food, clothing and household running costs.

'Do nothing': (in coastal management) an approach that allows natural processes to take their course without any intervention.

Dormitory town: a settlement dominated by commuters who work in a nearby city.

Drainage basin: the area of land drained by a river and its tributaries.

Eco footprint: a measure of how much land is needed to provide a place (e.g. a city) with all the energy, water and materials it needs, including how much is needed to absorb its pollution and waste.

Economic migrant: a person who moves in order to find employment.

Economic status: the position held by an individual, group or country in terms of their economic power.

E–G

Ecosystem: a community of plants and animals that interact with each other and their physical environment.

Emigrant: a person leaving a country or region to live somewhere else (for at least a year).

Employment structure: the proportions of people who work in primary, secondary, tertiary or quaternary jobs.

Energy consumption: the amount of energy used by individuals, groups or countries.

Enhanced greenhouse effect: the increased greenhouse effect resulting from human action (emission of greenhouse gases) and leading to global warming.

Environment Impact Assessment (EIA): a method of evaluating the effect of plans and policies on the environment.

Environmental degradation: negative impacts on the natural environment , generally through human action.

Environmental pollution: the degradation of the environment through the emission of toxic waste material.

Erosion: the wearing away and removal of material by a moving force, such as a breaking wave.

Estuary: a river mouth that is wide and experiences tidal conditions.

Eutrophication: the loss of oxygen in water after too much nutrient enrichment has taken place.

Evacuation: the removal of people from an area, generally in an attempt to avoid a threatened disaster (or escape from an actual one).

Expatriate communities: overseas communities made up of non-nationals, e.g. the British living in Spain.

Exploitation: making full use of something (often implying that the use is unfair and has a negative impact).

Extinction: the permanent loss of something, generally used with reference to species of plants or animals, when there are no living examples left.

Extreme climate: a climate that is unusually challenging, usually in terms of its temperature conditions or type and extent of precipitation.

Factory farming: a highly organised agricultural system with high inputs and high outputs.

Farmers' market: a set of stalls run by farmers and food growers from the local area.

Fauna: animals.

Fetch: the distance of sea over which winds blow and waves move towards the coastline.

Flood plain: the relatively flat area forming the valley floor on either side of a river channel, which is sometimes flooded.

Flood risk: the predicted frequency of inundation (floods) in an area.

Flora: plants.

Flows: (in migration) the movement of labour within and between countries to satisfy a changing pattern of demand and supply.

Food miles: the distance covered supplying food to consumers.

Food web: an illustration of the grouping of animals and plants found in an ecosystem, showing the sources of food for each organism.

Formal sector: (of the economy) work where the people are formally employed, with permanent jobs and regular pay (and they pay their taxes).

GDP per capita: Gross Domestic Product per person, is the total wealth created within a country, divided by its population.

Gene pool: the genetic material contained by a specific population.

Geological climate events: climate changes that result from major geological events such as volcanic eruptions.

Geological structure: the way in rocks are arranged, both vertically and horizontally.

Geology: the science and study of the Earth's crust and its components.

Glacial region: an area that is covered by ice (either a valley glacier or much larger ice sheets).

Global city: a major urban area that has a significant role in controlling the international flows of capital and trade.

Global hub: a major centre of global communications, such as an international airport.

Global shift: the movement of manufacturing from developed countries to cheaper production locations in developing countries.

Global warming: a trend whereby global temperatures rise over time, linked in modern times with the human production of greenhouse gases.

Globalisation: the process, led by transnational companies, whereby the world's countries are all becoming part of one vast global economy.

GNI per capita: a measurement of economic activity that is calculated by dividing the gross (total) national income by the size of the population.

Goods: produced items and materials.

Gradient: the slope of the land.

Grassroots scheme: a scheme that originates within a local community rather than being imposed from above.

Green consumerism: choosing to buy environmentally friendly products.

Green sector: that part of economic activity that pays attention to environmental issues.

G-M

Greenfield site: a piece of land that has not been built on before, but is now being considered for development.

Greenhouse gases: those gases in the atmosphere that absorb outgoing radiation, hence increasing the temperature of the atmosphere.

Groundwater: water contained beneath the surface, as a reserve.

Habitat: an animal or plant's natural home.

Hard engineering: using solid structures to resist forces of erosion.

Hard rock coast: a coastal region composed of resistant materials.

Headland: a part of the coastline that protrudes into the sea.

Holistic approach: an approach to environmental management that treats the whole area as an interrelated system.

Honeypot: a place of special interest or appeal that attracts large numbers of visitors and tends to become overcrowded at peak times.

Hot arid regions: parts of the world that have high average temperatures and very low precipitation.

Household waste: material produced by households that needs to be disposed of.

Human Development Index: a measure of development that uses four economic and social indicators to produce an index figure that allows comparison between countries.

Human resources: the skills and abilities of the population.

Hybrid cars: cars which use electric batteries as well as petrol engines.

Hydro-electric power (HEP): the use of fast flowing water to turn turbines which produce electricity.

Hydrograph: a graph which shows the discharge of a river, related to rainfall, over a period of time.

Hydrological cycle: the global stores of water and linking processes that connect them.

Ice age: a period in the Earth's past when the polar ice caps were much larger than today.

Immigrant: a person arriving in a country or region to live (for at least a year).

Impermeable: not allowing water to pass through.

Industrial stage: the economic stage when manufacturing industry develops.

Industrialisation: the process whereby industrial activity (particularly manufacturing) assumes a greater importance in the economy of a country or region.

Infant mortality rate: the number of deaths of children (under the age of one) per thousand live births a year.

Infiltration: the process whereby water soaks into the soil and rock.

Informal sector: (of the economy) forms of employment that are not officially recognised, e.g. people working for themselves on the streets of developing cities.

Inner city living space: residential areas within city centre areas.

Integrated Coastal Zone Management (ICZM): the system of dividing the UK coastline into zones that can be managed holistically.

Integrated river management: a holistic system of managing rivers that takes an overview of the whole river basin and the relationship between its different parts.

Interlocking spurs: areas of high land which stick out into a steep-sided valleys.

Intermediate technology: a technology that the local community is able to use relatively easily and without much cost.

Joints: lines of weakness in a rock that water can pass along.

Labour-intensive: needing a large workforce (or a large amount of work) and relatively little machinery/capital.

Land degradation: the declining quality and quantity of land, generally because of human action.

Landfill: disposal of rubbish by burying it and covering it over with soil.

Latitude: the position of a place north or south of the Equator, expressed in degrees.

Levees: natural embankments of sediment along the banks of a river.

Life expectancy: the average number of years a person might be expected to live.

Little Ice Age: a period of slight global cooling that lasted from around the mid-fifteenth century to the mid-nineteenth century.

Long profile: the gradient of a river, from its source to its mouth.

Long-term planning: planning that looks beyond immediate costs and benefits by exploring impacts in the future.

Longshore drift: the movement of material along a coast by breaking waves.

Lower course: that part of a river system that is close to the mouth of the river.

Magnitude: the size of something.

Malthusian theory: the view that population growth is the main reason why a society would collapse.

Mangrove swamp: a tidal swamp dominated by mangrove trees and shrubs that can survive in the salty and muddy conditions found along tropical coastlines.

Marine ecosystem: the web of organisms that live in the ocean or a part of an ocean.

Mass movement: the downslope movement, by gravity, of soil and/or rock by the processes of slumping, falling, sliding and flowing.

M-P

Material resource: a natural substance that humans choose to use.

Meanders: the bends formed in a river as it winds across the landscape.

Megafauna: very large mammals, such as those that lived during the last ice age.

Micro-hydro schemes: small-scale HEP systems that generate electricity locally.

Mid-course: the central section of a river's course.

Migration: the process of people changing their place of residence, either within or between countries.

Millennium Development Goals (MDGs): the development goals agreed by world governments at the UN summit in September 2000.

Natural causes: those processes and forces that are not controlled by humans.

Natural change: the change (an increase or a decrease) in population numbers resulting from the difference between the birth and death rates over one year.

Natural increase: the difference between birth rate and death rate.

Natural resources: those materials found in the natural world that are useful to man, and that we have the technology and willingness to use.

Net in-migration: the increase in a country's population as a result of more people arriving than leaving.

Net migration (migration balance): the difference between migrant arrivals and departures. If arrivals exceed departures, the balance is positive. If departures exceed arrivals, the balance is negative.

Network: a system of linkages between objects, places or individuals.

New economy: the emergence of new types of economic activity and employment in the last few decades.

Nomadic pastoralism: a type of farming where farmers have no permanent land and migrate with their cattle, etc. from one place to another.

Non-renewable resource: those resources – like coal or oil – that cannot be 'remade', because it would take millions of years for them to form again.

Nutrient cycle: a set of processes whereby organisms extract minerals necessary for growth from soil or water, before passing them on through the food chain – and ultimately back to the soil and water.

Oceanic crust: the part of the crust dominated by denser basaltic rocks.

Orbital changes: changes in the pathway of the Earth around the Sun and in its axial geometry.

Organic agriculture: farming systems that use no artificial chemicals.

Organic produce: food grown or produced without the use of chemicals.

Outsourcing: a process in which a company subcontracts part of its business to another company.

Over-abstraction: when water is being used more quickly than it is being replaced.

Overfishing: taking too many fish (or other organisms) from the water before they have had time to reproduce and replenish stocks for the next generation.

Ox-bow lake: an arc-shaped lake which has been cut off from a meandering river.

Park-and-ride scheme: a system whereby private vehicles are left on the edge of an urban area and people are then bused into the centre.

Periphery: the outer limits or edge of an area, often remote or isolated from its core.

Permafrost: permanently frozen ground, found in polar (glacial and tundra) regions.

Permeable: allowing water to pass through.

Plate margin: the boundary between two tectonic plates.

Players: individuals and groups who are interested in and affected by a decision-making process.

Polar: relating to the North or South Pole. In polar regions the land is covered with ice (glacial) or frozen (tundra).

Pollution: the presence of chemicals, noise, dirt or other substances which have harmful or poisonous effects on an environment.

Population pyramid: a diagrammatic way of showing the age and sex structure of a population.

Population structure: the composition of a population, usually in terms of its age and gender.

Pores: small air spaces found in a rock or other material that can also be filled with water.

Post-industrial stage: that period in the development of a society when manufacturing industry declines in importance, and is replaced by other forms of employment.

Poverty: a state of shortage of money and goods, usually measured in terms of average wealth and income in a society.

Poverty cycle: a set of processes that maintain a group or society in poverty.

Pre-industrial stage: that period in the development of a society when manufacturing industry has yet to develop.

P-S

Precipitation: when moisture falls from the atmosphere – as rain, hail, sleet or snow.

Prediction: forecasting future changes.

Preparation: the process of getting ready for an event.

Preserve: maintain (something) in its existing state.

Primary employment: working in the primary sector – extracting and exploiting raw materials.

Primary sector: the economic activities that involve the working of natural resources – agriculture, fishing, forestry, mining and quarrying.

Production chain: the sequence of activities needed to turn raw materials into a finished product.

Pull factor: something that attracts people to a location.

Push factor: something that make people wish to leave a location.

Quality of life: the degree of well-being (physical and psychological) felt by an individual or a group of people in a particular area. This can relate to their jobs, wages, food, amenities in their homes, and the services they have access to, such as schools, doctors and hospitals.

Quaternary employment: working in jobs that are related to ICT and research.

Quaternary Period: the most recent major geological period of Earth's history, consisting of the Pleistocene and the Holocene.

Quaternary sector: the economic activities that provide intellectual services – information gathering and processing, universities, and research and development.

Ramsar: The Ramsar Convention on Wetlands is an intergovernmental treaty for the conservation and wise use of wetlands.

Redevelopment: development that aims to stimulate growth in areas that have experienced decline.

Regeneration: growth in areas that have experienced decline in the past.

Regulated flow: the steady movement of water through a drainage basin that will not bring flash flooding.

Renewable resource: resources, such as forests, that can be maintained by management.

Remote countryside/rural area: rural areas that are distant from and thus little affected by urban areas and their populations.

Response: the way in which people react to a situation.

River cliff: steep outer edge of a meander where erosion is at its maximum.

River pollution: the emission of harmful or poisonous substances into river water (or their presence in the river).

Run-off: water that flows directly over the land towards rivers or the sea after heavy rainfall.

Rural depopulation: the decline of population in rural areas and regions.

Rural idyll: the common perception that rural areas are quiet and attractive – and therefore good places to live in.

Sea-level rise: the increase in the level of the sea, relative to the land.

Secondary employment: working in the secondary sector, making things.

Secondary sector: the economic activities that involve making things, either by manufacturing (TV, car, etc.) or construction (a house, road, etc.). The sector also includes public utilities, such as producing electricity and gas.

Sediment: usually sand, mud or pebbles deposited by a river.

Services: those things that are provided, bought and sold that are not tangible.

Short-term emergency relief: help and aid provided to an area to prevent immediate loss of life because of shortages of basics, such as water, food and shelter.

Siltation: the deposition of silt (sediment) in rivers and harbours.

Site of Special Scientific Interest (SSSI): a small area that has been officially designated for protection because of its wildlife or geology.

Slip-off slope: inner gentle slope of a meander where deposition takes place.

Soft rock coast: a coastal area made up of easily eroded materials.

Solar output: the energy emitted by the Sun.

Solifluction: the movement downhill of soggy soil when the ground layer beneath is frozen. It often occurs in tundra regions.

Spit: material deposited by the sea which grows across a bay or the mouth of a river.

Stack: a detached column of rock located just off-shore.

Strategic realignment: the reorganisation of coastal defences that is often part of managed retreat.

Stump: a stack that has collapsed, leaving a small area of rock above sea-level.

Sub-aerial processes: weathering and mass movement.

Superpower countries: the world's most powerful and influential nations – the USA and, increasingly, China and India.

Sustainability: the ability to keep something (such as the quality of life) going at the same rate or level. From this stems the idea that the current generation of people should not damage the environment in ways that will threaten future generations' environment (or quality of life).

T-Z

Sustainable cities: cities that have a number of policies that attempt to reduce their impact on the environment (including the surrounding regions).

Sustainable development: development that meets the needs of the present without compromising (limiting) the ability of future generations to meet their own needs.

Sustainable living space: living spaces that are designed in such a way as to have a small impact of the environment and are thus more durable than others.

Sustainable resources: resources – such as wood – that can be renewed if we act to replace them as we use them.

Swash: the forward movement of water up a beach after a wave has broken.

Tectonic hazards: threats posed by earthquakes, volcanoes and other events triggered by crustal processes.

Telecommuter: person who works away from the office (usually at home), through the use of the internet.

Telecottaging: working from a home in the country, using computer communication.

Teleworking: any form of work in which telecommunications replace work-related travel (commuting).

Temperate climate: a climate that is not extreme (in terms of heat, cold, dryness or wetness).

Tertiary employment: working in the service sector, producing 'intangible' products.

Tertiary sector: the economic activities that provide various services – commercial (shops and banks), professional (solicitors and dentists), social (schools and hospitals), entertainment (restaurants and cinemas) and personal (hairdressers and fitness trainers).

Throughflow: water that flows slowly through the soil until it reaches a river.

Throw-away society: a society with an attitude to consumption that pays little attention to the need to recycle.

Tipping point: the point at which the momentum of a change becomes unstoppable.

Top-down approach: approach in which projects are set up and organised by governments, often with little consultation with local communities.

Transnational company (TNC): a large company operating in several countries.

Tundra: the flat, treeless Arctic regions of Europe, Asia and North America, where the ground is permanently frozen.

Unsustainable: unable to be kept going at the same rate or level.

Upper course: the source area of a river, often in an upland or mountainous region.

Urban fringe: the countryside adjacent to or surrounding an urban area.

Urban sprawl: urban growth, usually weakly controlled, into surrounding rural and semi-rural areas.

Urbanisation: the development and growth of towns or cities.

Volcanic activity: the escape of molten rock, ash and gases from an opening in the Earth's surface (or when there is evidence that it is imminent).

Water flow: movement processes of the Earth's water, including evaporation, precipitation and overland flow.

Water harvesting: storing rainwater or used water ('grey water') for use in periods of drought.

Water management schemes: programmes to control rivers, generally organised by local or central government.

Water store: a build-up of water that has collected on or below the ground, or in the atmosphere.

Water table: the level in the soil or bedrock below which water is usually present.

Waterfall: sudden descent of a river or stream over a vertical or very steep slope in its bed.

Weathering: the breakdown and decay of rock by natural processes, without the involvement of any moving forces.

Wilderness: uncultivated, uninhabited and inhospitable regions.

Youthful population: a population in which there is a high percentage of people under the age of 16 (or sometimes 18).

Zero population growth: when natural change and migration change cancel each other out, and there is no change in the total population.

Index
A–C

E-H

M-R

R-S

T–Z

We are grateful to the following for permission to reproduce copyright material:

Figures

Figure 1.1 from "This dynamic earth", U.S. Geological Survey; Figure 1.2 'The convectional currents in the mantle', U.S. Geological Survey; Figure 1.6 from Figure 10i-2: "Creation of oceanic crust on the ocean floor", U.S. Geological Survey; Figure 1.6 from 'Diagram to show Continental-oceanic convergence boundary', U.S. Geological Survey; Figure 1.6 from 'Figure 10i-6: Collision of a oceanic plate with a continental plate', U.S. Geological Survey; Figures 1.7 and 1.8 from Figure 10i-2: Creation of oceanic crust on the ocean floor, U.S. Geological Survey; Figure 1.8 from 'Diagram to show an oceanic/oceanic convergence boundary', US Geological Survey; Figure 1.9 from "NASA's Cosmos are excerpted from Cambridge Guide to the Solar System" or "Sun, Earth, and Sky", http://ase.tufts.edu/cosmos/print_images.asp?id=4 (Professor Lang, K.) Tufts University and published by Cambridge University Press, copyright © Professor Lang; Figure 1.10 from http://earth.rice.edu/MTPE/geo/geosphere/hot/volcanoes/volcanoes_map.gif, copyright © Smithsonian Institution, Global Volcanism Program; Figure 1.12 from "The Hawaiian chain of islands", U.S. Geological Survey; Figure 1.14 from "A cut-away view of a composite volcano" U.S. Geological Survey; Figure 1.19 'The San Francisco seismic net on-line', U.S. Geological Survey; Figure 2.1 from "Changes in the Earth's average temperature during the last million years", http://www.seed.slb.com/subcontent.aspx?id=3750, data from NOAA http://www.ncdc.noaa.gov/oa/ncdc.html, copyright © 2009 Schlumberger Excellence in Educational Development, Inc. All rights reserved. Permission is granted to make copies of this document for educational purposes only, provided that this copyright notice is reproduced in full. For more information, visit our Web site, at www.seed.slb.com; Figure 2.2 from Forestry Commission, http://www.forestry.gov.uk/images/emissionspiechart.jpg/$File/emissionspiechart.jpg, Crown Copyright material is reproduced with permission under the terms of the Click-Use License; Figure 2.6 from http://www.worldmapper.org/, copyright © 2006 SASI Group (University of Sheffield) and Mark Newman (University of Michigan); Figure 2.8 from UKCIP 09: 'The climate of the UK and recent trends', http://www.ukcip.org.uk, copyright © Met Office 2007; Figure 2.9 from "Environmental and economic implications of rising sea level and subsiding deltas: the Nile and Bengal examples" AMBIO, 18(6), 340-5 (Milliman, J.D., Broadus, J.M. and Gable, F. 1989), Allen Press, Inc, Copyright © (1989) Royal Swedish Academy of Sciences. Reprinted by permission of Allen Press Publishing Services and Professor John Milliman; Figure 3.1 from Elements of Ecology, 6th ed. (Smith, T.M. and Smith, R.L.) Figs 23.4, p.500; 23.12, p.507 and 23.19, p.511, copyright © 2006 by Pearson Education, Inc. Reprinted by permission; Figure 3.8 adapted from National Geographic, 'http://ngm.nationalgeographic.com/2007/01/amazon-rain-forest/amazon-map-interactive, copyright © National Geographic Image Collection; Figure 4.9 'The Colorado River regime before and after construction of the Glen Canyon Dam', U.S. Geological Survey; Figure 5.4 adapted from "Landforms produced by the erosion of a headland", http://www.georesources.co.uk/sea6.gif copyright © David Rayner (Georesources); Figure 5.5 'The Swanage coast' copyright © www.collinsbartholomew.com Ltd, reproduced with kind permission of HarperCollins Publishers; Figure 5.7 'Destructive Waves' adapted from http://cgz.e2bn.net/e2bn/leas/c99/schools/cgz/accounts/staff/rchambers/GeoBytes%20GCSE%20Blog%20Resources/Images/Coasts/Destructive_Waves.jpg, copyright © Rob Chambers; Figure 5.8 adapted from 'Constructive Waves', http://cgz.e2bn.net/e2bn/leas/c99/schools/cgz/accounts/staff/rchambers/GeoBytes%20GCSE%20Blog%20Resources/Images/Coasts/Constructive_Waves.jpg, copyright © Rob Chambers; Figure 5.10 from "The process of Longshore drift", http://geographyfieldwork.com/LongshoreDrift.htm, reproduced by permission of Barcelona Field Studies Centre http://geographyfieldwork.com; Figure 5.12 from http://www.le.ac.uk/bl/gat/virtualfc/217/images/ley2.jpg, copyright © Ted Gaten; Figure 5.13 from www.tulane.edu/~sanelson/geol111/slump.gif, copyright © Professor Stephen A. Nelson; Figure 6.2 from Geography and Change, Hodder Arnold (Flint, D., Flint, C. and Punnett, N. 1996); Figure 6.6 'Formation of an ox-bow lake' adapted from http://cgz.e2bn.net/e2bn/leas/c99/schools/cgz/accounts/staff/rchambers/GeoBytes%20GCSE%20Blog%20Resources/Images/Rivers/ox-bow_lake.gif, copyright © Rob Chambers; Figure 6.9 from 'River Severn Hydrograph July 2007' Bewdley Case Study, copyright © Geographical Association and the Environment Agency; Figure 6.15 from Tomorrow's Geography John Murray, Fig 5.13 (Harcourt, M., Warren, S. and Warn, S. 2001), reproduced by permission of John Murray (Publishers) Ltd; Figure 6.16 from http://www.swenvo.org.uk/environment/images/Flooding_properties_1996_2006.gif, copyright © Environment Agency; Figure 7.4 adapted from "Aquatic food chain" Encyclopaedia Britannica, Reprinted with permission, copyright © 2006 by Encyclopaedia Britannica, Inc; Figure 7.5 from "De-oxygenated "dead spots": areas of water where eutrophication has occurred on a large scale" UNEP, United Nations Environment Programme, www.unep.org, copyright © UNEP; Figure 7.13 from "Out of sight, out of mind" by John Papastan and John Bradley, copyright © Greenpeace; Figure 7.14 from http://www.env.go.jp/en/wpaper/1994/eae230000000035.html, Source: The International Whaling Commission, 2008 compiled by Greg Donovan; Figure 8.10 adapted from "Native Peoples and Languages of Alaska Map" by Krauss, Michael E. 1974. Fairbanks, Alaska: Alaska Native Language Center. Revised 1982, http://www.uaf.edu/anlc/; Figure 9.1 from US Census Bureau, International Data Base, http://www.census.gov/ipc/prod/wp02/wp-02003.pdf, copyright © US Census Bureau; Figure 9.4 adapted from Population and Migration, Philip Allan Updates (Witherick, M. 2004) Figure 10, Reproduced by permission of Philip Allan Updates; Figure 9.5 adapted from AQA AS Geography, Philip Allan Updates (Barker, A. et al 2008) Figure 5.5, reproduced with permission of Philip Allan Updates; Figure 9.6 from 'Population pyramid for China, 2005' US Census Bureau, International Data Base, copyright © US Census Bureau; Figure 9.7 from 'Population pyramids for Indonesia, Mexico and the UK, 2005', US Census Bureau, International Data Base, copyright © US Census Bureau; Figure 9.8 from Population and Migration Philip Allan Updates (Witherick, M. 2004) Figure 32, Reproduced by permission of Philip Allan Updates; Figure 9.13 from 'UK residents born abroad (1951 – 2001)', Office for National Statistics, Crown copyright © 2002, Crown Copyright material is reproduced with permission under the terms of the Click-Use License; Figure 9.16 adapted from Economic migrants from Eastern Europe, Figure 2 (Witherick, M.) Geo Factsheet, 184 (Sept 2005), copyright © Curriculum Press; Figure 9.18 from Edexcel AS Geography, Figure 12.4, Philip Allan Updates (Warn, S. et al, 2008), reproduced by permission of Philip Allan Updates; Figure 10.1 from "New Reserves: As Prices Surge, Oil Giants Turn Sludge Into Gold - Total Leads Push in Canada To Process Tar-Like Sand; Toxic Lakes and More CO2 - Digging It Up, Steaming It Out", The Wall Street Journal, 27 March 2006 (Gold, R.), copyright © 2006, Dow Jones and Company, Inc; Figure 10.4 from 'The World's leading diamond producing nations', http://www.pangeadiamondfields.com/diamond-sector.htm, copyright © WWW International Diamond Consultants; Figure 10.5 'World wealth distribution in 2002 by country' from www.worldmapper.org/ copyright © 2006 SASI Group (University of Sheffield) and Mark Newman (University of Michigan); Figure 10.6 adapted from Global Footprint Network 2006 and United Nations Development Programme 2006, copyright © Global Footprint Network; Figures 10.7, 10.8 from BP Statistical Review of World Energy 2007 www.bp.com, copyright © BP plc; Figure 10.9 from "The growth of China and India", copyright © Goldman Sachs International (UK); Figure 10.11 from Beyond the Limits: Confronting Global Collapse, Envisioning a Sustainable Future (Meadows, et al 1992), copyright © Donella Meadows, with permission of Chelsea Green Publishing (www.chelseagreen.com);

Acknowledgements

Figure 11.1 from SE Urban Rural Classification - Scottish Executive 05-06, http://www.scotland.gov.uk, reproduced by permission of Ordnance Survey on behalf of HMSO © Crown copyright 2009. All rights reserved. Ordnance Survey Licence number 100030901; Figure 11.6 from 'Population pyramids of London and the London Borough of Camden', www.statistics.gov.uk, Crown Copyright material is reproduced with permission under the terms of the Click-Use License; Figure 11.9 from "Population, Highlights" North to south net migration, United Kingdom, Social Trends 38, http://www.statistics.gov.uk/cci/nugget.asp?id=1924, Crown Copyright © 2008, Crown Copyright material is reproduced with permission under the terms of the Click-Use License; Figure 11.12 'Singapore', source: Central Intelligence Agency; Figure 12.3 adapted from "The employment structure of Bangladesh, China and the USA, 2007" CIA Factbook of the World 2007, source: Central Intelligence Agency; Figure 12.9 'Sandwell in the West Midlands', www.laws.sandwell.gov.uk/ copyright © Sandwell Metropolitan Borough Council; Figure 12.10 adapted from 'The biomass-energy project in Berlin, New Hampshire' http://www.laidlawenergy.com/images/Berlin_Biomass-Energy_Q_and_A_9-14-07.pdf, copyright © Michael B. Bartoszek, Laidlaw Energy Group, Inc; Figure 13.2 adapted from "How York's eco-footprint is made up" The Eco Footprint of York, York lifestyles and their environmental impact, www.york.ac.uk, copyright © University of York; Figure 13.3 adapted from "The eco-footprints of different cities and countries", from Living Planet Report 2006 http://assets.panda.org/downloads/living_planet_report.pdf copyright © WWF; Figure 14.5 from "Complimentary Set Changing Environments for as" Changing Environments, Longman (Warn & Naish, 2000) Figure 1.8, copyright © Pearson Education Limited; Figure 14.7 'Employment in agriculture in the UK, 1978–2005' from Economic & Social Research Council ESRC; Figures 14.11, 14.12 from Essential AS Geography, Stanley Thornes (Ross, Morgan and Heelas), first printed in 2000, reproduced with the permission of Nelson Thornes; Figure 14.16 adapted from "The Polycentric Metropolis: Learning from Mega-City Regions in Europe" by Hall, P. and Pain, K. (2006) London: Earthscan, www.lboro.ac.uk/gawc/rb/images/rb225f3.jpg, granted with permission; Figure 14.22 adapted from http://www.new-forest-national-park.com/, copyright © Pete Carpenter, New Forest National Park; Figure 15.1 adapted from "Tax Regions Map", http://www.tra.go.tz/regions.htm, copyright © Tanzania Revenue Authority; Figure 15.6 adapted from "The Human Development Index for the UK, Brazil and Peru, 1975–2005" data from Human Development Report 2007/2008 published 2007, UNDP, reproduced with permission of Palgrave Macmillan; Figure 15.10 adapted from "Location of Chambamontera village in Peru" The Inventory of Conflict & Environment (ICE) https://www.cia.gov/library/publications/the-world-factbook/geos/pe.html, source: US Central Intelligence Agency; Figure 15.13 adapted from 'Cumulative change in protected areas', www.index.mundi.com source: Central Intelligence Agency; Figure 16.3 'Global distribution of development' from http://www.worldbank.org/, data copyright © The World Bank; Figure 16.10 from Global Challenge: A2 Level Geography for Edexcel B, Figure 4.21 Longman (McNaught, A. and Witherick, M. 2001) copyright © Pearson Education Limited; Figure 16.12 adapted from "Population change in the Western Isles, 1901 to 2001", copyright © GROS www.gro-scotland.gov.uk and Comhairle nan Eilean Siar; Figure 17.5 from Edexcel exam paper, 6472, May 2007, M26095A (Dunn, C), copyright © Edexcel Limited; Figure 17.6 from Edexcel exam paper, 647201, M30456A (Warn, S), copyright © Edexcel Limited; Figure 17.8 adapted from "Revealed: The 15 sites where controversial new 'eco' towns could be built", Daily Mail, 4 April 2008 , copyright © Solo Syndication 2008; Figure 17.9 'Sustainable features of eco-towns' from GeoFact Sheet 237, copyright © Curriculum Press;' and Figure 17.13 adapted from Strategic Vision, www.claycountryvision.imerys.com/content/1_StrategicVision.asp, copyright © Imerys' Minerals Ltd.

Tables

Table 15.2 from Tax income from the tax districts of Tanzania, 2005–06, http://www.tra.go.tz, copyright © Tanzania Revenue Authority.

Text

Extract on page 162 from "Why is striving for sustainability so important?", http://www.interfaceinc.com/goals/sustainability_overview.html, copyright © InterfaceFLOR www.interfaceflor.eu; Extract on page 163 from "goals to improve its environmental performance", http://www.interfacesustainability.com/seven.html, copyright © InterfaceFLOR www.interfaceflor.eu; Extract on page 176 adapted from Sustainable Cities, http://sustainablecities.dk/city-projects/cases, Sustainablecities.dk is sponsored by Realdania and owned and operated by DAC, Danish Architecture Centre, http://sustainablecities.dk; Extract on page 274 adapted from Eco-towns Prospectus, Department for Communities and Local Government: London, www.communities.gov.uk © Crown Copyright, 2007; and Extract on page 277 adapted from source www.claycountryvision.imerys.com, copyright © Imerys' Minerals Ltd

In some instances we have been unable to trace the owners of copyright material, and we would appreciate any information that would enable us to do so.

ON-Page Credits

Figures 14.11, 14.12 from Essential AS Geography, Stanley Thornes (Ross, Morgan and Heelas), first printed in 2000, Reproduced with the permission of Nelson Thornes